Autorenkollektiv

New Trends in Chemistry

Trends in Analytical Chemistry

D. Betteridge, T. J. Sly, Swansea (UK)

Topics in Photochemistry

H. E. Zimmerman, Wisconsin (USA)

New Developments in Polymer Synthesis

T. Saegusa, Kyoto (Japan)

Biochemical Engineering

H. Janshekar, A. Fiechter, Zürich (Switzerland)

Chemistry and Spectroscopy of Rare Earths

C. K. Jørgensen, Geneva (Switzerland)
R. Reisfeld, Jerusalem (Israel)

Systematization and Structures of the Boron Hydrides

L. Barton, St. Louis (USA)

AKADEMIE-VERLAG · BERLIN

Autorenkollektiv

New Trends in Chemistry

Autorenkollektiv

NEW TRENDS IN CHEMISTRY

With 95 Figures and 22 Tables

AKADEMIE-VERLAG · BERLIN
1982

Die Originalausgabe erscheint im Springer-Verlag
Berlin · Heidelberg · New York

Vertrieb ausschließlich für die DDR und die sozialistischen Länder

Erschienen im Akademie-Verlag, DDR-1086 Berlin, Leipziger Straße 3—4
Lizenznummer: 202 · 100/539/82
Gesamtherstellung: VEB Druckerei „Thomas Müntzer", 5820 Bad Langensalza
Umschlaggestaltung: Karl Salzbrunn
Bestellnummer: 763 120 5 (6710) · LSV 1205
Printed in GDR
DDR 98,— M

Table of Contents

Trends in Analytical Chemistry

D. Betteridge and T. J. Sly

Chemistry Department, University College of Swansea, Singleton Park, Swansea SA2 8PP, UK

Table of Contents

1 Introduction

The last two decades have seen some spectacular achievements in analytical science: the placing of the environmental revolution on a sound basis by the routine determination of p.p.m. or p.p.b. levels of pollutants in the atmosphere, hydrosphere and biosphere; the routine testing of athletes and race horses for traces of stimulants; the remote analysis of the surface of the Moon and Mars and the atmosphere of Venus, etc. It has also been a period when the normal criteria for acceptable limits of impurities has dropped from the level of per cent to p.p.b., when non-destructive testing has become routine and when samples can be so small that even destructive methods of analysis scarcely have a deleterious effect on bulk of the material from which the sample is taken. In short, the nature of analysis has changed greatly.

The way in which the changes have affected a well equipped R & D laboratory situated in the Plastics Division of I.C.I. is shown in Fig. 1. This demonstrates the impact of instrumentation on sample throughput. It is interesting to note however that the growth has not been linear. The period of rapid improvement in efficiency coincided with the time when mechanical forms of semi-automation, which were easy to develop, install, operate and maintain were in vogue. With the introduction of more sophisticated instrumentation, the per capita output has levelled off, but the complexity of the analysis has increased: For example, with ^{13}C NMR as well as ^{1}H NMR it is possible to make structural assignments to polymer chains routinely.

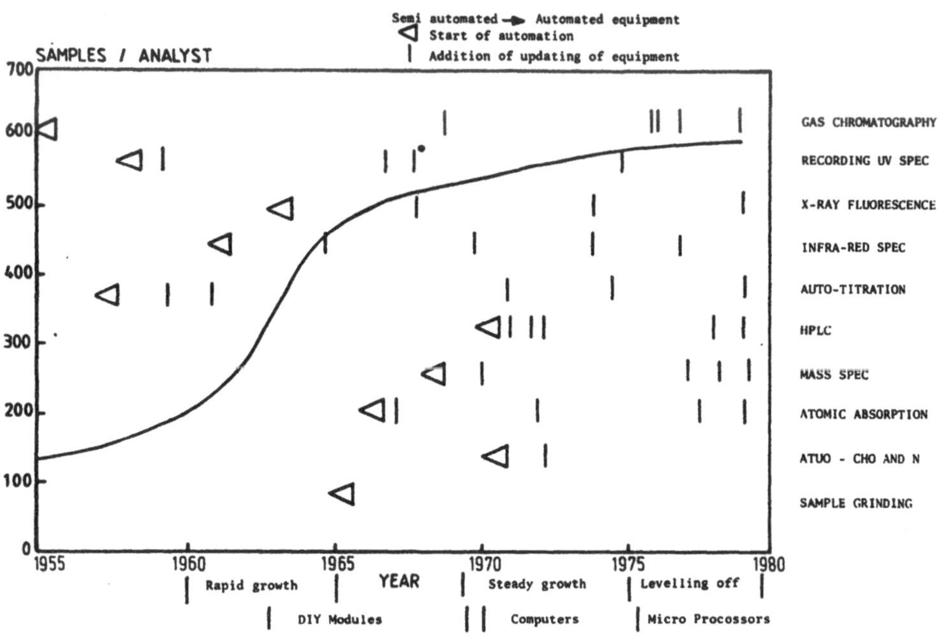

Fig. 1. Productivity of the analyst in a major industrial laboratory 1955–1980, and the influence of new instrumentation and automation. (By courtesy D. C. M. Squirrell, ICI Plastics Ltd.)

This information, which is of the greatest value and has contributed to the marketing of very well characterised polymers, would have been impossible to obtain 10 yrs ago, but it may take a 10 hr run to obtain the ^{13}C NMR spectrum and this reduces the analyst's output as measured in numbers of samples. The experience of the plastics industry is paralleled in many other laboratories — analyses are becoming much more complex and the analyst is becoming more involved with the scientists on the project who request the analysis. The days of what have been called "hole-in-the-wall analysis" where a sample was passed through a hole in the lab wall for an isolated and impartial assay are drawing to a close.

It has been argued elsewhere that the development of analysis is an interplay between the trinity of contemporary scientific theory, analytical technique and problems to be solved [1)]. For many, technique is the most important of these, and the advance of analysis is dominated by technique. We feel, however, that it is all too easy to overemphasise this aspect of the subject, and that it is dangerous to make any forecast about specific techniques. It is certain that over the next 50 years, existing techniques will improve and new ones will be discovered, but it is not possible to predict what these will be. (The record of previous pundits is not good in this respect). In this article, we shall concentrate on the total analytical problem, and especially on the impact on analysis of microelectronics, computers, automation and mathematical methods. As noted above, the analyst is already being set increasingly complex problems and the trend will continue as improved methods of obtaining and processing analytical data become available. This will require a change in the analyst, who as well as being an all round chemist will also have to have a familiarity with modern electronics, computers and management science and quite likely physics or biochemistry as well, depending on his industrial location. Given the scarcity of polymaths, it is likely that interdisciplinary analytical teams will become the norm, and the outcome should be some very exciting science.

In the following pages we first give brief accounts of the developments in microelectronics, computers and data processing which underpin virtually all modern analytical methods. Secondly some novel methods, which indicate the breadth of the subject and the trends towards high sample throughput and/or complexity of analysis will be described. Finally a major specific analytical problem, represenative of many likely to be encountered in the foreseeable future will be discussed. In the space available, selectivity and brevity is essential. Our object is to indicate trends, not to be comprehensive.

2 Developments in Electronics and Instrumentation

2.1 Introduction

Over the last twenty five years, the instrumental techniques available to the analyst have become far more varied and of much better quality. This is to a large extent due to the advances which have been made in the field of electronics over the same period, with the development firstly of transistors, then simple integrated circuits, and more recently such complex and versatile devices as microprocessors.

The impact of these developments on the instrument manufacturers has been twofold: firstly, by incorporating these advances as they occurred into successive generations of instruments, manufacturers have been able to lower production costs and increase the accuracy and reliability of their instruments, thus making them available to a wider market. Secondly, it has led to the development of increasingly sophisticated instrumental techniques which were not previously feasible for either economic or technical reasons. In order to view this process of evolution in perspective, it is wortwhile to consider briefly some of the developments in electronics which have helped it to occur.

2.2 The Transistor

In 1947, the first experimental transistors were produced by a team led by Shockley at Bell Laboratories in the USA. These were unreliable devices, expensive to produce and frequently short-lived. When compared to the well proven thermionic valves then in widespread use, which were cheap and relatively reliable, the transistor appeared to most people as nothing more than an interesting novelty of little practical value, and apart from the work at Bell Labs little attempt was made initially to develop transistor technology or to find applications for the new devices. By the early 1950s they were available in production quantities, but it was some years before they became widely accepted. This was in part due to the poor reliability of the early devices, but mainly because after nearly fifty years of working with valves, the electronics industry found it hard to come to terms with the new technology; indeed, many early transistor-based designs are closer in concept to the valve designs which they replaced than to a modern transistor circuit.

As far as we can tell, no one in the early 1950s thought of applying transistor circuits to an analytical instrument. This is hardly surprising in view of the extreme simplicity of the electronics in most instruments then available. Leaving aside for the moment the 'large' instruments (such as mass spectrometers and NMR spectrometers) which were then just beginning to come into experimental use, the most sophisticated piece of instrumentation likely to be found in routine use in most laboratories was a UV spectrophotometer, which contained many precision made mechanical and optical parts, but probably no more than two valves and a handful of other electronic components. The electronics were there almost as an afterthought, simply a convenient means of quantifying a light level. In such a context, the familiar valve technology was able to provide all that was required. It was to be another decade before increased sophistication in instrument design, coupled with a dramatic fall in the price of transistors, led to their widespread usage in instrumentation.

Not surprisingly, the first major application for transistors turned out to be one for which they were inherently better suited than valves due to their small size and low power consumption: the computer.

In the early 1950s several groups built experimental electronic calculators based on transistors, and in the UK a team at Manchester University built an experimental transistor computer between 1953 and 1955. This was at a time when new manufacturing techniques were being introduced which overcame many of the reliability problems which had plagued the early transistors, and in 1956 an adaptation of the Manchester experimental computer was marketed as the Metropolitan Vickers

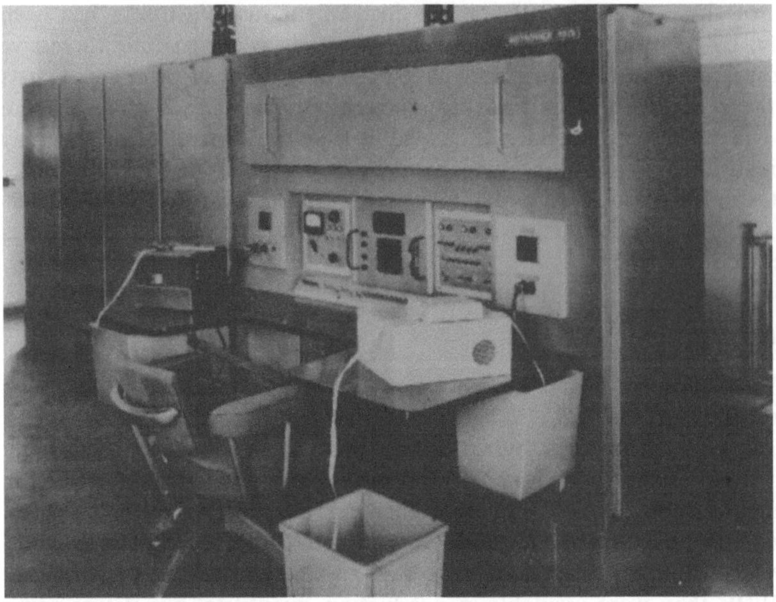

Fig. 2. The Metropolitan — Vickers MV950, believed to be the first commercially available transistor computer. (Reproduced by courtesy of Dr. S. Lavington)

MV950 (Fig. 2), the world's first commercially available transistor computer [2]. Although it had a somewhat limited performance, even by the standards of the day, it marks the point at which the transistor first became accepted as a serious tool.

2.3 Integrated Circuits

Shortly after the first commercial transistors were produced, there was speculation that it might one day be possible to fabricate an entire electronic circuit on a single piece of semiconductor material, and thus greatly reduce the size of electronic circuits. In 1957 this goal was realised with the production of the first experimental integrated circuit (IC) by a British team. By the early 1960s these were available commercially in the form of simple amplifiers, comparators, and logic gates, which contained the equivalent of about 20 transistors. Their high cost meant that, as in the case of the transistor some years before, their initial uses were in those applications where they showed a decisive advantage over valves or dicrete transistors, such as computers, aircraft, and military equipment.

Meanwhile, the price of transistors had fallen sharply as they started to become more widely used in domestic applications, and this, coupled with the desire to produce more sophisticated instruments, led instrument manufacturers to start incorporating them into their new products. The principle advantage in doing this was that the new circuits could be built on a modular basis using printed circuit boards, allowing considerable savings in production costs and at the same time making the instruments simpler to service. The accompanying reduction in circuit size and power comsumption was an added bonus to the manufacturers, as it enabled a considerable increase in circuit complexity without any increase in the overall size of the instruments, compared to previous models.

It was, however, not until the late 1960s that valve circuits disappeared entirely from new analytical instruments; when they finally did, they were replaced by solid state circuits utilising both transistors and ICs, as both came into widespread use for instrumentation at much the same point in time. By now, there had been a fundamental change of attitude on the part of the instrument manufacturers, who now started to look upon electronics as an integral and important part of the overall system, rather than simply an accessory to the mechanical or optical parts of the instrument.

With this new attitude came a realisation of the potential offered by automatic instrumentation, with many of the functions of the machine under electronic control. From the late 1960s through to the mid 70s instruments appeared which incorporated many extra features designed to make them easy to use. This was especially true of the 'large' instruments, such as mass spectrometers, many of which incorporated minicomputers for the first time to handle functions which had previously been performed manually, such as peak assignments, etc. (techniques of this type will be discussed more fully in the following sections). For some time however, innovations such as the use of on-board computers were strictly confined to the large instruments (MS, NMR, etc.), for reasons of cost. It required another breakthrough in electronics and computer technology to bring about the cost reductions neccessary to allow such facilities to be incorporated into simpler instruments.

2.4 Large Scale Integration and Microprocessors

As the use of integrated circuits expanded throughout the 1960s, semiconductor manufacturers were working to develop new fabrication techniques which would allow the equivalent of ten thousand or so transistors to be etched onto a single chip, rather than the hundred or so then possible. They realised the enormous cost reductions (and consequent increases in sales) which such a process would allow, and as these so — called Large Scale Integration (LSI) devices came into widespread use in the early 1970s they caused huge price reductions for many types of equipment. Probably the most obvious example of this is the pocket calculator. In 1970 a pocket calculator was a rarity: those that did exist were comparatively bulky and cost about £ 100. By 1975 the size had shrunk to about half that of the 1970 model, whilst the price had dropped tenfold. Such reductions were made possible largely by the enormous market for devices such as calculators, for although LSI devices are extremely cheap to manufacture, their design and development costs are quite astronomical, and hence must be spread over a large number of units to enable a reasonable selling price to be achieved. Thus, for many small-quantity applications, such as analytical instruments, the cheapest method of construction remained the discrete transistor and Small Scale Integration (SSI) IC.

The problem of these high development costs was foreseen by the American computer giants IBM when, in 1971, they filed a patent application for a "reconfigurable circuit" — in other words, a single circuit which could be programmed to perform many different functions to suit different applications, avoiding the cost of developing a specialised or 'dedicated' chip for each application. These devices, known as microprocessors, began to appear commercially shortly afterwards, and have had a profound effect on almost all areas of applied electronics, not least analytical instruments. Indeed, it can be argued that they have affected instrument design to a greater extent than any other previous development in electronics, by bringing the benefits of computerisation to low cost routine laboratory instruments.

In the following section we will consider some of the applications of microprocessors in laboratory microcomputers, but we conclude this section with an illustration of the effects which electronic developments (and in particular microprocessors) have had upon analytical instruments by comparing an infrared spectrometer produced two decades ago with an equivalent model manufactured today.

2.5 Effect of Advances in Electronics upon Instrumentation

Figure 3 shows two infrared spectrophotometers, both manufactured by Perkin-Elmer. The upper one (model 237) was introduced in 1961, whilst the lower (model 1330) was introduced twenty years later in 1981. The most obvious difference between the instruments, apart from the differences of styling, is the absence of any mechanical

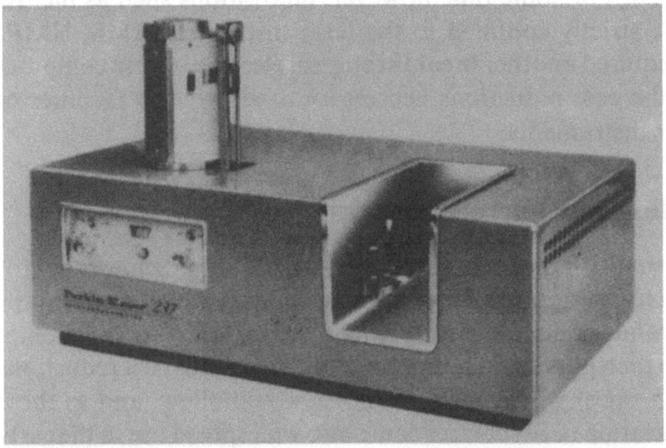

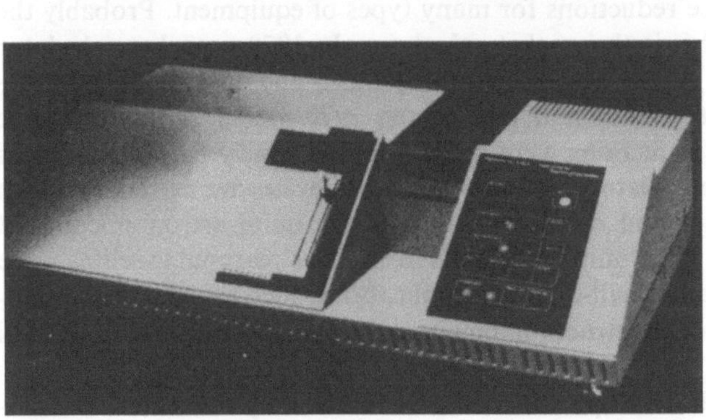

Fig. 3. Two infrared spectrometers produced by Perkin-Elmer. Model 237 (top) introduced in 1961, and model 1330 (bottom) introduced some twenty years later in 1981. (Reproduced by courtesy of Perkin-Elmer Ltd.)

controls from front panel of the newer instrument, since virtually all mechanical and optical parameters are adjusted by motors controlled from a small microcomputer built into the instrument. This facility prevents the operator from selecting inappropriate combinations of parameters [3], but more significantly actually reduces production costs, since this control approach is nowadays cheaper to produce than a system of complicated precision made mechanical linkages.

Without doubt, however, the greatest difference between the two is that the sophisticated electronics of the newer instrument enable it to be directly connected to an 'Infrared Data Station', a microcomputer system which can smooth spectra, produce an average spectrum from a number of scans, subtract one spectrum from another, or reformat a spectrum. In addition, it is possible to identify possible structural features of an unknown compound and to match the spectrum against a reference library stored on magnetic disks, to facilitate a rapid identification of unknowns [3]. It is interesting to note that the basic optical specifications of the two instruments are not all that dissimilar!

It is clear that the growth of electronics in instrumentation has been dramatic, but it has not been linear. By far the greatest advances have occurred within the last five years, and the pace of change appears to be increasing, making it difficult to make any firm predictions of the likely developments over the next two decades or so. We may be certain, however, that considerable progress will be made in the field of electronics over this period, and that it will have a substantial impact on the type of instrumentation available to the analyst.

3 Developments in Laboratory Computers

3.1 Introduction

In the preceding section we mentioned the MV 950, the world's first commercial transistor computer. This machine was produced at a time when the total number of computers in use was a tiny fraction of todays figure, and when the major scientific application for computers was in solving the extremely complex equations of theoretical physics and chemistry. Today, many mainframe computers throughout the world are employed to solve problems in these fields, but they are tiny in number compared with the vast quantitites of small and medium sized computers used to solve problems of a much more everyday and routine nature, but which are nevertheless important. In this section we will attempt briefly to review the many developments in small computers over the last twenty years, by reference to specific examples, and consider the various ways in which these developments have broadened the role of small computers in the analytical laboratory.

3.2 Minicomputer Developments

The first computer to be designed specifically for use in comparatively small-scale laboratory applications was the LINC (Laboratory Instrument Computer), produced in 1962 by a team at Massachusetts Institute of Technology in the USA,

Fig. 4. Two generations of laboratory computers: on the left, the LINC from 1962 and on the right the MINC from 1978. Although the MINC is about a quarter of the size of the LINC it contains sixteen times as much memory, as well as a far more powerful processor. (Reproduced by courtesy of Digital Equipment Co.)

and subsequently manufactured by Digital Equipment Corporation (DEC). This system, which was based on transistors, sold for the exceptionally low price (for 1962) of $ 43.600 and included such facilities as a cathode-ray tube display, magnetic tape storage, and a 2K magnetic core memory [4]. It is shown in Fig. 4 together with its present-day counterpart, the DEC MINC.

What really made the LINC different from the other small computers then available was the ability to be linked on-line to an experiment, to collect and process data and perform control functions. Up to this point, scientific computing had been performed almost exclusively off-line; that is to say, readings taken from an experiment were recorded onto paper tape or punch cards, either manually or directly by the instrument, and then fed into a mainframe computer for processing at the end of the experiment (much data is still processed in this fashion today). Savitzky [5] gives a good account of the state of the art at this time. With the LINC, data could be collected in analogue form from the experiment, digitised, stored and then processed immediately, resulting in much greater convenience for the user, and allowing signals from the experiment to provide a feedback loop for control functions. The main disadvantage of the LINC was that the user was required to program it in assembly language, i.e. using the instruction level of the processor itself (so-called 'machine code'). Anyone who has ever attempted it will know that this is an extremely laborious and time consuming task, requiring detailed knowledge of the actual computer hardware if the resultant program is to run efficiently (or at all!). Equally, to edit or modify an assembly language program is not a task for the faint-hearted, and for this reason many scientists who were not computer specialists found this type of programming language difficult to use.

One of the first modern minicomputers, the DEC PDP-8, was introduced in 1965 at a basic price of $ 18,000. It was constructed from transistors and integrated circuits and overcame the problems of assembly language programming. For the first time it provided in a small computer many of the facilities conventionally found in mainframe machines, the most significant of which was the facility to write programs in high level languages such as FORTRAN, which were translated or 'compiled' by the computer into a form which it was capable of executing (object code). This development greatly eased the task of the programmer, who was now able to write:

```
A = B + C
```

if he wished to add two numbers together, instead of having to write a complicated series of instructions such as:

```
ADDIT LDA FIRSTNUM
      ADC SECNUM
      STA RESULT
      RTS
```

which would be required to achieve the same result with a typical assembly language. Not surprisingly, machines of this type proved extremely popular for many business and scientific applications where the use of a computer had not previously been feasible for financial reasons.

DEC produced several derivatives of the PDP-8 aimed specifically at laboratory applications, and several thousand of these were sold altogether. Many are still in use today, and minicomputers in general are still widely sold for larger analytical applications. One consequence of the large sales of these and similar machines was the emergence of software libraries — collections of frequently used algorithms for many different mathematical and data processing applications, written both by the computer manufacturers themselves and also by users. In effect, ready-made programs were now available for a great many applications, thus simplifying the task of the user still further.

3.3 Microcomputers in the Laboratory

With the arrival of microprocessors in the mid 1970s the emphasis in lab computing has shifted somewhat. Minicomputers now tend to be used for applications involving a lot of iterative computation, owing to their speed, whilst the more powerful microcomputers have taken over from mincomputers as the general laboratory workhorse due to their lower price. The DEC MINC (Fig. 4) is a good example of this type of machine. It has a 16-bit processor, the LSI-11, and is software compatible with the DEC PDP-11 range of minicomputers. In terms of computing power, it is on a par with a low-powered minocomputer, i.e. about an order of magnitude more powerful than a small 8 bit microcomputer. It has optional facilities for analogue inputs and outputs, to allow direct connection to an experiment, and can be programmed in either BASIC or FORTRAN.

Another important feature of the MINC is the inclusion of the IEEE-488 interface bus. As its name implies, this is a bus designed for communication between one or more computers and a number of instruments or peripherals, according to a standard format, thus allowing various pieces of equipment produced by different manufatcurers to be linked together without any of the compatability problems usually encountered [6]. This bus was originally introduced by Hewlett-Packard, but has since been adopted by a large number of instrument, peripheral, and computer manufacturers. Many analytical instruments introduced during the last few years incorporate it, making it much easier than ever before to interface analytical instruments to laboratory computers to provide sophisticated computer control over the instrument and to collect data for processing. The extreme flexibility of the IEEE-488 bus, coupled with its high rate of data transfer, mean that it will probably become the most widely used method for interfacing between various laboratory instruments and computers over the next few years.

At the bottom end of the scale, many cheap 8-bit microcomputers are finding their way into analytical applications. These include machines such as the Apple and the Commodore PET (which incorporates the IEEE-488 bus) as well as many others. Although computers such as these lack the computing power and sophisticated interfacing abilities of the MINC, there are many applications for which they are quite powerful enough, whilst their low prices enable them to be used in a wide range of situations where computers have not previously been employed for reasons of cost. The impact which microcomputers and microprocessors have had upon analytical instrumentation is reviewed in Ref. [7].

The trend, therefore, is towards large numbers of small computers, rather than small number of larger machines. A recent development in this respect has been the emergence of networks of small computers, which can replace a much larger single computer whilst offering greatly increased flexibility and lower cost [8]. The next twenty years will see an increasing reliance by the analyst on computers, to increase both the number and quality of analyses. We shall consider next some of the mathematical techniques employed in conjunction with computers to achieve these ends.

4 Mathematical Methods for Chemical Analysis

4.1 Introduction

The availability of fast, cheap and accessible data processing is encouraging analysts to employ standard statistical evaluation of results, which previously have been very time-consuming to undertake, and to explore new ways of extracting the maximum amount of chemical information from the analytical data. This last is a rapidly expanding branch of the subject and has been termed "chemometrics"; it is some aspects of this that are discussed below.

The spread of the methods is being assisted by the availability of computer programs. Kowalski has assembled and generously made available a package known as ARTHUR which contains many routines of value to the analyst. Another which has been developed for social scientists but has much of interest to the analyst is the SSPS. In addition, standard statistical programs, suitable for specific micro and mini-computers are being marketed. The situation is analogous to that with instrumentation; powerful methods may be deployed with ease and by those with little knowledge of them. When used correctly they can greatly enhance the analysis, but with inadequate or poor data misleading conclusions will be generated with facility.

The range of methods and their scope is shown by the symposium arranged by Kowalski [9)] and the books of Massart, Dijkstra and Kaufman [10)], and Kateman and Pijpers [11)]. For the purpose of discussion they may be grouped as (i) statistical — quality control; (ii) improvement in quality of signal (iii) improvement in quality of the method (iv) extraction of analytical information.

4.2 Consideration of Selected Methods and Strategies

4.2.1 Statistical Methods

Many of the standard methods of statistics have been applied for quality control for many years. The computer has merely made their routine application easier, and enables the results to be presented in a more useful form. For example, it is possible to program the computer to take the analytical output, to combine it with other data so that the output is in the form of a decision e.g. "this sample is acceptable" or "this sample is suitable for internal use, but not for outside customers" or "this sample is not to specification, being deficient in the following respects . . ." This is time-saving and reduces error at the decision-making stage.

Because analytical results can be accumulated over long periods on easily accessible disks — a big improvement on laboratory notebooks — it is easier to evaluate the performance of a method or process over a period of time. It is easy to determine the frequency of defects in performance, both by inspection and by more formal methods. One of these which is being used more widely is the cusum procedure. A reference value, *k*, is selected; typically it is the mean value of the expected result. Then sequentially the cumulative sum, S_i, is calculated

$$S_i = \sum_{j=1}^{i} (x_j - k)$$

where x_j is the jth result. Since some results are low and some high, a perfect process would result in a plot of S_i *vs* j being a horizontal line with S_i being approximately zero. If, however there is a positive or negative bias, the cusum curve will deviate from the horizontal, and the slope will be indicative of the extent of malfunction. This is an example of a simple form of statistical analysis, which gives improved information, compared to standard control charts, but which is tedious to carry out by hand.

It is also a form of time series analysis, which is a growth point in statistics. It is to be expected that as improved and novel statistical methods are developed, accumulated data will be used increasingly for the purposes of quality control and forecasting performance.

4.2.2 Improvement in Quality of Signal

There are few things more satisfying to the experimentalist than seeing the improved quality of signal which is obtained with modern instrumentation compared to the old. Spectral bands, which previously were bumps on the side of a peak are clearly resolved; signals formerly buried in background noise are extracted and evaluated with certainty. Some of this improvement is due to the developments in

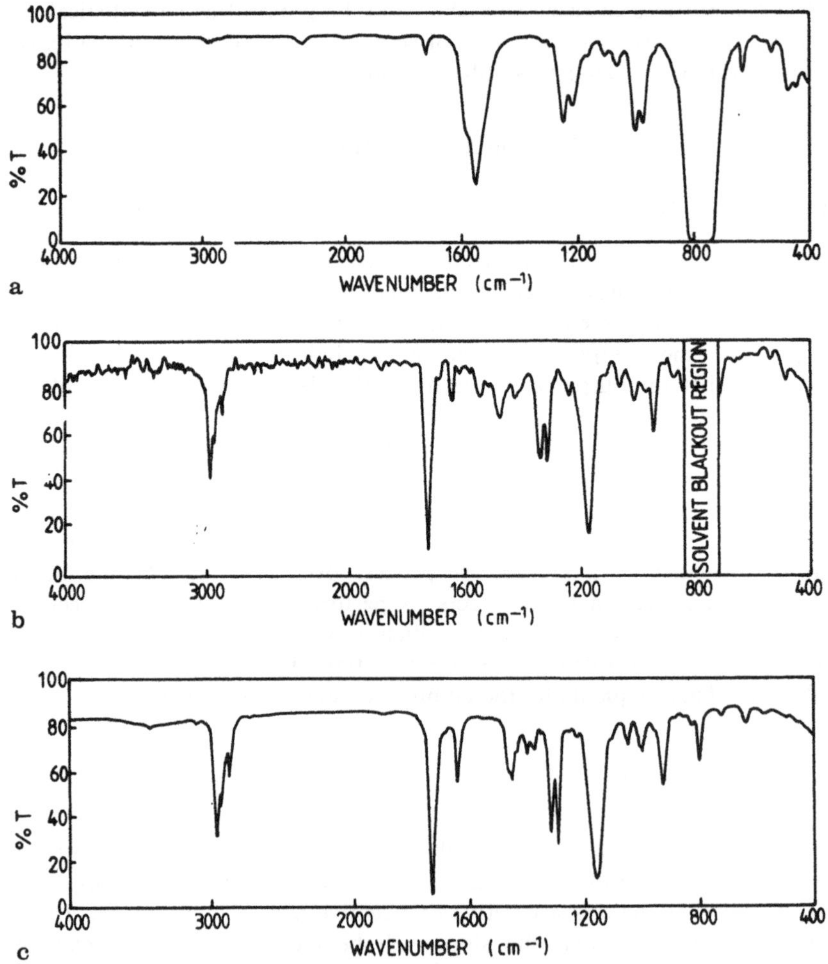

Fig. 5. Digital Spectrum Stripping of an Infrared Spectrum. **a** 0.05% butyl methacrylate in carbon tetrachloride; **b** Residual spectrum after stripping that of carbon tetrachloride; **c** Spectrum of pure butyl methacrylate in 0.1 mm path length cell for comparison. (Adapted from Ref. [12])

electronics described above, but much is due to the application of mathematical methods.

The result of collecting data in digitised form and processing it with very basic methods is shown in the example of the infrared spectrum of butyl methacrylate in carbon tetrachloride Fig. 5 [12]. Horlick and Hieftje have reviewed a number of methods, especially autocorrelation, and demonstrated how effectively signal can be extracted from noise [13]. All methods depending on the supposition that noise is generated randomly whereas the signal is regular. Thus it is possible to categorise regularly generated noise, e.g. mains intereference, as signal, but elementary checks should prevent misleading conclusions being drawn.

Perhaps the most dramatic example of the improvement in signal brought about by mathematical means is the application of the Fourier transform to infrared and NMR spectroscopy [14–16]. In the former instance, interferometry has become a practical possibility and in the latter it has permitted carbon-13 NMR to be developed.

4.2.3 Improvement in the Quality of the Method — Optimisation

Everyone believes, or at least hopes, that their experiment is carried out under "optimum conditions". The time-honoured way of doing this is to vary one of the perceived variables in the experiments whilst keeping all others constant in order to establish the point or range of maximum benefit. Then that is fixed and another is varied and so on until the optimum conditions are established. This is a satisfactory procedure for most classical analytical methods, but it presupposes that the variables are independent of each other, which is not always the case, and it is a lengthy process if several variables have to be investigated.

Even experimentally simple procedures may be complex. For example, if one were to develop continuous flow spectrophotometric methods of flow-injection analysis (such as described below) the following variables are relevant: concentrations of reactants; pH; physical dispersion of sample, which in turn is dependent on flow rate, tube diameters and size; configuration of the apparatus, whether tubes are coiled or straight, the presence, by design or accident, of mixing chambers; the rate of chemical reaction and the temperature. Because of the flexibility of the method, it is easy to "optimise" in the classical manner, but the variations in published versions of essentially the same procedure suggest that more often "acceptable" rather than "optimum" conditions have been established. It is obvious that many of the variables interact, for example the concentration of reactants affects both the thermodynamic equilibrium and the rate of reaction. Furthermore, there is a tracit assumption that in optimising, one is aiming for the most sensitive, reproducible and accurate analytical method. This is not necessarily so, for it may be acceptable to trade off accuracy and precision in order to maximise economy of operation e.g. minimum reagent consumption or speed.

There is every advantage in understanding the system thoroughly so that the various interactions can be recognised and the best conditions be established. This type of problem is commonplace in engineering design and management science, and numerous strategies have been examined. These fall into the following broad categories:

(i) *Simultaneous experimental designs*. Such variables, or factors, as are perceived to be important are determined before — hand and a series of experiments are carried out in which these variables are altered simultaneously according to some fixed procedure. Then by a statistical analysis of results the effects of each variable and interactions between them can be evaluated and the optimum conditions determined.

(ii) *Linear programming*. This is a method which is widely used in engineering and management science, and for which standard computer programs are available [17]. It has not been widely used so far for chemical analysis [18]. The principle of the method is that the relationships between many of the variables in any given problem may be represented by a series of linear equations whose range of applicability are limited by well defined constraints. In the ideal situation these would be a set of simultaneous equations which could be solved to give the optimum set of parameters. In reality there is no exact solution, but by application of linear programming, normally the simplex method, the best compromise can be established. Further, the effect of varying the boundary conditions is easily established and it is also possible to optimise for different objectives e.g. the most sensitive method or the most economic.

To illustrate the method we may consider a standard spectrophotometric method, in which the basic relationships at equilibria between absorbance, A, and reagent concentration, [R], metal ion concentration [M], and pH, are defined as a series of linear equations within definable boundary conditions (Table 1. Eqns. 1–6 and Fig. 6). By inspection the conditions defined by equation 2, would appear to be optimum and these are the ones conventionally recommended. However, if there was a desire to minimise reagent cost, the conditions given in Equation 5 might be preferable provided the metal ion concentration range were acceptable and the pH could be controlled over a narrow range. If time, t, is taken into account there is a direct relationship between sensitivity and reagent concentration, Eqn. 7. At this stage the problem is so trivial that it would be inappropriate to apply formal mathematical procedures to its solution. However, of one then considers other parameters. The procedures to its solution. However, if one then considers other parameters such as the cost of the determination (reagents + instrument, time + labour) or restrictions imposed by the sample, the relationship between the variables is too

Table 1. Basic relationships in a hypothetical complex formation reaction

Equation	Boundary Conditions		
	[R], *M*	[M], *M*	pH
1. $A = k_1 [M] pH + C_1$	10^{-4}	$10^{-6} - 5 \times 10^{-5}$	3–5
2. $A = k_2 [M]$	10^{-4}	$10^{-6} - 5 \times 10^{-5}$	5–8
3. $A = -k_3 [M] pH + C_2$	10^{-4}	$10^{-6} - 5 \times 10^{-5}$	8–9
4. $A = k_1 [M] pH + C_3$	10^{-5}	$10^{-6} - 5 \times 10^{-6}$	1–5
5. $A = k_2 [M]$	10^{-5}	$10^{-6} - 5 \times 10^{-6}$	6.8–7.2
6. $A = k_3 [M] pH + C_4$	10^{-5}	$10^{-6} - 5 \times 10^{-6}$	7.2–9
7. $A = k_6 t[R] + C_5$	10^{-4}	$10^{-6} - 5 \times 10^{-5}$	5–8

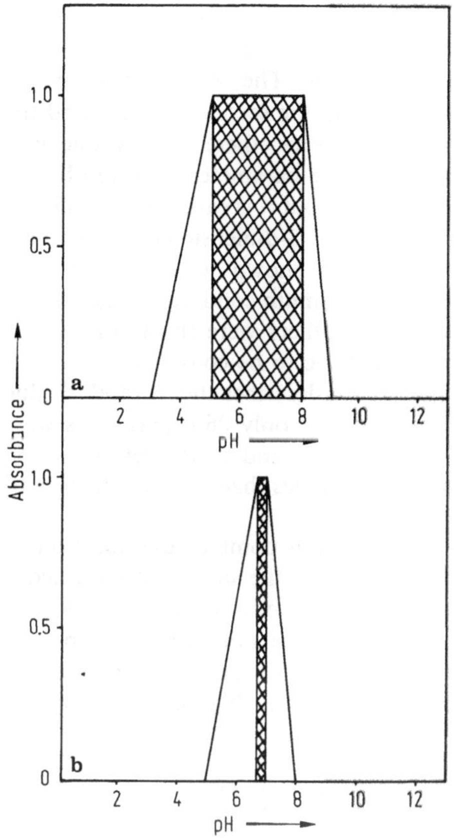

Fig. 6a and b. Classical "optimisation" procedure using data from table 4.1. Note one variable, [M], remains fixed at a predetermined value. **a** [R] = 10^{-4} M; **b** [R] = 10^{-5} M, optimum pH region shaded

complex to be revealed by simple graphical procedures. If this is true for one of the simplest of analytical methods, how much truer it must be for automated systems where in addition to the chemistry, one has to allow for physical factors such as flow and where marginal savings in the cost of an individual analysis may have an impact on the overall budget because so many determinations are performed. The disadvantage of the method is that the system has to be fairly well understood before it can be applied, so it is preeminently suited for managerial assessment of major analytical procedures. A modification of the simplex method applicable to the optimisation of small systems is now described.

(iii) *Sequential methods of experimental design — modified simplex.* This method has some similarities with the classical optimisation procedure, in that the number of experiments required to determine the optimum conditions, is not predetermined. Experiments are performed sequentially until it has been decided that an optimum has been reached, however the procedure is very different and much more efficient.

First, the objective is defined e.g. maximise absorbance, and so are the number of variables, n, to be investigated. Then n + 1 combinations of variables are selected as a trial and n + 1 experiments are carried out. This combination is the first *simplex* which is defined as a geometrical figure with (n + 1) vertices. The results at each point are ranked in order of best to worst. The worst result, w_1, is reflected through the centroid of the simplex, to set up the second simplex, and the procedure is repeated. For the second simplex, however, there is only one new experiment to be performed, that with the values of the parameters as given by the reflected w_1. The procedure is applicable to n variables, but it is easiest to visualise in 2-dimensional space, where the simplex is a triangle, or 3-dimensional space where the simplex is a tetrahedron. The idealised case is shown in Figs. 7 and 8. Obviously, fewer experiments are required to reach optimum conditions, than would be the case if each variable were altered whilst the others were fixed, and this is especially significant if a large number of variables have to be considered.

Morgan and Deeming, in a benchmark paper which describes the method, [19] due in the first instance to Nelder and Mead [20], found that only 26 experiments were required to optimise a complicated spectrophotometric method for the determination of cholesterol in blood serum, for which five variables had to be taken into account.

In reality, an even-paced series of steps from starting point to optimisation, is neither to be expected nor desired, so the extent to which the simplex is modified is governed by a set of rules, which are shown in algorithmic form in Fig. 9, and whose operation is illustrated in Fig. 10. Even these are not sufficient, and the basic procedure has been modified by Denton [21] to give a super modified simplex, in which it (a) is easier to adjust the size of the simplex, to take big steps to begin with

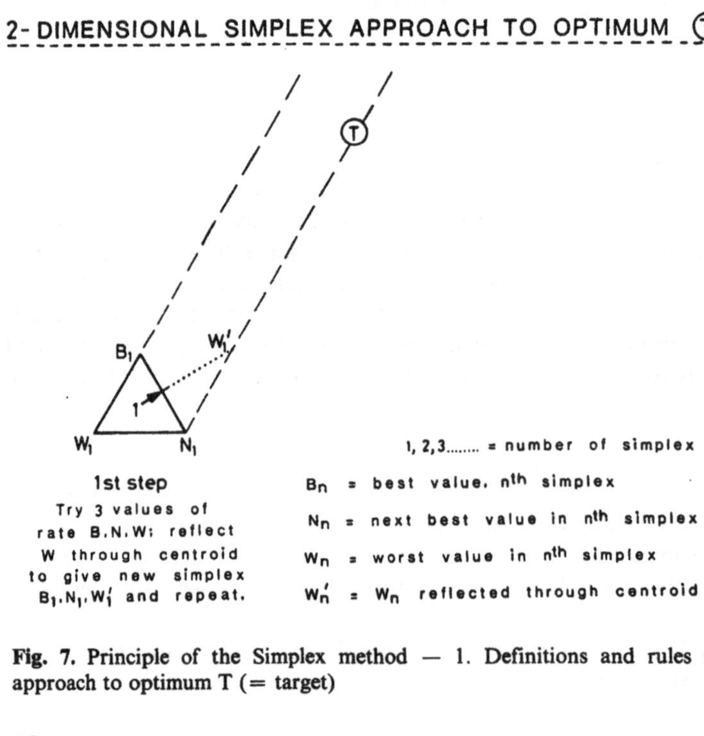

Fig. 7. Principle of the Simplex method — 1. Definitions and rules for a 2 — dimensional approach to optimum T (= target)

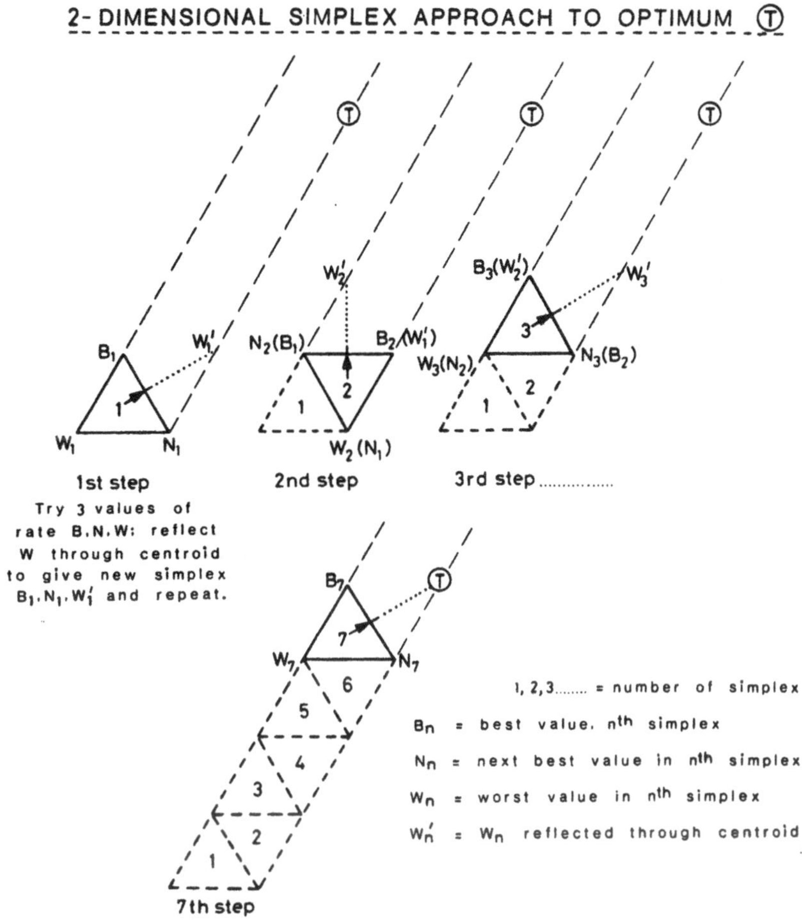

Fig. 8. Principle of the Simplex method — 2. Application of the basic procedure in a 2 — dimensional representation of finding the optimum T. Eight simplices are required

and small ones close to the optimum and (b) takes into account reflections through a boundary condition, which would otherwise demand an impossible experiment. Other variations have also been proposed. All of these are different in concept from the simplex procedure used in linear programming.

One of the attractions of the method is that it is eminently computerisable, so that for an automated system in which the results of the simplex can be processed by a computer, which is then able to reset the apparatus for the next simplex, it is possible for a system to optimise itself. It is also, possible to detect different patterns of interaction, in the mountaineering terminology which pervades the discipline, one may detect "ridges" of optimisation of different "peaks".

These facets are well illustrated by the papers of Denton [21] on atomic emission spectrometry and Steig and Nieman [22] on a chemiluminescence reaction. It was found, in the former, that different reactions were going on at different heights above

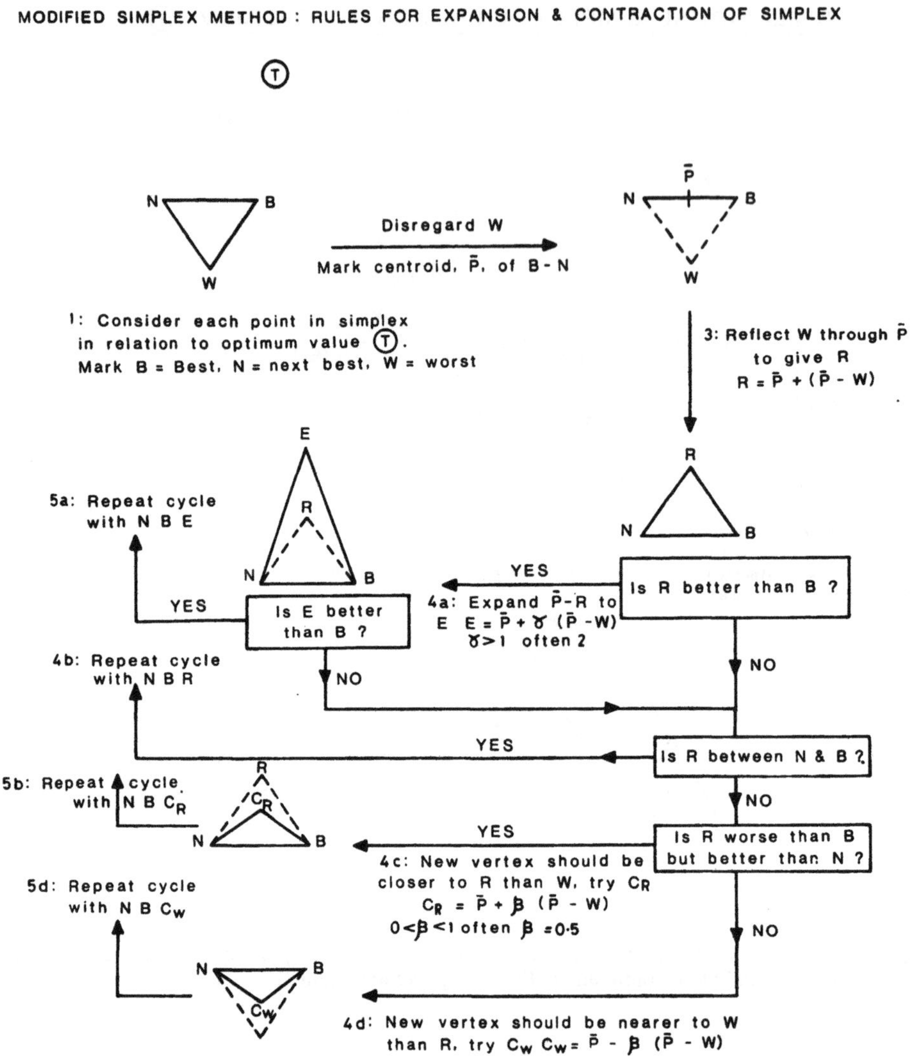

Fig. 9. Principle of the Modified Simplex method — 1. Schematic representation of the rules for expansion and contraction of the simplex

the plasma, and that each could be optimised by well-defined but different experimental conditions. This study rationalised the apparently contradictory observations of earlier groups of workers, who independently, by the classical approach, had each found just one optimum set of conditions. In the investigation of Steig and Neiman, a microcomputer was used to control the optimisation process and again different

---

Fig. 10a and b. Principle of the Modified Simplex method — 2. Steps in the 2 — dimensional ▶ approach to T, showing (**a**) expansion (simplices 2–5) and (**b**) contraction (simplices 5 and 6)

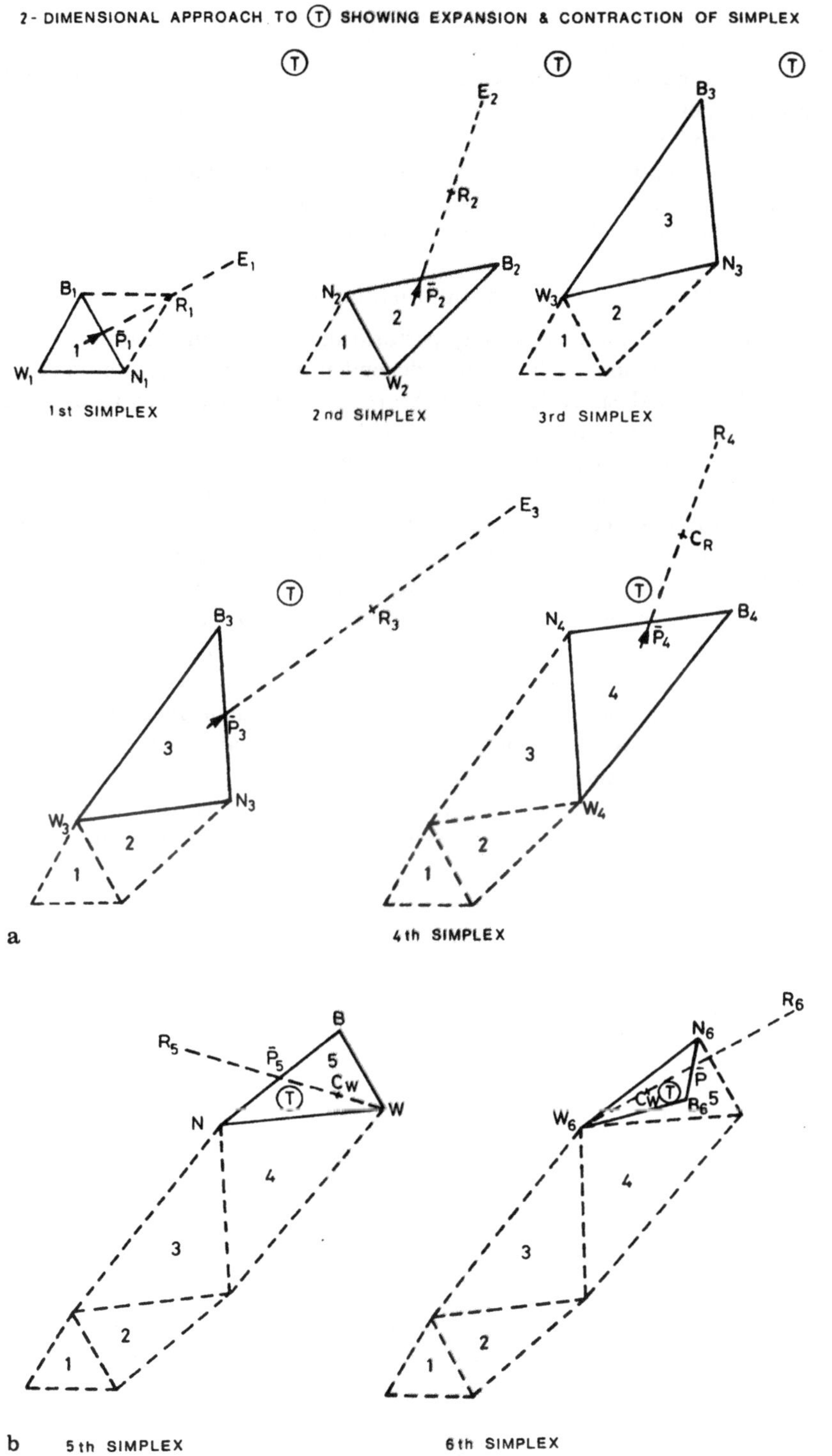
2-DIMENSIONAL APPROACH TO (T) SHOWING EXPANSION & CONTRACTION OF SIMPLEX
1st SIMPLEX
2nd SIMPLEX
3rd SIMPLEX
a
4th SIMPLEX
b
5th SIMPLEX
6th SIMPLEX

optimum conditions were found to pertain for the determination of cobalt(II) and silver(I) with gallic acid and hydrogen peroxide. The optimisation procedure resulted in improved methods for the selective determination of both elements, and reduced the experimental investigation of the search for the optimum condition from about 1 week to 2 hours.

This is a powerful empirical method which is likely to find widespread chemical application, but a fuller understanding of a particular system is likely to accrue from use of all three of the approaches described in this section.

4.2.4 Extraction of Maximum Analytical Information from the Data

Conventionally at present, the analyst may perform many different determinations on one sample, but then use each result independently for the decision-making process, for which the analysis is demanded. More recently, it has been recognised that the results used in totality may yield much more information about the sample. Three such methods will be taken as representative.

Factor analysis. It is a truism that any observation may be a compound of several factors. In analysis one often seeks to hold all but one factor constant and measure the one against some standard. This approach has much to commend it; a result is obtained and the complexities involved in complete understanding of the system are avoided. In other disciplines it has not proven so easy to avoid the complications e.g. in psychology and social science the relative importance of hereditary and environmental characteristics or the relationship between race and intelligence are matters of controversey, and so the field of *factor analysis* has been developed. The objective is to extract from the data available, the number of independent factors present and their relative importance.

A chemical example is the mass spectrum of a mixture of oils. It is possible to construct a matrix which relates relative intensity of a component A at an m/e value n, a_n etc.:

component \ m/e	1	2	3 ... n
A	a_1	a_2	$a_3 \dots a_n$
B	b_1	b_2	$b_3 \dots b_n$
C	c_1	c_2	$c_3 \dots c_n$
.	.	.	. ... .

The observed signal at m/e of n is given by $X_A a_n + X_B b_n + X\,C_n \dots$ where X_A is the fraction of component A in the mixture. By appropriate matrix manipulation, which constitutes the factor analysis, it is possible to calculate X_A, X_B etc. Interestingly, it is also possible to deduce the number of independent factors, so that if an unexpected oil were present in the mixture this would be shown up.

In a practical example adapted from the early work of Gallegos et al. [23)], the various fractions of Kuwait waxy distillate oil are routinely analysed at BP Sunbury Research Laboratories by high resolution mass spectrometry. Drs. Maddens and Meade have written about this work as follows [24]:

"The coupling of on-line computer to the mass spectrometer has permitted the gathering of data on a much larger scale. It is now possible to measure the atomic mass of *all* the ions in a high resolution mass spectrum to within a few ppm. ... The complexity of petroleum products precludes analysis of higher boiling fractions in terms of individual compounds. Type analyses are widely used; the '19 × 19' method [23]. This involves mass measurement and peak intensity measurement of over a hundred peaks. The peaks are summed into groups and multiplied by a 19 × 19 matrix. The results are normalised, averaged and printed. By hand this operation, per spectrum, takes 2–3 hr. By on-line computer 16 spectra can be analysed in less than 2 min."

There are a number of methods closely related to factor analysis and some of these exploit its capability of being able to abstract from the data the number of independent factors which are operating. This is clearly of value for elucidating complex systems, for spectral analysis of mixtures and may be employed for pattern recognition analysis.

The book by Malinowski and Howery [25] concisely describes the method and reviews its chemical applications, that by Cattell [26] is more expansive. Both serve to suggest that the method will be increasingly applied to chemical analysis.

Pattern recognition and cluster analysis. In many current applications of large analyses, the results are viewed on a one-by-one basis. One may routinely measure 8 components of blood sera and then see if any single one is beyond acceptable limits. It may make more sense, however, to view the eight results as one group for it is well established that there are interactions which lead to the likelihood that a diagnosis based on a single limit measurement will be erroneous [27]. What is really needed is a way in which to recognise distinctive patterns within large amounts of data. This is the object of "pattern recognition analysis" which contains within it as a sub-group "cluster analysis" [28–30].

The terminology is a little obscure for the historic reason that much of the early development was carried out as an investigation of machine intelligence. This has given rise to the two major classes of the method being termed "supervised learning" and "unsupervised learning". These are most simply defined by example.

"Supervised learning" techniques are applied when one is seeking to classify a set of data with the aid of previously defined patterns. In analysis one of the most investigated problems of this sort has been the interpretation of mass spectra, both with respect to identifying components of a mixture and for naming an unknown compound on the basis of fragmentation patterns.

"Unsupervised learning" techniques are used when the data is being examined to see if any patterns can be recognised within it. A striking example is provided by Batchelor (Fig. 11), who, by using three parameters from the analysis of urine, showed that useful diagnostic information could be obtained about an individual patient [28]. Having achieved both a separation of data into clusters and an association of a cluster with some medical effect, the next stage is to convert the pattern recognition procedure to "supervised learning" and to set up a procedure whereby the analytical information can directly assist medical diagnosis.

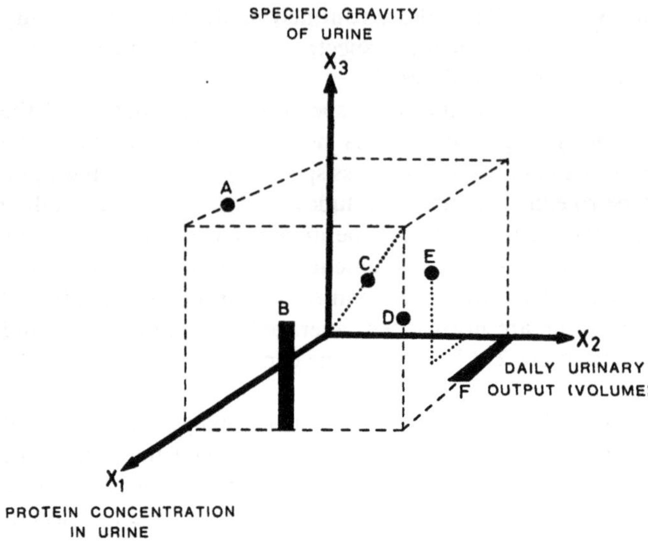

Fig. 11. Cluster analysis used to assist in diagnosis of kidney diseases (adapted from Batchelor [4.18]). (A) acute nephritis, (B) nephrotic syndrome, (C) normal, (D) acute renal infection, (E) essential hypertension, and (F) chronic renal failure

It is easy to be deceived by such a example. There is no guarantee that clusters will be found within a given set of data or that any physical meaning attaches to those that are. There is no fixed procedure which is applicable to all types of data nor one that is completely automatic. What is available are a number of procedures for the classification which at crucial stages involve the experimenter exercising his own judgment. They are a sophisticated aid to the analysis of complex data. One of the methods available, the compound classifier method detailed by Batchelor, will be outlined, as being representative of the style of cluster analysis.

Consider the original points a, b, ... j as shown in Fig. 12.

They are shown in 2 dimensions, but can be in any number of dimensions since each data point can be represented by a vector x_1, x_2, x_3 ... x_n. The data is searched and the two nearest points are found, b and c. These are then eliminated and replaced by one point at their centroid, 1. The next closest pair is a and 1 and these in turn are replaced by their centroid, 2. The procedure continues until all of the data is reduced to one point, in this case, 9. It will be noted that joining the two points 2 and 7 to give centroid 8, and 8 and 5 to give centroid 9 involve appreciably greater distances, than the earlier steps, b–c, h–i etc. All of this can be done quite automatically and rapidly by a computer, but the next stage involves some judgment. The results of the above procedure can be plotted out as a dendogram, which shows the branching points. The experimenter now has to decide which of these are important. In this instance it is likely that points 2, 7 and 5 would be taken as locates, i.e. centre of the cluster for a–c, d–g and h–j respectively. Then the program continues by defining limits to the clusters and establishes intercluster distances and sets up various checks as to the independence of a cluster. The elements of judgment

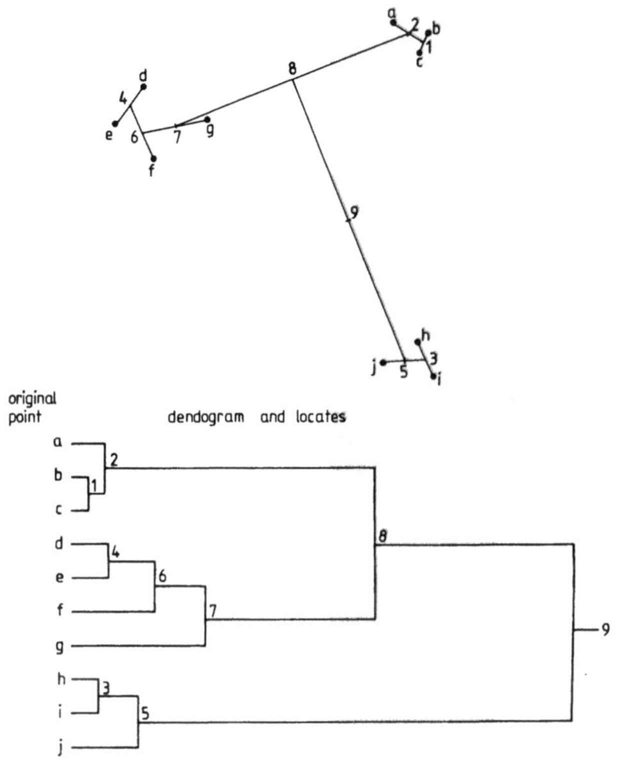

Fig. 12. Schematic procedure for cluster analysis. Original data points are shown as letters a–j; the sequence of joining nearest neighbours and replacing them with the centroid is shown numerically. The dendogram is shown below, the lengths of each branch being approximately proportional to the distance between the points or locates being joined

may be seen in this stylised example. Is point g, really associated with d, e and f or is it best left on its own? Is it better to consider a, b, c, d, e, f and g as one large group rather than two smaller ones? In real samples with a large amount of data it is not always so easy, and there is the fundamental point that any classification is going to contain subgroups which are sometimes best considered separately and at other times are more conveniently placed within the parent group. An example of classification of acoustic emissions is given in Fig. 13.

From the chemists' viewpoint the technique has been pioneered by Kowalski, who has described it very clearly in benchmark papers [30,31] reviewed it [32], demonstrated that it is equally applicable to the classification of the clays used in ancient pottery and to the identification of oils from spillage incidents [33], and has made his set of computer programs ARTHUR available to the scientific community. Many others, have been active within the area and their work has been comprehensively reviewed by Kryger [34] and Varmuza [35] who has also described virtually all of the techniques which have been applied. All of these have important features in common:

(i) n parameters or n dimensional space may be used.

(ii) The parameters may be cardinal or ordinal, i.e. exact measure with units, e.g. nm, s, v or a value judgement which may be placed on a simple scale, eg. red, green, blue being equivalent to 1, 2, 3 or very slow, slow, fairly slow, fast, very fast being equivalent to 1, 2, 3, 4, 5.

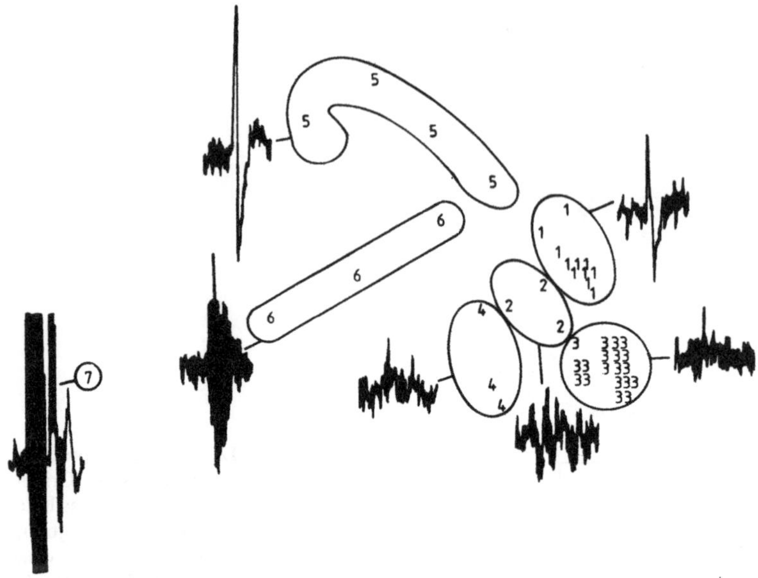

Fig. 13. Cluster analysis of a batch of 64 acoustic emissions obtained from a stressed sample of polypropylene + 40% calcium carbonate. Parameters: maximum amplitude, variance, and median frequency. The bounds of each cluster are indicated in the mapped projection

(iii) A measure of similarity between two points X_i and X_j is given by

$$d_{ij} = \left(\sum_{k=1}^{n} (X_{ik} - X_{jk})^2\right)^{\frac{1}{2}}$$

where the summation is over the measurements, or by

$$S_{ij} = 1 - d_{ij}/MAX(d_{ij})$$

where MAX d_{ij} is the largest inter point distance. For identical points $S_{ij} = 1$ for completely dissimilar points $S_{ij} = 0$.

(iv) Scaling is important because (i) distortions occur if one of the parameters is numerically dominant; one is left with the pattern recognition equivalent of differentiating small differences between large numbers (ii) the scale must match the problem eg. if one were using data, one would use $1700 \pm 100\ cm^{-1}$ to sort out carbonyls from other organics but a much finer scale would be required to group different carbonyls.

(v) The program is developed through interaction with the problem. A program is written and tested on some suitable data. If satisfactory, it is tried out on more test data until the operator is satisfied that the programme can be relied upon to solve the problem i.e. there is no black box solution, the chemist must be in full control of the data analysis.

(vi) It helps to display the results in two dimensions by non-linear mapping. By this means the clusters are often easily recognised, whereas in n-dimensional space they are virtually beyond conception.

Because of the flexibility of the approach, especially the possibility of linking quantitative chemical data with non-chemical data it is a most valuable addition to the armoury of the analytical chemist. By the same token it is a double-edged sword which must be treated with care.

5 New Techniques

5.1 Introduction

Despite the reservations noted above, it would be improper if no comment at all were made on the future of new techniques. In general, the trends of the last few decades suggest that developments in spectroscopy, chromatography and biological methods will continue to take place. Some of these will be in the direction of improving the speed, sensitivity and resolution of existing techniques, others may be the establishment of entirely new methods.

Of the other established methods one can be less certain. Electrochemical methods, other than ion selective electrodes, seem to be practised more in academic laboratories than industrial, and are prone to fundamental problems relating to electrode contamination and chemical interferences. Nuclear methods would seem to have reached their apogee (if one classifies radio immuno assay as biological rather than nuclear) and the same seems to be true of thermal methods. This is not to say that these methods will not continue to be used and to be important. It is a comment that despite the missionary work of numerous adherents of these methods, one notes in the large industrial laboratories much more application of and enthusiasm for spectroscopy, chromatography and biological methods.

It is even more difficult to predict the future of kinetic methods. There is scope here for considerable development and they have many advantages; they can be used for "difficult" mixtures, they are well suited for automation and kinetics is an integral part of every course in chemistry, so its principles are well understood. However, as yet, these methods have not become as widespread as one might expect.[1]

Three new techniques will now be described to give a "flavour" of growth areas. The first, flow injection analysis, demonstrates how chemical methods may be reinvigorated by adoption to continuous flow analysis. The next, thin film analysis, is being hailed as a major breakthrough in clinical analysis. It is an excellent example of the imaginative science and ingenuity which typifies many biological methods and clinical analysis. The last, analysis based on acoustic emissions, is an

1 Six years ago one of us vividly remembers a very distinguished analytical chemist predicting, during the course of a brilliant lecture on novel kinetic methods, that within 4–5 years a kinetic method would be first choice for the majority of analytical determinations.

example of the applications of a novel physical method in which acoustic energy is exploited for analytical purposes; one which may or may not prove of value in the long run.

5.2 Flow Injection Analysis

Skeggs in 1957 made a breakthrough in automated chemical analysis, by demonstrating how chemical procedures could be carried out on a continuous flow basis. His ideas were exploited by Technicon who developed the Autoanalyser, which in one form or another has become the main stay of clinical analysis. The key step as Skeggs saw it was to introduce bubbles of air into the sample and reagent streams so that each sample-reagent mixture was separated from the preceding and following by a barrier of air. This procedure, although extremely effective, does require the introduction of air, its removal before measurement and a careful proportioning of the streams of reagents and sample.

Several workers [36] have questioned the importance of air-segmentation, which was Skeggs' breakthrough and which has been held as an item of faith over 20 years of continuous flow analysis (CFA). They have all shown that satisfactory analyses without carryover of sample are feasible without air segmentation, but there is a split of opinion as to whether a mixing chamber is desirable or not. Pungor's group in Hungary have favoured a mixing chamber on the grounds that it results in good precision and is especially suitable for the electrochemical sensors which they prefer [37]. The thinking behind this approach is in the mainstream of CFA, in that it seeks to put a standard chemical procedure onto a continuous flow basis. The more radical approach is that which dispenses with the mixing chamber, and it is this which is described briefly below. K. K. Stewart in the US and Ruzicka and Hansen in Denmark have been the pioneers, the latter being responsible for coining the term Flow Injection Analysis, FIA, and for many innovations in the field [38,39].

A diagram of one of the simplest form of FIA is shown in Fig. 14a. It consists simply of a carrier stream of reagent propelled by either a pump or a constant head device, and injection system, a detector and a recorder. Typically the carrier stream is a solution of organic reagent, buffered to a suitable pH and containing any appropriate masking agents and the sample is a solution containing a metal ion which is to be determined. The sample is injected directly into the carrier stream. The analysis time is ca 15 s, the sample throughout may be up to 200 h^{-1}, the sample volume is 5–300 μl and the precision is 1–2 per cent. For typical results see Fig. 14b.

On the practical side, non-segmented systems have much more tolerance with respect to pumping and sample injection than the air-segmented system. This enables one to create a working analytical system from parts available in most laboratories. The exercise makes for a good Friday afternoon experiment or student project. Not surprisingly there are almost as many systems as there are workers, as is evident from the papers presented at a conference in Amsterdam, published as a special issue of *Anal. Chim. Acta* [40]. Systems developed from the cheap modular system of Ruzicka and Hansen and from the HPLC based system of Stewart, are available commercially. The choice between DIY and a commercial model is straightforward

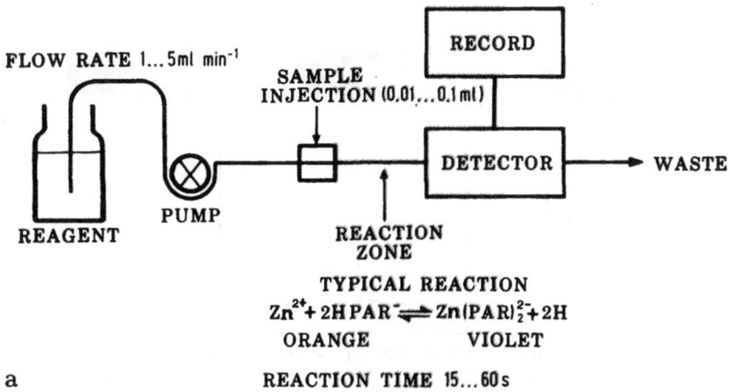

a

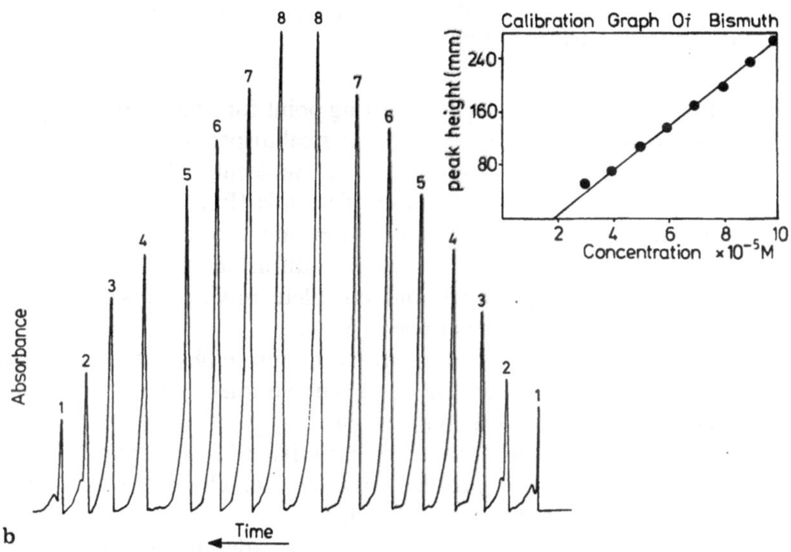

b

Fig. 14a. Schematic arrangement for Flow Injection Analysis for a simple spectrophotometric determination. **b** A typical experimental output from such a system, showing lack of carry over even with large samples. (Metal ion: Bismuth; Reagent: Pyrocatechol violet, 5×10^{-3} M; pH: 2–4; Sample size: 200 μl; Analysis rate: 80 hr^{-1})

and depends on the individual's needs and resources. However, it is worth noting that the rudimentary 'LEGO' system of Ruzicka and Hansen is surprisingly robust; as developed by Bergamin's group in Brazil it is used routinely for the 1000 determinations per day of diverse samples of soils and waters from Amazonia.

The flexibility of the apparatus stems directly from the avoidance of air segmentation, and for systems without mixing chambers. There are, in addition to the mechanical benefits, some interesting chemical possibilities.

The physical dispersion of a plug of sample in a liquid flowing through a pipe was thoroughly discussed by Taylor in 1953 [41]. His theory, although modified in some

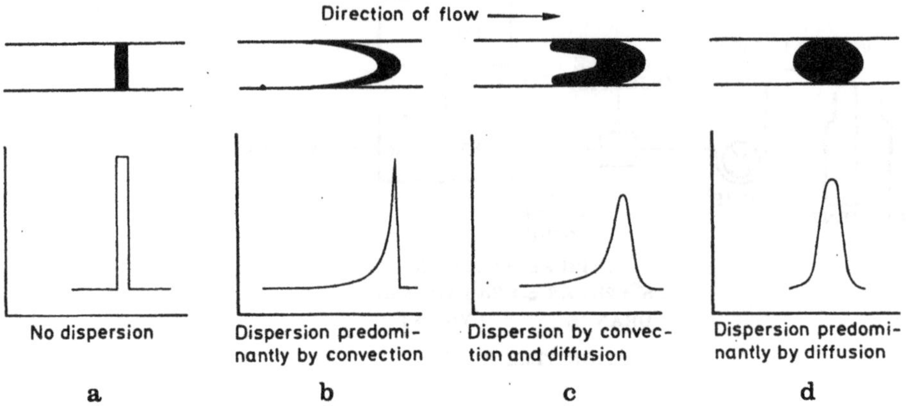

Fig. 15a–d. Concentration profiles of samples monitored downstream from the point of injection and their relationship to the mode of mixing (schematic)

detail by subsequent workers, provides a good starting point for understanding FIA. In essence, it is that there are two major modes of physical dispersion (i) convection, which gives rise to the parabolic plug and (ii) radial molecular diffusion, which results in a much more symmetrical distribution of sample (Fig. 15). The relative importance of these modes of dispersion depends on flow rate, tube length and radius. Under Taylorian conditions, it is radial molecular diffusion that is the most important, and which brings about rapid and complete mixing of sample and carrier. Thus, there are three basic options open in FIA:

(i) To use short tubes and relatively high flow rate to minimise physical dispersion. This is useful if the FIA system is to be used as a sample inlet system for atomic absorption spectroscopy, pH or pM measurements etc.

(ii) To achieve a medium degree of dispersion so that mixing of carrier and sample plug is complete but the concentration profile is asymmetric. This is a widely used option being suitable for spectrophotometric procedures.

(iii) To use relatively long tubes and slow flow rate in order to achieve a wide dispersion of sample and near Gaussian concentration profiles. Under these conditions, the mixing of carrier and sample is largely by radial molecular diffusion. At any point in time the concentration of sample in a given section of the tube is given, according to Taylor, by the normal error curve, and the concentration of carrier in the same section is obtained by difference. Thus, if for example, the carrier stream contains a weak base and the sample a strong acid, a definable and reproducible pH gradient exists across the sample bolus, as illustrated by the titration curve of phosphoric acid (Fig. 16). If the carrier also contains a reagent which forms coloured complexes with a number of metal ions, and if the sample contains a mixture of these ions, the absorbance vs time curve reflects the sequential, pH dependant reactions. Consequently, mixtures of metal ions may be determined by one injection of sample [42,43]. Of course, there is a time dependance as well, and the sample bolus moving down the tube resembles a mobile chemical reactor in which parameters such as pH and concentration of reactants vary with time in a reproducible way. Given the concurrent advances which are being made in image sensing

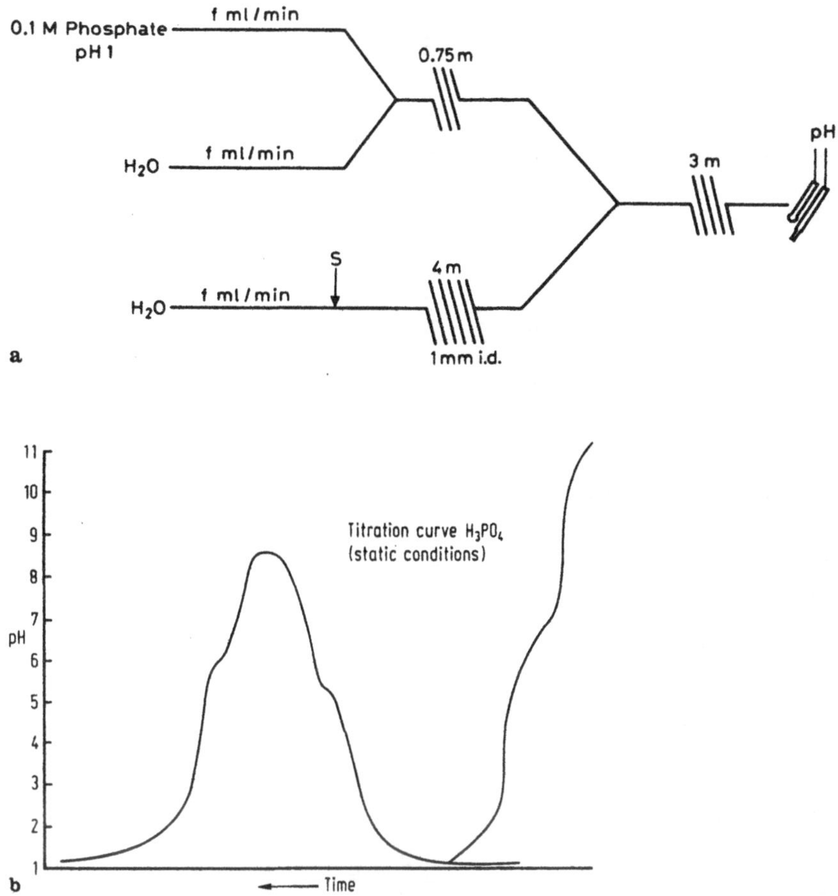

Fig. 16a. Experimental arrangement to establish a pH gradient by FIA with phosphate buffer. Sodium hydroxide is injected at S, disperses as shown in Fig. 15, and then interacts with phosphate at point of confluence; **b** Reaction between phosphoric acid and sodium hydroxide superimposed on the sodium hydroxide peak, i.e. the 'dynamic' titration curve of phosphoric acid, which may be compared with that obtained under static conditions. (Flowrate = 1.55 ml · min^{-1}, sample size = 0.5 ml and [NaOH] = 0.75 M; different pH gradients may be obtained by varying these conditions)

detectors and data handling, there is great potential for analysis of mixtures and the study of chemical reactions.

The analytical applications of FIA are too numerous to detail, but Table 2 indicates the scope of the method. It is one which is proving of great interest to analysts, but it has much to offer physical chemists studying dispersion phenomena and solution kinetics as well. Both air segmented and non-segmented continuous flow analysis systems are capable of providing high sample throughput and analytical reliability. The selection of one system as opposed to another may be made sometimes on purely chemical grounds, but there is the interesting possibility that the introduction

Table 2. Examples of application of flow injection analysis[a]

Application	Procedure	Comment
Determination of trace elements or anions.	(i) Direct injection of sample into carrier stream of reagent.	Effectively modification of Autoanalyser procedures, sample throughput 70–200 h^{-1} depending on the chemistry used. Has been applied to diverse samples.
	(ii) Merging zones: sample and reagent are injected simultaneously but in separate carrier streams of H_2O. They merge and reaction product is measured.	Economical of reagent.
Inlet system for pH meters, aas etc.	Use short length of tubing to minimise dispersion.	Convenient inlet system, but needs calibration.
Kinetic determination of mixtures of metal ions or enzymes etc.	(i) Use two detectors. (ii) Sense peak, stop the flow and measure the change of signal with time.	A variation on stop flow, suitable for reaction times of 0.2–5 min.
Determination of mixtures of metal ions or organic compounds by exploiting pH gradients.	Use a reagent which forms similarly coloured complexes with a number of metal ions. Establish a pH gradient over the sample and monitor the absorbance.	So far only simple mixtures have been analysed, but there is potential for a more complex system. Kinetic factors may be exploited.
Measurement of viscosity and diffusion coefficient.	(i) Measure the time taken for the sample to travel from point of injection to the detector under constant pressure.	(i) The system functions as a modified Ostwald viscometer.
	(ii) Measure the spread of the peak under Taylorian conditions.	(ii) Taylor's theory is assumed to be applicable.

[a] For details see Refs. [38–40] and bibliographies supplied by the manufacturers BIFOK and Fiatron.

of non-segmented systems will affect the strategy of CFA. The approach of Technicon to dealing with the problem of determining several substances in one sample is to use a computerised system, SMAC. It permits 40 simultaneous assays at a rate of 200 samples per hr, but costs ca £ 200,000. It exploits the fact that with air segmented systems, long reaction times are permissible. This approach for non-segmented systems, is not very attractive — the very speed of analysis makes it very difficult to proportion a sample into many parts and to perform parallel determinations.

On the other hand, the apparatus is so cheap and simple to use that it may be feasible to run several independent systems in parallel. The obvious analogy is between a hierarchical computer system linked to a mainframe and a series of dedicated microcomputers. Also the cheapness, speed and simplicity of non-segmented systems make them attractive for a small number of determinations, less than would be considered viable with Autoanalysers.

5.3 Thin Film Analysis

This is becoming known by the above name but the name used in the seminal papers of Curme et al. and Spayd et al. of Eastman Kodak Company is "multi layer film elements for clinical analysis" [44,45]. It is a technique which unites two aspects of clinical analysis. One is the procedure for carrying out quantitative tests on fibre strips. These strips may be dipped into a sample of urine and instantly reveal the concentration of up to 8 basic components; it is an elegant variation of indicator paper. The other aspect is the increasing use being made of biological methods, particularly enzyme methods for clinical analysis.

The achievement at Kodak is to design films, in which a transparent base is successively covered with layers of dry reactants and to top it with a layer onto which the sample is spotted. This last is known as the spreading layer. This is shown schematically in Figure 17. For the determination of glucose the chemistry layer consists of glucose oxidase, 1,7-dihydroxynaphthalene and peroxidase bound with a gelatine binder. This is for the reaction sequence, which the formation of a coloured dye (λ_{max}, 490 nm) being formed in proportion to the concentration of added glucose.

Approximately 10 µl of sample is required and after development, the density of the colour is measured via the transparent bottom layer by reflectance spectrophotometry. There is a relatively complex relationship between the signal and concentration, but with suitable transformations a perfectly acceptable calibration curve is obtained.

The spotting may be done automatically, and the whole placed on a type of conveyor belt operation, so that it is possible to complete 3 analyses per min with a 7-min reaction time.

The technical difficulties of making films which, both work and stay in place are immense. But now the breakthrough has happened, many other and more complex chemistries have been fixed on film. Film is available for urea, triglycerides and anylase in blood sera, and even ion selective electrodes have been adapted for the detection of metal ions such as Na^+, K^+, etc. in a single step.

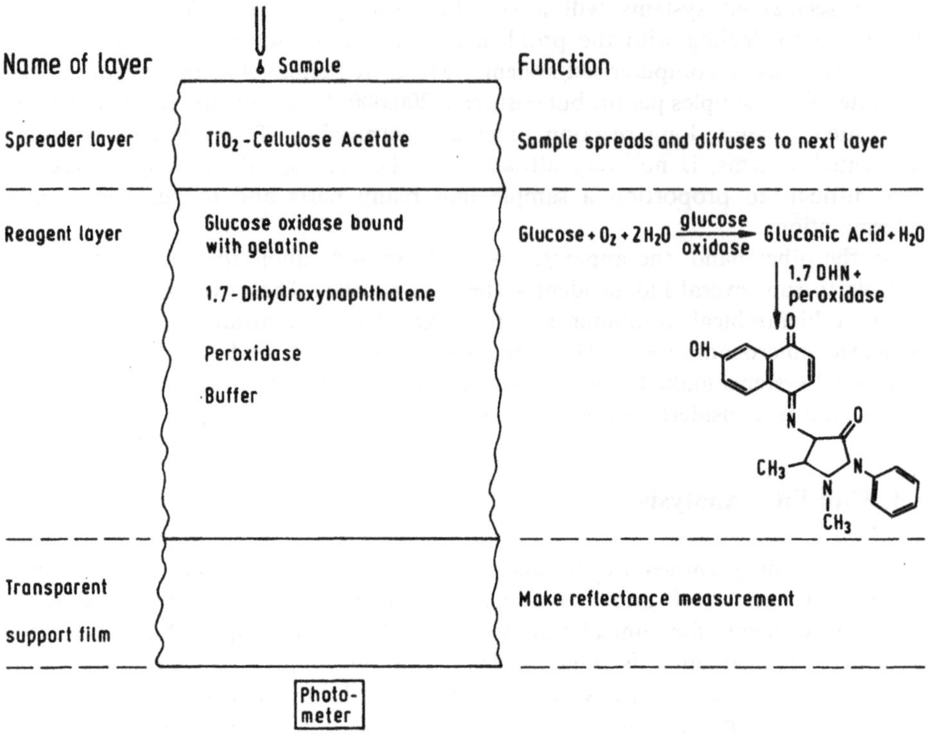

Fig. 17. Schematic arrangement for the 'thin film' analysis of glucose

By virtue of its simplicity of operation and great chemical and biochemical adaptability it is coming into widespread use.

5.4 Acoustic Emissions

Methods of analysis based on sound were investigated in some detail during the 19th century, but the primitive forms of detection then available made them impracticable for routine analysis. Now, with the dramatic improvement in the detection, recording and analysis of acoustic signals, a reevaluation of these methods is taking place. There has been a revival of the photoacoustic method of detecting the absorption of light and extensive use has been made of ultrasonics for the examination of materials. One acoustic method which has recently been examined is that in which the sound emitted by materials under stress or from chemical reactions is used for analytical purposes [46,47]. There have been some traditional examples of the technique; there is a 14th Century drawing of an analyst listening to a lump of sulphur to check its purity, and "singing tin" is similarly used. Nevertheless, it is only in the last decade that the phenomenon has received serious attention.

At first the measurement process amounts to counting the number of acoustic emissions given out as a function of time. A microphone is placed on the sample and the signals it detects are passed through some amplification and recording system set so that the arrival of signals greater than a preset threshold value are recorded. It was found that for solids the acoustic emission pattern reflects the physical and mechanical condition of the sample. The details of the origin of acoustic emissions is still being investigated, but there is no doubt that they are associated with microcrack formation. In solids the onset of this form of mechanical deterioration is more readily detected by the acoustic signal than by microscopic examination. Thus metal fatigue may be recognised by changes in the acoustic trail before there is serious deterioration in mechanical performance. This is obviously an important piece of analytical information, and it is becoming quite common for structures under stress, such as large chemical reactors, to have microphones positioned around them so that the onset of mechanical failure may be detected in time for remedial action to be taken.

It has now been shown that sound is given out during the course of many chemical reactions, and that the acoustic energy released is possibly a measure of the extent of the reaction. If this link is proved, then acoustic monitoring of chemical reactions may well become widespread since the detector (the microphone) is attached to the outside of the vessel in which the reaction takes place and sound is generated by reactions which are difficult to monitor by conventional means, such as gel formation.

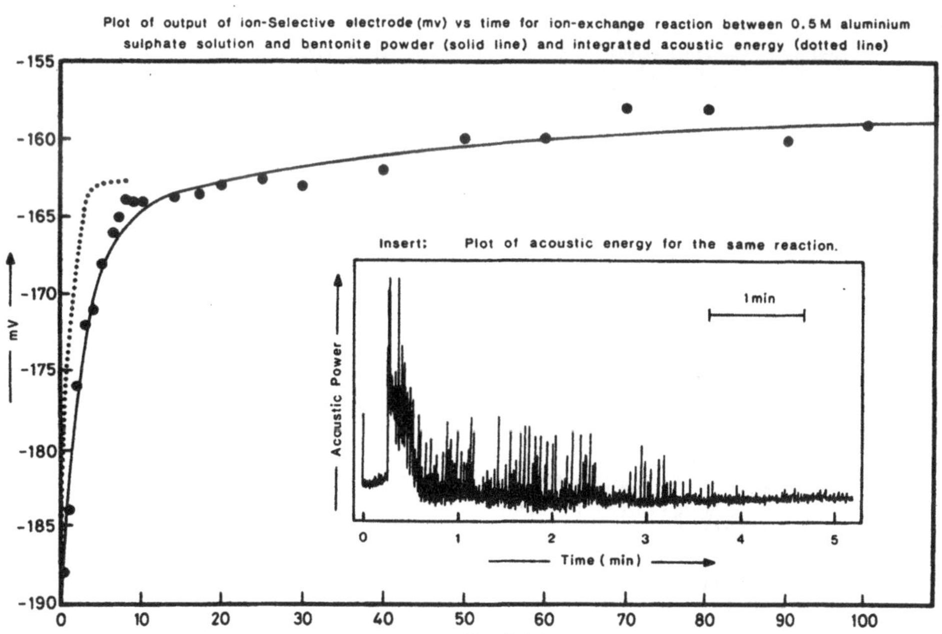

Fig. 18. An ion exchange reaction followed acoustically via a microphone attached to the outside of the column and potentiometrically via an ion-selective electrode in the effluent (solid line — electrode response; dotted line — integrated acoustic energy). Inset: The raw plot of acoustic power versus time

There is also the possibility of following reactions in vivo, something of obvious interest to clinicians.

In addition to detecting the rate of acoustic events, it is possible to measure acoustic power as a function of time or to detect individual events. The time scale for the former is of the order of minutes, sometimes hours, that for the latter is micro seconds. Examples of both are shown in Figures 18 and 19.

There is still much that remains to be understood about this phenomenon. Pattern recognition analysis of the individual systems indicates that the signals fall into well defined clusters and that these signals are characteristic for a given system (Fig. 13). There are various models for associating acoustic emission from solids with crack propogation. Still, it may turn out in the long run that the data becomes intractable, or uninterpretable, but even so the present studies serve to illustrate that there is still much to learn and exploit.

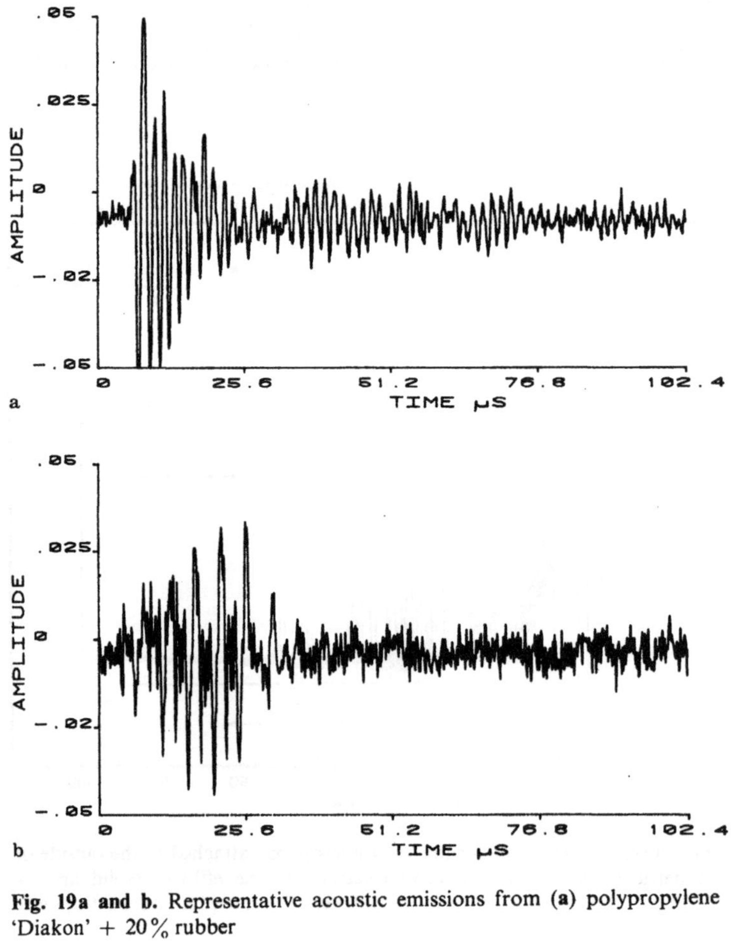

Fig. 19a and b. Representative acoustic emissions from (**a**) polypropylene + 30% glass and (**b**) 'Diakon' + 20% rubber

6 Total Systems Automation of a Major Analytical Project

6.1 Introduction

The final section of this chapter deals with a description of total systems automation in the analysis of tobacco smoke. This is used as an example partly because it is one of the few instances of large-scale laboratory automation of which full details have been published, but more importantly because it provides an example of both the benefits and the pitfalls which can be encountered when all the areas discussed previously are brought together for the solution of a specific problem. Here we give only a brief outline of the significant points; a full and detailed discussion of the project can be found in Ref. [48].

The laboratory of the Government Chemist in London undertakes the analysis of tobacco smokes in order to produce 'league tables' (Fig. 20) of the tar and nicotine content of all cigarette brands available in the UK, on behalf of the British government. In this context 'tar' is defined as the total particulate matter present in the main-stream smoke (inhaled smoke), adjusted for its content of water and nicotine alkaloids. 'Nicotine' is defined as the total nicotine-type alkaloids present in the mainstream smoke; both are expressed in mg per cigarette. The tar and nicotine 'league tables' are published every six months, and include details of approximately 130 brands of cigarette.

6.2 Sampling Procedures

The extremely large numbers of cigarettes involved in the production of a table (approximately 20,000) require a rigorously controlled sampling procedure and standardised methods of analysis to be applied if the results are to show any statistically significant difference between brands, as well as a true average yield for each brand over the sampling period. Variations may occur due to the types of tobacco used, changes in the types of filters, etc., used by the manufacturer, the age of the cigarettes at the time of analysis, and several other factors.

At the start of the project in 1972, samples were purchased from retail shops, as it was thought this would give results which most closely reflected the tar and nicotine yields of cigarettes actually being consumed. Accordingly, one packet of each brand was bought from a large retailer in each of five regions for each of the six months of the survey, giving a total of thirty packets of each brand. It was found, however, that the samples obtained in this manner spanned a very wide age range, as a result of poor stock rotation, low demand for certain brands in particular areas, and patchy distribution. Since most cigarettes are smoked within a few months of manufacture, it was obviously unrepresentative to include in the survey results from samples which were several years old; sampling at retail outlets was therefore discontinued in favour of obtaining samples directly from the manufacturers, and for subsequent surveys five packets of each brand were drawn directly from the factory warehouse each month. This ensured that the published tar and nicotine figures fairly represent the content of the cigarettes being sold at the time of publication of the table.

TAR & NICOTINE YIELDS OF CIGARETTES

As determined by the Government Chemist from samples obtained during the period November 1979 to April 1980

As 'tar' is regarded as a greater danger to health than nicotine, the brands are listed in 'tar' yield order. Brands with the same figure for 'tar' yield are listed in alphabetical order. The 'tar' groups are LOW TAR (0-10 mg/cig), LOW TO MIDDLE TAR (11-16 mg/cig), MIDDLE TAR (17-22 mg/cig), MIDDLE TO HIGH TAR (23-28 mg/cig), and HIGH TAR (29 and over mg/cig).

'Tar' yield groups shown on cigarette packets and in brand advertisements may be for current production and could differ from those given in this table.

Differences between brands of up to 2 mg of 'tar' can generally be ignored.

(P) indicates plain cigarettes. All other brands have filters.

LOW TAR

Tar Yield mg/cig	Brand	Nicotine Yield mg/cig
Under 4	Embassy Ultra Mild	Under 0·3
Under 4	John Player King Size Ultra Mild	Under 0·3
Under 4	Silk Cut Ultra Mild with Substitute	Under 0·3
6	Silk Cut Extra Mild	0·7
7	Silk Cut King Size with Substitute	0·7
8	Embassy Premier King Size	0·6
8	John Player King Size with NSM	0·7
8	Piccadilly Mild	0·5
9	Consulate Menthol	0·6
9	Craven A King Size Special Mild	0·7
9	Dunhill International Superior Mild	0·8
9	Dunhill King Size Superior Mild	0·7
9	Embassy No. 5 Extra Mild	0·7
9	John Player King Size Extra Mild	0·7
9	Lambert & Butler King Size Mild	0·7
9	Peer Special Extra Mild King Size (with Cytrel)	0·7
9	Silk Cut	0·8
9	Silk Cut International	0·9
9	Silk Cut No. 5	0·7
10	Belair Menthol Kings	0·6
10	Consulate No. 2	0·6
10	Embassy Extra Mild	0·8
10	Embassy No. 1 Extra Mild	0·8
10	Peter Stuyvesant Extra Mild King Size	0·8
10	Player's No. 10 with NSM	0·7
10	Player's No. 6 Extra Mild	0·7
10	Silk Cut King Size	0·9
10	Silk Cut No. 3	0·8

LOW TO MIDDLE TAR

Tar Yield mg/cig	Brand	Nicotine Yield mg/cig
11	Gauloises Longues	0·6
12	Gauloises Filter Mild	0·6
12	Gitanes International	0·9
12	Rothmans King Size Mild	0·9
12	St. Moritz	0·9
13	Gitanes Caporal Filter	0·7
13	John Player Carlton King Size	1·5
13	John Player Carlton Long Size	1·4
13	John Player Carlton Premium	1·2
13	Player's No. 10 Extra Mild	0·8
14	Black Cat No. 9	1·0
14	Dunhill International Menthol	1·0
14	Gauloises Disque Bleu	0·6
14	Kensitas Mild King Size	1·4
14	Kent	1·0
14	Lark Filter Tip	1·1
14	Peer Special Mild King Size (with Cytrel)	1·2
14	Peter Stuyvesant King Size	1·2
14	Piccadilly No. 7	1·1
15	Benson & Hedges Sovereign Mild	1·2
15	Chesterfield King Size Filter	1·1
15	Gauloises Caporal Filter	0·7
15	Kensitas Club Mild King Size	1·4
15	Kensitas Corsair Mild	1·1
15	L & M Filter	1·1
15	Merit	1·2
15	More	1·5
15	More Menthol	1·5
15	Pall Mall Filter	1·4
15	Three Fives Medium Mild King Size	1·3
15	Three Castles Filter	1·0
16	Cadets	1·2
16	Camel Filter Tip	1·1
16	Carroll's No. 1	1·2
16	Carroll's No. 1 King Size	1·3
16	Dunhill International	1·4
16	Dunhill King Size	1·3
16	Guards	1·1
16	Kensitas Club Mild	1·3
16	Marlboro	1·3
16	Phillip Morris International	1·4
16	Piccadilly Filter De Luxe	1·4
16	Piccadilly King Size	1·3
16	Rothmans International	1·5

MIDDLE TAR

Tar Yield mg/cig	Brand	Nicotine Yield mg/cig
17	Craven 'A' Cork Tipped (P)	1·3
17	Craven 'A' King Size	1·3
17	Kent De Luxe Length	1·2
17	Lucky Strike King Size Filter	1·4
17	MS Filter	1·3
17	Rothmans Royals	1·5
17	Silva Thins	1·5
17	Sobranie Virginia International	1·5
18	Benson & Hedges Gold Bond	1·5
18	Benson & Hedges Gold Bond King Size	1·5
18	Benson & Hedges King Size	1·6
18	Benson & Hedges Supreme	1·7
18	Du Maurier	1·5
18	Embassy Envoy	1·2
18	Embassy Filter	1·4
18	Embassy Gold	1·3
18	Embassy King Size	1·4
18	Embassy No. 3 Standard Size	1·4
18	Embassy Regal	1·4
18	Emperor King Size	1·4
18	Fribourg & Treyer No. 1 Filter De Luxe	1·5
18	Imperial International	1·6
18	John Player Special	1·5
18	Kensitás Club King Size	1·5
18	Kensitas Corsair	1·3
18	Kensitas Tipped King Size	1·5
18	Piccadilly No. 1 (P)	1·4
18	Player's Medium Navy Cut (P)	1·4
18	Rothmans King Size	1·4
18	Slim Kings	1·4
18	Three Fives Filter Kings	1·5
18	Weights Plain (P)	1·2
18	Winston King Size	1·4
18	Woodbine Plain (P)	1·5
19	Benson & Hedges Sovereign	1·4
19	Capstan Medium (P)	1·3
19	Embassy No. 1 King Size	1·4
19	Embassy Plain	1·3
19	Gallaher's De Luxe Green (P)	1·5
19	Gold Flake (P)	1·3
19	John Player King Size	1·5
19	Kensitas Club	1·5
19	Lambert & Butler International Size	1·5
19	Lambert & Butler King Size	1·5
19	Major Extra Size	1·3
19	Park Drive Plain (P)	1·5
19	Park Drive Tipped	1·5
19	Player's Filter Virginia	1·4
19	Player's Gold Leaf	1·5
19	Player's No. 6 Filter	1·3
19	Player's No. 6 King Size	1·6
19	Player's No. 6 Plain (P)	1·2
19	Player's No. 10	1·3
19	Regal King Size	1·4
19	Senior Service Plain (P)	1·6
19	Sterling	1·6
19	Woodbine Filter	1·3
20	Kensitas Plain (P)	1·6
20	Nerit	1·5
20	Three Fives International	1·7
20	Three Fives Selected Virginia (P)	1·4
22	Sweet Afton Bank Size Plain (P)	1·9

MIDDLE TO HIGH TAR

Tar Yield mg/cig	Brand	Nicotine Yield mg/cig
24	Lucky Strike Plain (P)	1·8
25	Gallaher's De Luxe Blue (P)	2·1
25	Gauloises Caporal Plain (P)	1·2
25	Gitanes Caporal Plain (P)	1·4
26	Capstan Full Stength (P)	2·7
27	Pall Mall King Size (P)	2·2

New Brands introduced during Nov. 79 to Apr. 80 not yet analysed by the Government Chemist for a period of 6 months.

Tar Yield mg/cig	Brand	Nicotine Yield mg/cig
14	Du Maurier King Size	1·2
15	Benson & Hedges Sovereign King Size	1·3
18	Benson & Hedges Academy	1·8
19	John Player Special King Size	1·4

6.3 Analytical Procedures

The first stage of the analysis involves smoking the cigarettes. It was obviously important that this should be done in a standardised and controlled fashion, in order to produce reproducible results, whilst at the same time mimicking as closely as possible the way in which cigarettes are smoked by humans, i.e. short 'puffs' interspersed by much longer periods of smouldering. This is achieved by the use of a commercially produced 20 channel smoking machine which smokes 20 samples simultaneously and which is capable of reproducibly maintaining the volume, frequency, and duration of puffs, as well as the point at which smoking is terminated (the butt). The number of puffs required to smoke each cigarette is also recorded for each channel and sent to the central computer. Five cigarettes of each type are smoked consecutively, and smoke from the cigarettes is drawn through a filter holder containing a glass fibre filter disc on which all particulate matter above 0.3 μm in diameter is trapped. The filter disc is then weighed using an electronic balance interfaced to a central computer, the difference in weight of the filter before and after smoking giving the total particulate matter (TPM) present in the smoke of the five cigarettes. The carbon monoxide present in the smoke is also determined using

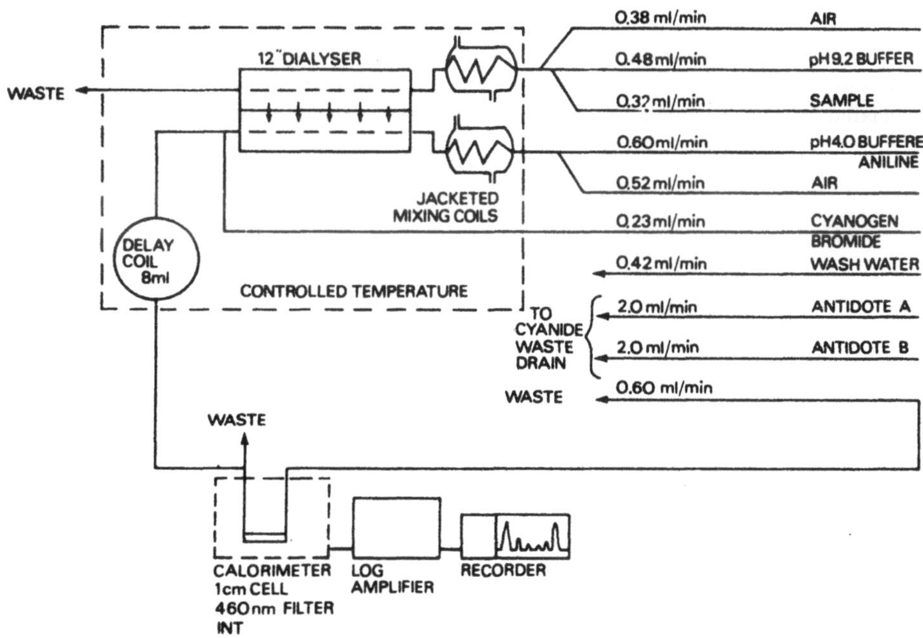

Fig. 21. Schematic diagram of the autoanalyser method used for determining nicotine alkaloids. (Reproduced by courtesy of Dr. P. B. Stockwell)

◀ **Fig. 20.** 'League table' published by the British povernment's health departments, showing tar and nicotine yields for cigarettes. (Reproduced with acknowledgement to the copyright owner)

a selective non-dispersive infrared (NDIR) detector, although at present these results are not published in the tables.

The filter disc is transferred to a stopperred flask containing ethanol and propan-2-ol, and shaken to extract the water present which is then determined by gas-liquid chromatography; the quantitity of water present is calculated from the ratio of the areas of the peaks for water (unknown) and ethanol (internal standard). The alkaloids are extracted from the filter disc using sulphuric acid and determined by a specially developed autoanalyser procedure based on the Koenig reaction; this is shown schematically in Fig. 21. Since the procedure was developed specially for this application, results obtained from it were compared closely with those produced by the traditional manual method, a steam distillation technique, before it was adopted as a standard method.

6.4 Role of the Computer

At the centre of the system is a Rank Xerox RX 530 computer, which is used not only to acquire and process data, but also to generate a statistically designed testing sequence in the form of randomised smoking plans covering the entire period of a survey, to minimise the incidence of systematic errors. The analyst allocates a code to each packet of cigarettes according to manufacturer, brand, type of cigarette (plain, tipped, king size, etc.), and the factory of origin, as well as various other information. This code is used in the generation of the smoking plans and is associated with a particular computer file access code, allocated by the computer

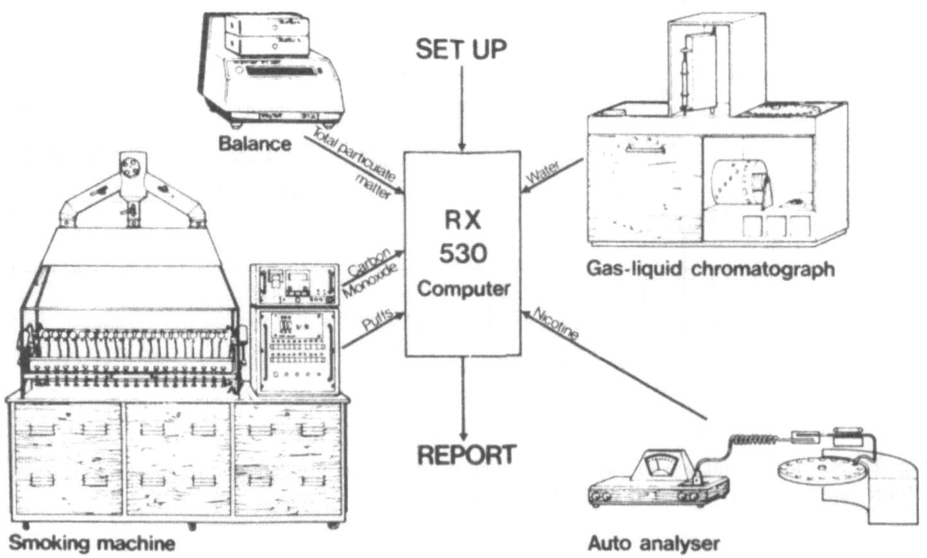

Fig. 22. Relationship between the major parts of the total automation system. (Reproduced by courtesy of Dr. P. B. Stockwell)

from a pool, to 'tag' the results from each stage of the analysis and direct them to the appropriate file.

All the instruments are connected on-line to the computer, as shown in Fig. 22. The electronic balance and the puff counter send a direct digital output to the computer, whilst for the other instruments signals are sent to the computer in analogue form and subsequently digitised. Details of the samples and standards being analysed can be entered on a data terminal adjacent to each instrument, which may also be used to view or manipulate data held on the computer files. In the case of instruments such as the gas chromatograph and the autoanalyser, the data capture procedures are synchronised to the injection of each sample by means of a simple control interface known as a 'commbox', developed specifically for this application. This unit also gives the analyst an audible warning if the computer fails to acknowledge receipt of data.

In addition to the functions outlined above, the computer is also used to generate requests for new samples, validate results, perform statistical analyses of experimental data, and finally produce a report sheet setting out all the relevant information in tabular form. The fact that the computer is required to handle such a diversity of different tasks, many of which must be performed simultaneously, necessitates the use of sophisticated software. The relationship between the analytical procedures, the computer files, and the control programme BRANDER is shown in Fig. 23.

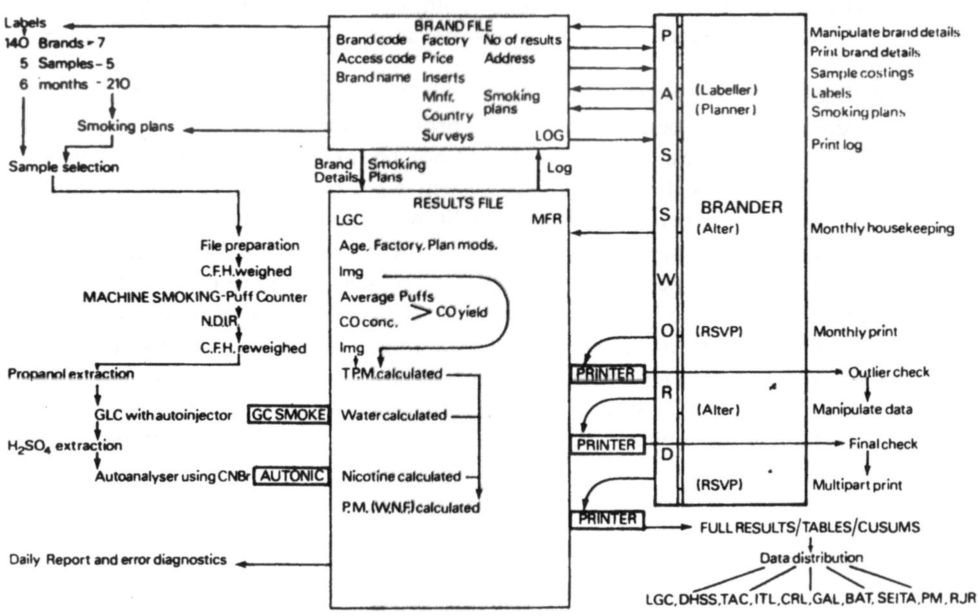

Fig. 23. Schematic diagram showing the relationship between the automated analytical procedures, the data files, and the interactive control program (BRANDER). (Reproduced by courtesy of Dr. P. B. Stockwell)

6.5 Discussion

The total automation system was introduced gradually over several years, via intermediate stages of semi-automation, and represented a major project which required the coordinated skills of a large number of specialists in such fields as chemistry, statistics, computer science, and electronics. The major benefits resulting from the use of this system in place of the manual and semi-automated systems which preceded it have been a substantial improvement in productivity and an overall saving in manpower, coupled with a simpler procedure from the analyst's point of view. In addition, it is now far simpler, thanks to the computer, to check the validity and statistical significance of the results obtained, and since the computer also handles much of the administration associated with the analyses, such as the generation of smoking plans and requests for samples, a lot of tedious paperwork has been eliminated.

However, the system also has its disadvantages, the greatest of which appears to be the use of a large central computer to perform all the computation required. This tends to be an inefficient solution because with a large number of instruments and data terminals connected to it, the machine will spend a large proportion of its time just performing the imput/output routines necessary to send data to and from the peripherals. If too many terminals, etc. attempt to communicate with the computer at once, a position is reached where the computer is fully occupied with the input/output routines, and has no time left for computation. At this point the system will crash, with the result that data from the various instruments connected to it is lost. The same situation can occur if the computer develops a fault, and so it is necessary to keep a manual record of results as an insurance against a crash occurring.

At the time that this system was being planned and assembled, a large central computer represented the best solution to the problem, as microcomputers were not then available, and so these comments should not be taken as criticism of the way in which the system was designed. If the system were being planned for the first time today, however, a better approach might be to use a small microcomputer attached to each instrument which could control the instrument, acquire the experimental data from it, and perform the initial data reduction before transmitting the partially processed data to a central computer. This could now be much smaller, since it would no longer need to perform the data aquisition and control functions, nor would it need to be coupled on-line to the instruments continuously. The risk of data being lost when a crash occurred would be eliminated, since the data would be stored in the microcomputer and could thus be re-transmitted at a suitable time later.

6.6 Conclusion

In many ways, this example is typical of modern analytical problems, where increasingly people are looking to automated analytical systems to provide the answer. It also demonstrates that with the increasing complexity of such systems, it is no longer sufficient for the analyst to rely on his knowledge of chemistry alone. In the future, therefore, we may expect to see analysts who are 'all-rounders' with a background in mathematics, computer science and electronics in addition to their chemical knowledge, as well as analytical teams of perhaps three or four individuals, each of whom is a specialist in one of the above fields.

7 Acknowledgements

We are grateful to D. C. M. Squirrell of the Plastics Division of ICI, and Drs. W. H. Maddens and W. L. Meade of BP Research Centre, Sunbury, for the provision of material for this article, and to the Department of Industry for a studentship for T.J.S.

8 References

1. Betteridge, D.: Anal. Chem. *48*, 1034A (1976)
2. Lavington, S.: Early British Computers. Manchester, Manchester University Press 1980, p. 48
3. Instrument News (Perkin-Elmer Ltd.) *31* (1), 12 (1981)
4. Henderson, K. F.: MINC Newsletter (Digital Equipment Corp.) *1* (4), 6 (1979)
5. Savitzky, A.: Anal. Chem. *33* (13), 25A (1961)
6. IEEE Standard Digital Interface for Programmable Instrumentation: IEEE Std. 488–1975. New York: IEEE 1975
7. Betteridge, D., Goad, T. B.: Analyst *106*, 257–282 (1981)
8. Levy, G. C., Terpstra, D. (Eds.): Computer Networks in the Chemical Laboratory. New York, Wiley 1981
9. Kowalski, B. R. (Ed.): Chemometrics: Theory and Application. Washington, Amer. Chem. Soc. 1977
10. Massart, D. L., Dijkstra, A., Kaufman, L.: Evaluation and optimisation of Analytical Procedures. Amsterdam, Elsevier 1978
11. Kateman, G., Pijpers, F. W.: Quality Control in Anal. Chem. New York, Wiley 1981
12. Hannah, R. W., Coates, J. P.: Europ. Spectroscopy News *32* (1980)
13. Horlick, G., Hieftje, G. M.: Contemporary topics in Anal. a. Clin. Chem., Hercules, D. M., et al. (Eds.) Vol. 3, New York, Plenum 1978, pp. 153
14. Griffiths, P. R.: Chemical Infrared Fourier Transform Spectroscopy. New York, Wiley 1975
15. Griffiths, P. R. (Ed): Transform Techniques in Chemistry. London, Heyden 1978
16. Shaw, D.: Fourier Transform NMR Spectroscopy. Amsterdam, Elsevier 1976
17. Wagner, H. M.: Principles of Operations Research, with Applications to Managerial Decisions. Englebrook, Prentice-Hall 1969
18. Massart, D. L., et al.: op. cit. p. 435
19. Morgan, S. L., Deming, S. N.: Anal. Chem. *46*, 1170 (1974)
20. Nelder, J. A., Mead, R.: Computer J. *7*, 308 (1965)
21. Routh, M. W., Swartz, P. A., Denton, M. B.: Anal. Chem. *49*, 1422 (1977)
22. Stieg, S., Nieman, T. A.: Anal. Chem. *52*, 800 (1980)
23. Gallegos, E. J., et al.: Anal. Chem. *39*, 1833 (1967)
24. Maddens, W. H., Mead, W. L.: Private communication, March 1981
25. Malinowski, E. R., Howery, D. G.: Factor Analysis in Chemistry. New York, Wiley 1980
26. Catell, R. B.: The Scientific Use of Factor Analysis in Behavioural and Life Sciences. New York, Plenum 1980
27. Massart, D. L., et al.: Op. cit., p. 309
28. Batchelor, B. G.: Practical Approach to Pattern Recognition. London, Plenum 1974
29. Everitt, B.: Cluster Analysis. London, Heinemann 1974
30. Kowalski, B. R., Bender, C. F.: J. Amer. Chem. Soc. *94*, 5632 (1972)
31. Idem, ibid, *96*, 916 (1974)

32. Kowalski, B. R.: Anal. Chem. *47*, 1152A (1975)
33. Duewer, D. L., Kowalski, B. R., Shatzki, J. F.: Anal. Chem. *47*, 1573 (1975)
34. Kryger, L.: Talanta *28*, 000 (1981)
35. Varmuza, K.: Pattern Recognition in Chemistry. Heidelberg, Springer-Verlag 1980
36. Stewart, K. K.: Talanta *28*, 000 (1981)
37. Nagy, G., Feher, Zs., Pungor, E.: Anal. Chim. Acta. *52*, 47 (1970)
38. Betteridge, D.: Anal. Chem. *50*, 832A (1978)
39. Ruzicka, J., Hansen, E. H.: Flow Injection Analysis. New York, Wiley 1981
40. Anal. Chim. Acta *114*, 1 (1980)
41. Taylor, G.: Proc. Roy. Soc. London Ser. A. *219*, 186 (1953)
42. Betteridge, D., Fields, B.: Anal. Chem. *50*, 654 (1978)
43. Idem, Anal. Chim. Acta (in press)
44. Curme, H. G. et al.: Clinical Chem. *24*, 1335 (1978)
45. Spayd, R. W. et al.: Clinical Chem. *24*, 1343 (1978)
46. Betteridge, D., Joslin, M. T., Lilley, T.: Anal. Chem. *53*, 1064 (1981)
47. Betteridge, D. et al.: Polymer, in press
48. Copeland, G. et al. (Eds.): Topics in Automatic Chem. Anal., Vol. 1. Chichester, UK, Horwood 1979

Topics in Photochemistry

Howard E. Zimmerman

Chemistry Department, University of Wisconsin, Madison Wisconsin, USA

Table of Contents

1 Introduction

Until 1960 organic photochemistry was an only occasionally studied field. Part of the disinterest in photochemistry arose from the feeling by organic chemists that photochemical reactions were random and unpredictable. It was often stated that reactions occurred by virtue of the high energy imparted to reacting molecules by light photons absorbed.

In 1960, the author [1-7] suggested that despite the high energy of excited states, these molecules did not react indiscriminately but, rather, transformed themselves by continuous electron redistribution. It was further suggested by the author that excited state molecules react by seeking out low energy pathways and avoiding high energy routes. It was proposed that excited state molecules are controlled by the same forces as ground state reacting species. Similarly, it was noted that the energy of excitation of molecules in solution is not directly available for random reaction. Beyond this, it was postulated that, given an excited state structure, organic chemical intuition (e.g. electron pushing) would suffice to rationalize, if not predict, photochemical reactions. In the intervening two decades, photochemical research has dramatically increased in activity. Photochemical mechanisms have been intensively investigated, new reactions have accumulated in considerable number, and excited state structures have been elucidated.

It is clear that the very nature of organic photochemistry has changed in the last two decades. A variety of methods has been developed for describing electronically excited organic molecules, a variety of methods for describing the nature of excited state reactions has evolved, and a number of useful generalizations for predicting and understanding photochemical reactions has ensued.

The purpose of this article is to describe the author's thinking about photochemical reactions and their mechanisms as well as about methods of dealing with organic photochemistry. The emphasis is very heavily on the contributions of the author.

2 Use of Excited State Properties in Photochemistry

2.1 Electron Densities

2.1.1 General Philosophy

One approach of use in ground state organic chemistry is a static one. This assumes that one can predict the reactivity of a molecule from a description of the starting material itself. Thus, very electron rich centers are subject to electrophilic attack, electron poor sites in the molecules are expected to be susceptible to facile nucleophilic attack, weak bonds are subject to scission, etc. This approach is imperfect, in that a reaction course is really determined by a preference for the lowest energy transition state. Nevertheless, this starting state reasoning is quite useful, since most often it is, indeed, the predicted site of attack which affords the preferred transition state.

The same approach is useful in organic photochemistry. For this method, however, one needs a description of the excited state of the reactant. For carbonyl compounds, such descriptions are readily obtained from simple considerations. Because of their uniqueness, carbonyl excited states are discussed separately (vide infra).

For hydrocarbon excited states, there are simple resonance structures which may be written. More precise, however, are descriptions based on molecular orbital methods. One example of such a treatment is given in the following section.

2.1.2 Example of meta-Electron Transmission

The author's interest in this problem derived from an intriguing result reported by Havinga [8] in 1959 of the photochemical solvolyses of the isomeric nitrophenylphosphates and sulfates. It was observed by Havinga that the meta-isomers were more reactive than the para compounds, a result he noted was in contrast to ground state expectation and not explicable in terms of ordinary ground state qualitative resonance reasoning.

This intriguing observation led the present author to investigate whether or not the phenomenon was more general and to search for a quantum mechanical basis for the chemistry. Two studies were carried out [1,10,11]. In the first [1,10] a photochemical solvolysis of substituted-phenyl trityl ethers was encountered. Here p-nitrophenyl and p-cyanophenyl trityl ethers were found to solvolyze thermally in the dark faster than the meta-isomers as expected. However, the meta-isomers solvolyzed more readily on irradiation, and the quantum yields were higher for the meta-isomers. Note Equations 1.

Still another example of the phenomenon was encountered in a study of the photochemical solvolysis of 4-methoxybenzyl, 3-methoxybenzyl, and 3,5-dimethoxybenzyl acetates and chlorides [1,11]. Note Equation 2.

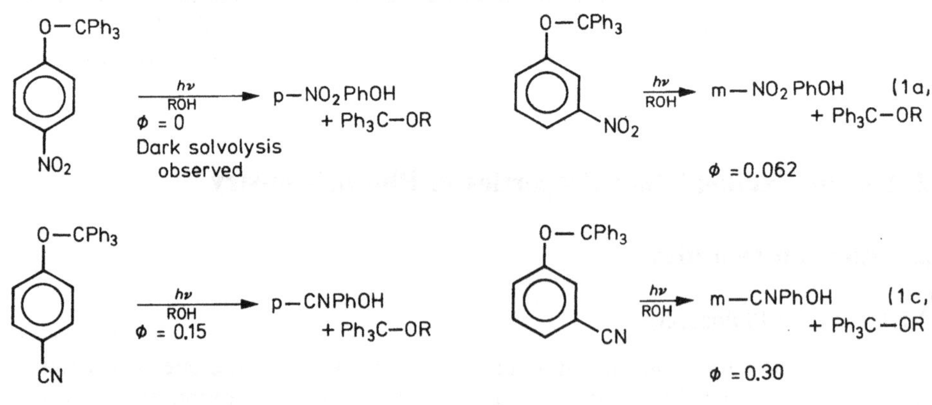

(1a, b)

(1c, d)

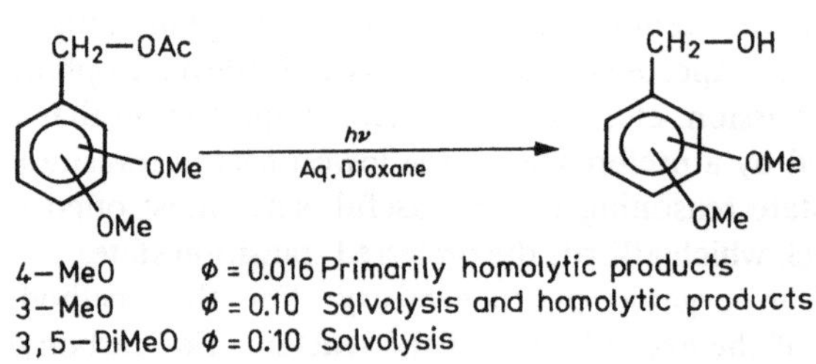

(2)

In 1961 the present author suggested [1,5,10,11] that the anomalous photochemical behavior was due to the different properties of the excited states involved which contrasted in electron densities from the ground states. More explicitly, it was suggested that in the excited state, meta-transmission of electron density could be expected rather than ortho-para transmission characteristic of ground state aromatic species. The molecular orbital rationale shown in Fig. 1 was presented [1,5,10,11]. Here, a generalized electron donating group D was simulated by a carbon bearing two unshared electrons. Then one has the $—CH_2^-$: substituent and thus the benzyl carbanion as a model. For the situation where one has an electron withdrawing group, a $—CH_2^+$ moiety was used to simulate a generalized electron withdrawing group. One can see that electronic excitation in the benzyl anion leads to enhancement of electron density meta to the donor. Similarly, electronic excitation in the benzyl cation leads to decreased electron density at the meta position.

Explicit MO calculations [1,5] on the various methoxy benzenes led to the same conclusion. Note Fig. 2. It also was suggested [1,5,10,11] that one could write convenient resonance structures to represent the excited state electron transmission. The solvo-

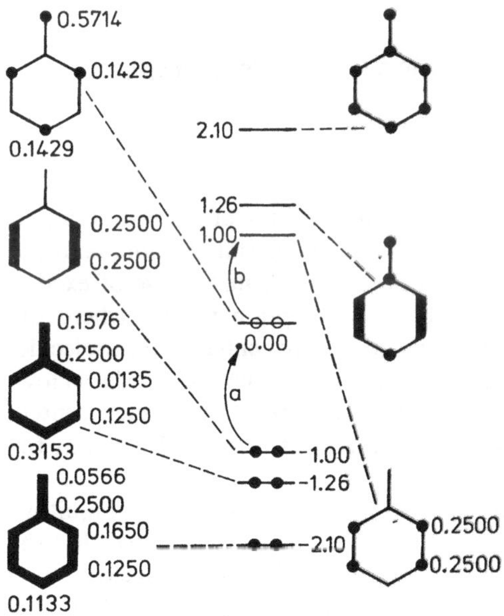

Fig. 1. Benzyl MO's and excitation processes for the 6 and 8 electron species. Blacked portions qualitatively represent electron densities. a. Excitation of 6 electron system simulating an electron withdrawing group being present. Promotion from an MO with heavy meta density to an MO with no meta density. b. Excitation of 8 electron system simulating a donor group being present. Promotion from an MO with no meta density to an MO with heavy meta density

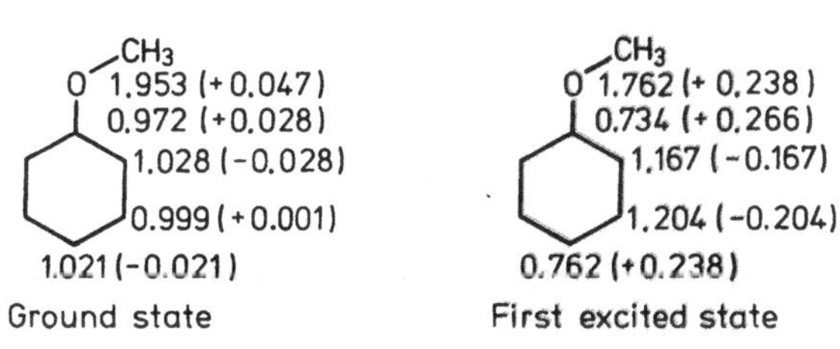

Fig. 2. Ground and excited state electron densities for methoxy aromatics from simple Hückel calculations

lysis can then be pictured in traditional organic chemical terms. For example, note Equation 3.

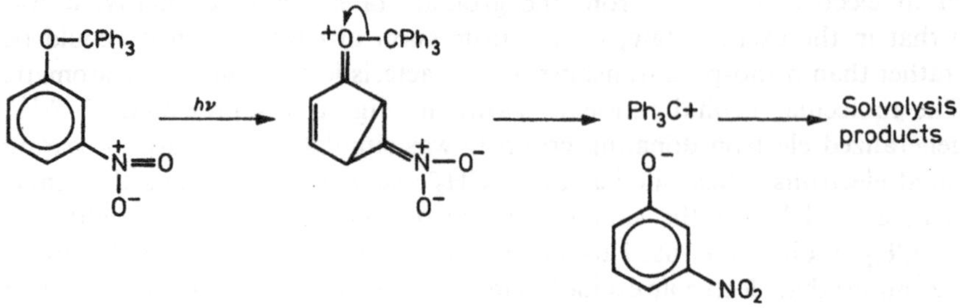

(3)

Since this early work, the phenomenon of meta-electron transmission has been extended very broadly in the very elegant and thorough research of Havinga and Coworkers [9]. For example, the concept is seen to extend to a photochemical preference for nucleophilic attack meta to an electron withdrawing group.

2.2 The Role of Bond Orders in Mechanistic Organic Photochemistry

2.2.1 Rationale

In the preceding, we considered the use of electron densities of excited states in predicting photochemical reactivity. This is the static, starting state technique of the type considered in the introduction. Another static approach makes use of excited state bond orders to predict reactivity. The basic assumption is that where the excited state has a high bond order between two orbitals at two centers of the molecule, there will be a tendency for these two centers to bond. Where the bond order is low, or especially where negative, the two centers will tend to repel. Two such centers certainly will not tend to form a bond. If already attached, the bond between the centers will tend to break.

This reasoning was used by the author in 1961 [1] to rationalize the ubiquitous photochemical cyclization of butadienes to cyclobutenes; here it was noted that the excited state has a high 1,4-bond order. The same reasoning was applied [6,12] to understanding the key step of cyclohexadienone rearrangements (vide infra). Still another example is the decreased central bond order in the excited state of stilbene which, as Daudel has noted [13], is in accord with photochemical cis-trans interconversion.

2.2.2 Di-π-Methane Reactivity as an Example of Bond Order Control

One interesting example derived from the Di-π-Methane rearrangement [14-16]. The overall mechanism of the rearrangement is depicted in Scheme 1.

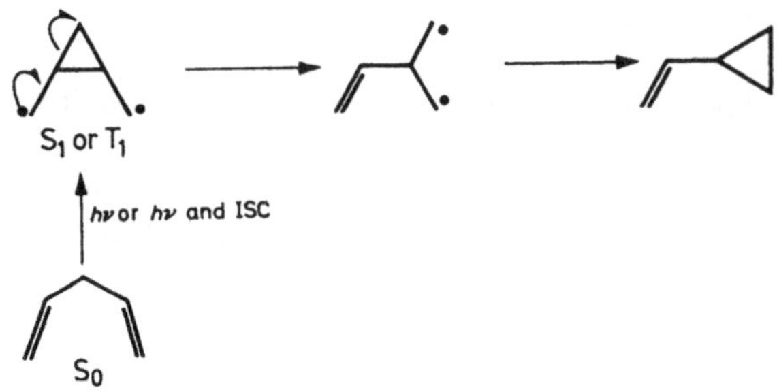

Scheme 1. Generalized Mechanism for the Di-π-Methane Rearrangement.

The primary step of the rearrangement involves vinyl-vinyl (or more generally) π-π bridging in the excited state. It is readily seen that such bridging of two vinyl (or π) moieties in proximity to one another corresponds to a locus with a high bond order. This is depicted in Equation 4.

The MO basis for this effect is seen quite simply in Fig. 3. Here it is seen that electron promotion is from MO 2 which is antibonding (having a negative bond order) between the orbitals at atoms 2 and 3; and, this promotion is to MO 3 which is bonding (i.e. with a positive and large bond order) between atoms 2 and 3. The net effect of excitation then is enhancement of the 2,3-bond order upon electronic excitation.

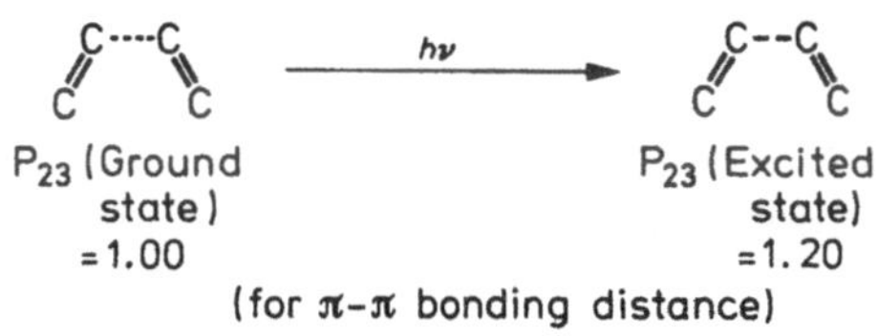

(4)

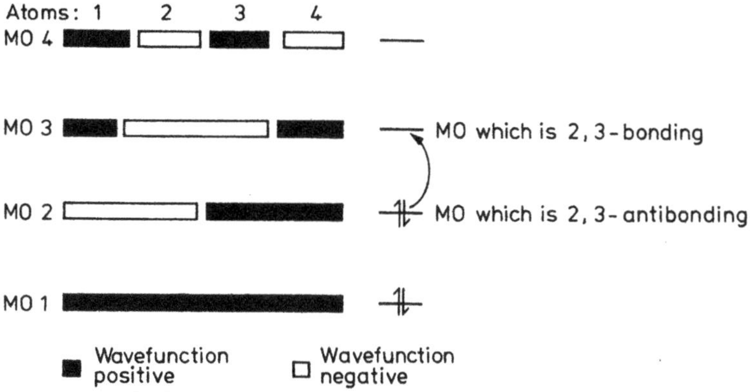

Fig. 3. Excitation of two contiguous and overlapping vinyl groups, top view of MO's

2.2.3 Bond Orders and Type A Cyclohexadienone Reactivity

Still another example of bond order control of photochemical reactivity is found in the photochemistry of 2,5-cyclohexadienones. Here, strong evidence suggests that the n-π* excited triplet is the species responsible for the rearrangement [6,12,17–23]. It is seen that the critical step is β,β-bonding in the excited state. The reaction mechanism is depicted in Scheme 2.

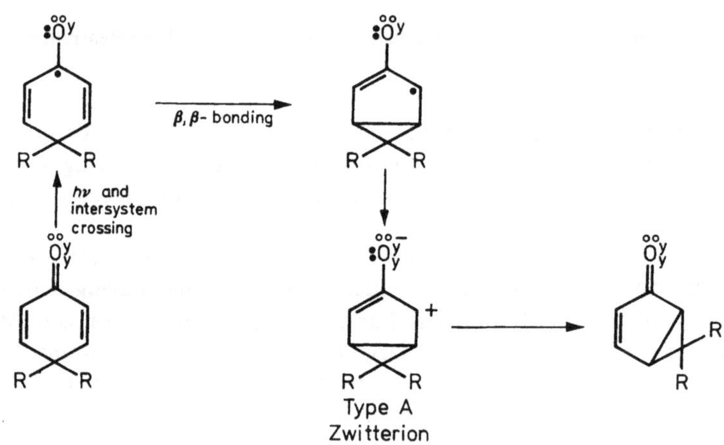

Scheme 2. Overall Mechanism of the Type A Rearrangement.

One of the most interesting theoretical results derives from configuration interaction calculations [19]. It was observed that of the triplets of photochemical interest, that is the n-π* and π-π* states, only the n-π* triplet reveals a high β,β-bond order. The π-π* triplet possesses a negative bond order, thus suggesting that this triplet should exhibit a reluctance to β,β-bond. Interestingly, the ground state of 4,4-diphenylcyclohexadienone also is 3,5-antibonding. Figure 4 summarizes this bond order information.

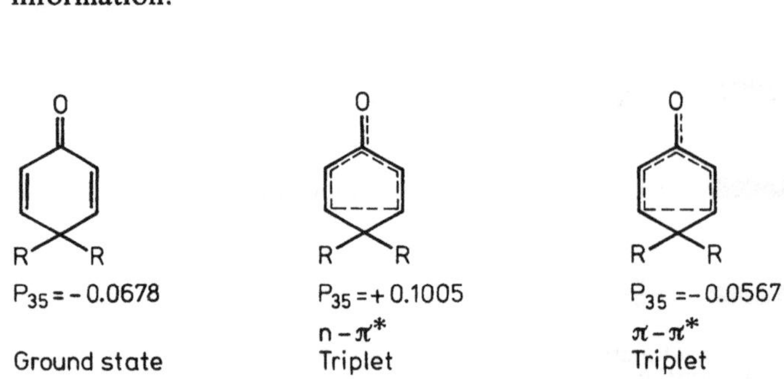

Fig. 4. β,β-bond Orders of the Triplets and Ground State of 3,5-cyclohexadienones

This bond order control is seen even in Hückel [17,18] and SCF [23] calculations on the 3,5-cyclohexadienone system. Although there are definite quantitative and other advantages to the use of SCF and SCF-CI calculations in organic chemistry, Hückel theory very frequently provides a reward in simplicity and in allowing one to understand the basic factors giving rise to some phenomenon. In the present instance, reference to Fig. 5 [12,17,18,20] allows ready understanding of the n-π* β,β-bond order and its contrast with the π-π* excited state.

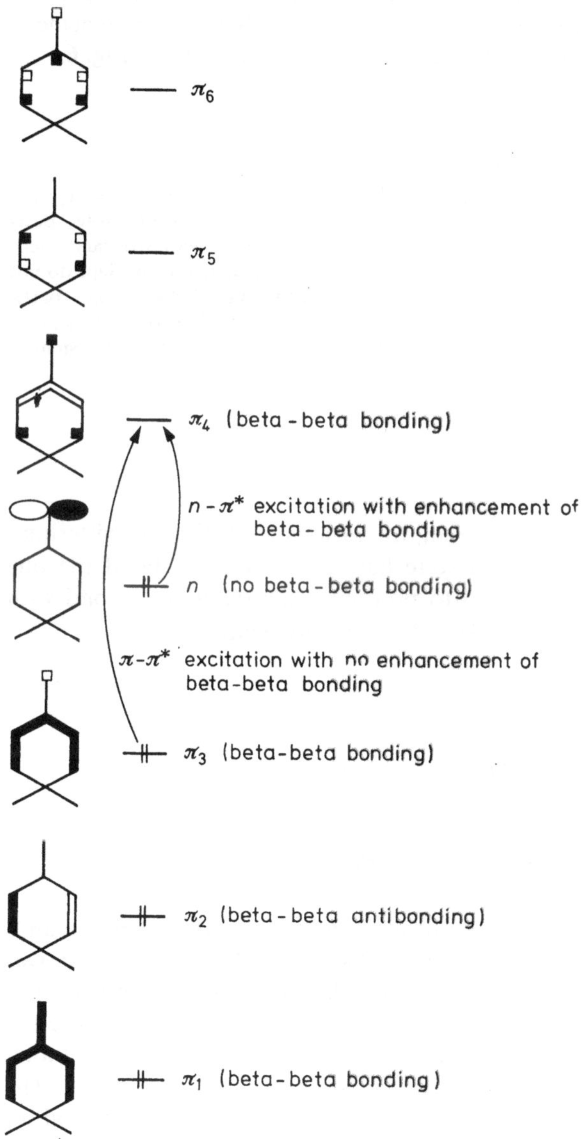

Fig. 5. MO Representation of n-π* and π-π* Excitation Processes

2.2.4 Application of Bond Order Considerations to the Type A Zwitterion Rearrangement

The Type A zwitterion (see Scheme 2) is a ubiquitous intermediate in a large variety of photochemical rearrangements. This intermediate was originally proposed by the author [1-4,7] as part of the photochemical mechanism to account for dienone photochemistry (vide supra). Since that time, it has proven possible to generate such zwitterions without the use of light by reaction of the alpha-bromo bicyclo[3.1.0]-hexan-2-ones with bases and the alpha-, alpha'-dibromo ketones with reducing agents [22-26]. The mechanism of zwitterion rearrangement is included in Scheme 2. The transition state for the rearrangement may be envisaged as shown in Fig. 6.

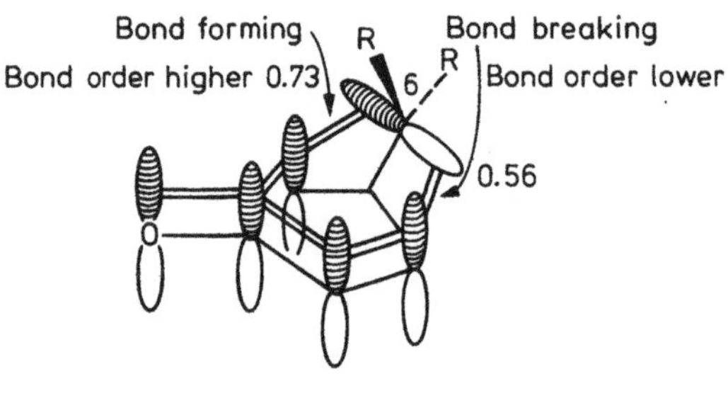

Fig. 6. Bond Order Effects in the Type A Zwitterion Rearrangement. Note: Basis orbitals are shown with arbitrary orientation, hence plus-minus overlaps do not imply antibonding between any particular orbital pair. However, with an odd number of such overlaps, the system is Möbius

The intriguing result is encountered wherein, for variously substituted zwitterions, carbon-6 in the rearrangement transition state has as the stronger bond the one which is being formed to proceed onward to product while the weaker bond with lower bond order is the one which is in the process of breaking.

2.3 Reactivity of the Carbonyl Moiety

2.3.1 Use of a Three Dimensional n-π* Model to Predict Reactivity

Our starting point in dealing with photochemical reactivity of carbonyl compounds is a three-dimensional structure of the n-π* excited state. Thus, in 1961 it was first suggested by Zimmerman [1-4,7] that essentially all of the known photochemical reactions of ketones were explicable on the basis of the three dimensional structure of the n-π* excited state of the carbonyl group as shown in Fig. 7. In parallel studies Kasha dealt [27,28] with the hydrogen abstraction ability of the n-π* state. The Kasha and Zimmerman picture of hydrogen abstraction by the p_y orbital of n-π* states of a ketone or aldehyde paralleled the proclivity [29,30] of t-butoxyl radical's oxygen atom to abstract hydrogen atoms.

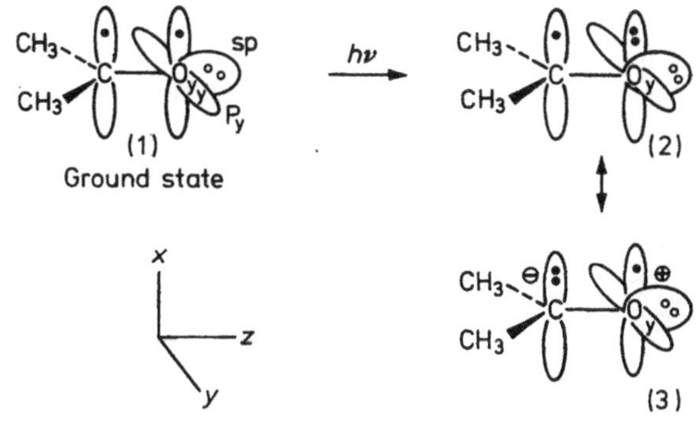

(5)

Fig. 7. Three Dimensional Representation of n-π* Excitation

A convenient short-hand representation of such three dimensional structures introduced by Zimmerman [1-4,7] is often convenient when one is following complex photochemical reactions. This two dimensional notation is shown in Equation 6 [31]. Thus Equation 6 in two dimensions represents the three dimensions of Equation 5.

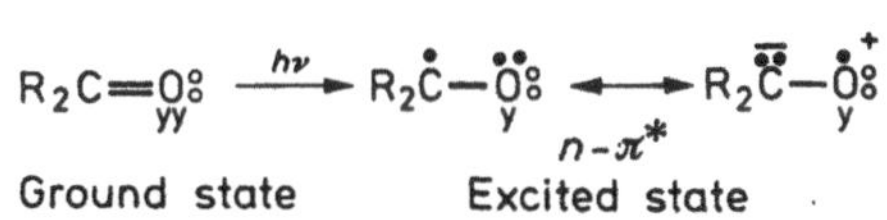

(6)

It is noted that the above n-π* model is an approximate one which provides a qualitative description of both singlet and triplet excited states.

Hydrogen abstraction is discussed again below in connection with MO Following; there we complete the discussion, obtaining a correlation diagram and showing that the reaction is, indeed, continuous to ground state and allowed as shown.

2.3.2 A Simple Type of Behavior: p_y-Orbital Reactivity and Some Examples of p_y-Orbital Behavior

The special case of hydrogen abstraction has been discussed in connection with the simple model for the n-π* excited state. However, it was noted in 1959–62 by Zimmerman [1-3,7] that a major fraction of the organic photochemistry of carbonyl compounds was explicable on the basis of simple basic principles and this simple model for the carbonyl group.

For example, oxetane formation was envisaged [12] as involving a parallel attack of the electrophilic p_y orbital on the π system of a double bond rather than on a hydrogen atom of a hydrogen donor.

Similarly, the Norrish Type I acyl fission process was depicted [1,3-5] as proceeding by gradual disengagement of the alkyl moiety initially bonded to the carbonyl carbon thus leaving an acyl radical. This process was noted as arising by overlap of the electron deficient p_y orbital with weakening of the alkyl to carbonyl carbon bond. Thus, the p_y orbital of the n-π^* state has only one electron and an electron deficiency consisting of an "electron hole". This positive "hole" can be pictured as being delocalized into the alkyl to carbonyl carbon bond, thus weakening the bond. Note Equation 7.

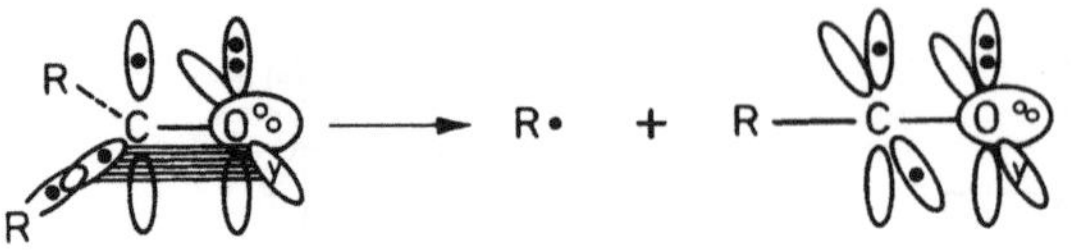

(7)

Also, beginning in 1961 it was noted by the author [1,3,4] that there are two general kinds of n-π^* reactions, those leading directly to ground state of photoproduct and those involving several steps before ground state is reached. The present example of the Norrish Type I (i.e. acyl fission) reaction is one of several of the first variety.

During the disengagement process there are three orbitals which overlap in a linear array: the p_y oxygen, the carbonyl carbon hybrid orbital bonding to the alkyl group, and the orbital centered on the alkyl group being disengaged. Thus, somewhere near "half reaction", one has an allyl-like array of orbitals with a total of three electrons. If we were to assume equal electronegativity everywhere, clearly a very approximate assumption, we would predict the distribution of the allyl free radical with 1.5 electrons in the sigma bond being broken. This type of delocalization involving sigma bonds has been discussed in many different instances by R. Hoffmann [32] and extended Hückel theory could be applied here to demonstrate the onset of this delocalization in the n-π^* ketone prior to bond weakening.

Still another reaction which is readily susceptible to our n-π^* model is the Norrish Type II with concomitant cyclobutanol formation (i.e. the Yang reaction [33]). This mechanism was described in detail by the author [1,3,12], again in those early papers. Here the two dimensional "circle-dot-y" notation suffices and is convenient. Note Equation 8.

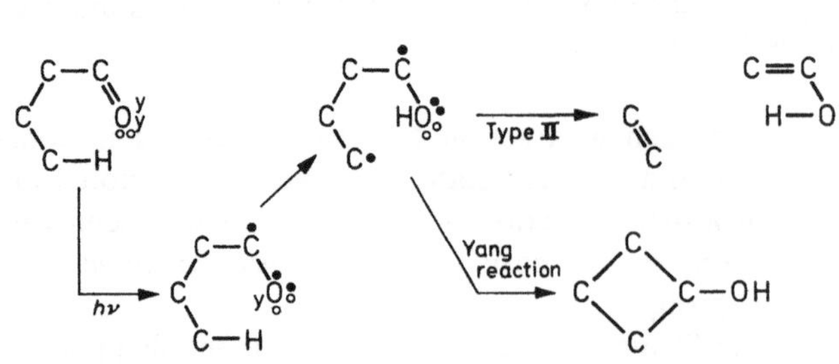

(8)

2.3.3 The Second Type of Reactivity: π^* Photochemical Effects

Also in those early years the present author [1-5] noted that photochemical reactions of the n-π^* states might arise by virtue of the changed π system which had one extra electron relative to ground state.

One interesting example cited [1,4,5] was alpha-expulsion. Here a moiety, X, alpha to the carbonyl group was noted to be subject to expulsion as an anionic species or as an odd-electron fragment, the mode depending on whether the solvent is ionic (favoring ionic fission) or non-polar (favoring radical fission). This is not surprising to the organic chemist adept at electron pushing when he views the structure of the n-π* excited state as in Scheme 3.

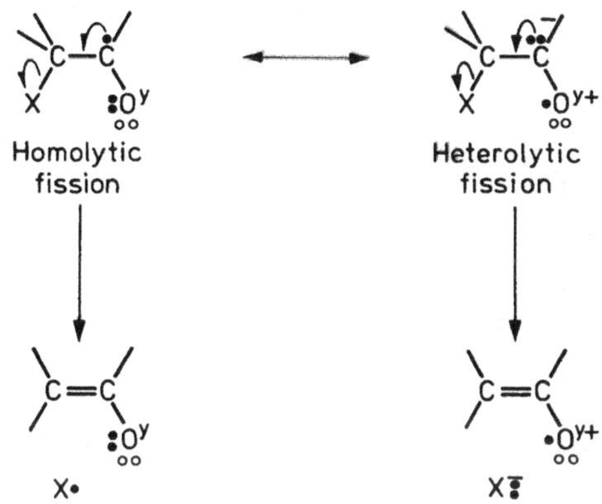

Scheme 3. Mechanism for alpha-Expulsion.

The alpha expulsion of chlorine atom from chloroacetone in the gas phase was cited [1,4,5] as one example of the homolytic fission while formation of chloride anion and alpha-hydroxyacetic acid from chloroacetic acid in aqueous solution was given as an example of anionic explusion.

Still further examples were found [1,34] in the photochemical rearrangement of α,β-epoxyketones to beta-diketones. Here the alpha C-O bond is broken homolytically in the primary photochemical process.

3 The Möbius-Hückel Concept and Photochemistry

3.1 Application to Correlation Diagrams

We now turn to a different type of photochemistry and different mechanistic questions, namely photochemical pericyclic reactions and the utility of the Möbius-Hückel treatment of these transformations.

The Möbius-Hückel concept was introduced by Zimmerman in 1966 [35]. It was suggested that each cyclic array of orbitals in a reacting system may be categorized as a "Hückel type" or a "Möbius type", depending on the number of plus-minus overlaps between adjacent orbitals. With zero or an even number of such sign inversions, the system is a Hückel variety array while with one or some other odd number the system is a Möbius system.

In selecting orbitals to constitute a cyclic array, one needs to select those orbitals involved in bonding changes during a reaction, photochemical or ground state. The orbitals used are the so-called basis orbitals, i.e. the set of atomic or hybrid orbitals used in a molecular orbital calculation and are not to be confused with final MO's. Note the unrealistic but instructive two arrays given in Figure 8.

The Möbius-Hückel approach now has become common to a large number of undergraduate textbooks because of its facile application to ground state chemistry. The method does not really differ in conclusions and derived rules from the Dewar method presented a year later [36].

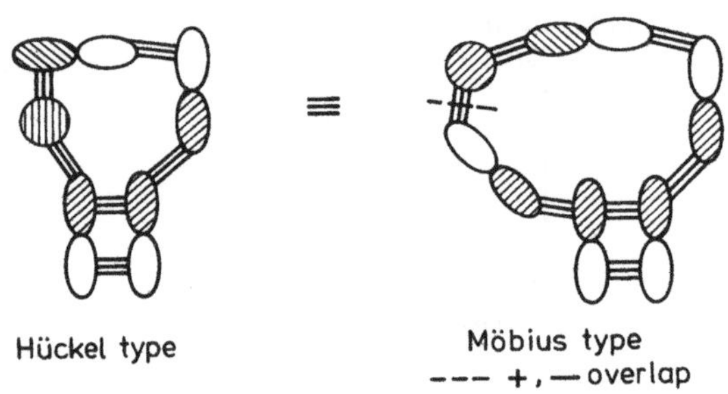

Fig. 8. Examples of Möbius and Hückel Orbital Arrays

The relevance of Möbius-Hückel theory [35] to organic photochemistry is seen first in its ability to allow facile construction of correlation diagrams and hence to predict reaction forbiddeness and allowedness.

The starting point are two mnemonics, one due to Frost [37] and one due to Zimmerman [35,38]. The Frost mnemonic analytically gives the MO energies of a Hückel cyclic array of equi-energetic orbitals while the Zimmerman mnemonic similarly gives the MO energies of a Möbius array. For each of the two mnemonics, a circle of radius $2\,|\beta|$ is drawn with its center at 0 on a vertical energy axis. The energy units are "beta" (or approx. 18 kcal/mole in Hückel theory). The energy zero is taken as the energy of the basis atomic or hybrid orbitals making up the array. A regular polygon is inscribed in the circle with as many vertices as there are basis orbitals in the cyclic array of interest.

Whereas in the Frost mnemonic for Hückel systems the polygon is inscribed with a vertex down, in the Zimmerman mnemonic for Möbius systems the inscription is with the polygon side down. Three examples of each type are shown in Figure 9. Note that each intersection of the polygon with the inscribed circle corresponds to an MO and that the vertical positioning of the intersection gives the MO energy analytically. Thus, all of the Hückel systems, with one vertex at the bottom, have in common one MO at $-2|\beta|$. Also the odd-sized arrays have their Hückel and Möbius relatives turned upside down from one another, while in the even series there is no such relationship.

In a pericyclic reaction the array of basis orbitals of the reacting molecule is cyclic halfway along the reaction coordinate. The array will either be of the Möbius or the Hückel type. Since the circle mnemonics give the distribution of MO energies

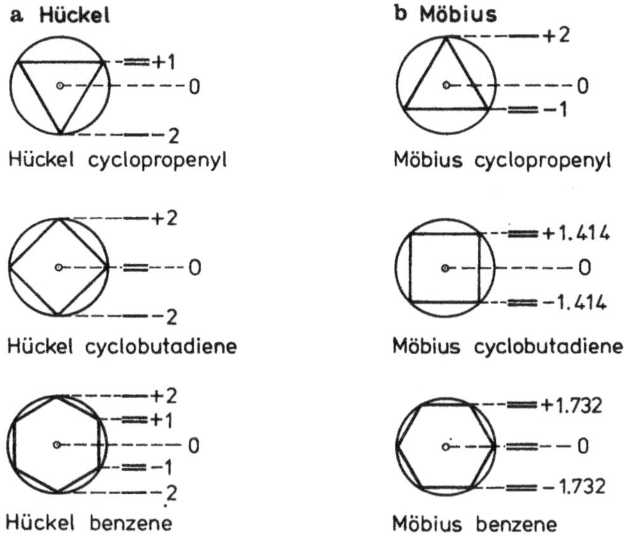

Fig. 9. The Frost and Zimmerman Mnemonics for Hückel and Möbius Cyclic Arrays, Frost Mnemonics on the left and Zimmerman Mnemonics on the right

and also the degeneracies, we can readily construct the correlation diagram by "tying" together MO's of the cyclic array (half-reacted species) with the MO's of starting material and also with product.

Most importantly, for every degeneracy, there is a pair of MO's crossing. Hence without the use of symmetry one can readily decide which reactant MO's cross and which do not. A simple example is seen in the correlation diagram for the ${}_{\pi}2_s + {}_{\pi}2_s$ cycloaddition of two ethylenes as in Fig. 10.

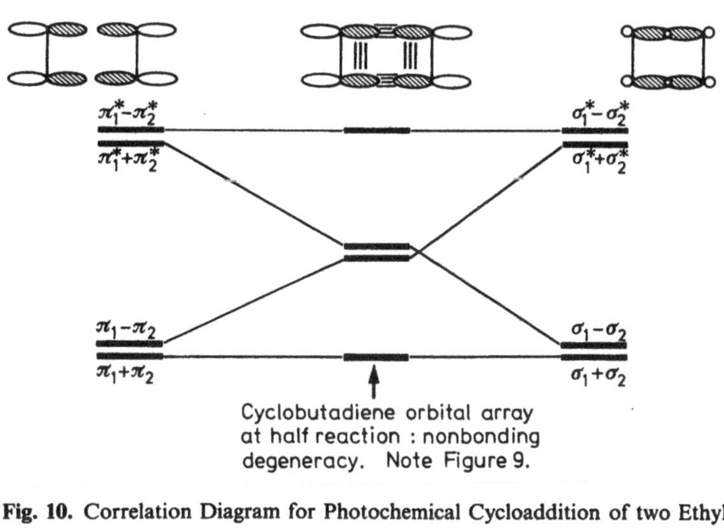

Fig. 10. Correlation Diagram for Photochemical Cycloaddition of two Ethylenic Components

It should be added that for the method to be analytically exact, the basis orbitals would have to all be identical which is generally not the case. Thus the method provides the behavior of an idealized array which then defines, qualitatively and semiquantitatively but not analytically, the behavior of an array of real chemical interest.

3.2 Degeneracies and Enhancement of Internal Conversion

In a related publication, the present author [39] presented a more general method of locating degeneracies. More importantly, he suggested that degeneracies between HOMO and LUMO along the reaction coordinate are particularly important to photochemistry because at the point of such degeneracies molecular vibrations are most likely to enhance internal conversion to ground state.

We note that it is the Hückel systems with 4N electrons and the Möbius systems with 4N + 2 electrons which have a nonbonding degenerate pair of MO's and thus a facile mode of converting starting excited state to ground state of product. Finally, it should be noted the the Möbius-Hückel method is fully consistent with the Woodward-Hoffmann treatment, both for ground state and for photochemical reactions [40].

4 More General Use of MO and State Correlation Diagrams and Surfaces

4.1 MO Following Applied to Photochemistry

Where symmetry is lacking, MO Following [41] has proven to be of value in obtaining correlation diagrams and in assessing the change in the MO's as a reaction proceeds. The emphasis of our previous presentation of MO Following was on ground state examples. However, the correlation diagrams thus obtained may be used with an excited configuration.

One example of interest is the hydrogen abstraction process by the triplet n-π^* excited state of a ketone. The correlation diagram is shown in Fig. 11.

The correlations are made simple once one recognizes that at half reaction the hydrogen dohor to hydrogen bond is stretched and that there is a new bond to the hydrogen formed by the p_y oxygen orbital. This forms an allyl-like array of three orbitals in a linear sequence (i.e. 1--2--3; note Fig. 11). In such a linear array, the lowest energy MO har the general form 1 + 2 + 3, the second has the form 1 — 3 or 3 — 1 and the highest energy MO has the form 1 — 2 + 3. In addition there are the two carbonyl MO's: π_{45} and its antibonding counterpart π^*_{45}.

It can be seen that, as a result of the allyl nature of the 1--2--3 array, there is no degeneracy along the reaction coordinate. With the excited state of reactant affording the lowest states of the two free radical primary products, the reaction is allowed.

Surprisingly, the correlation diagram differs from that written by Turro [42] in which MO's p_y and σ_{C-H} were thought to cross. A similar discrepancy in the alpha fission MO diagram arises, where again MO Following, similarly applied, leads to no MO crossing while Ref. [42] finds a crossing.

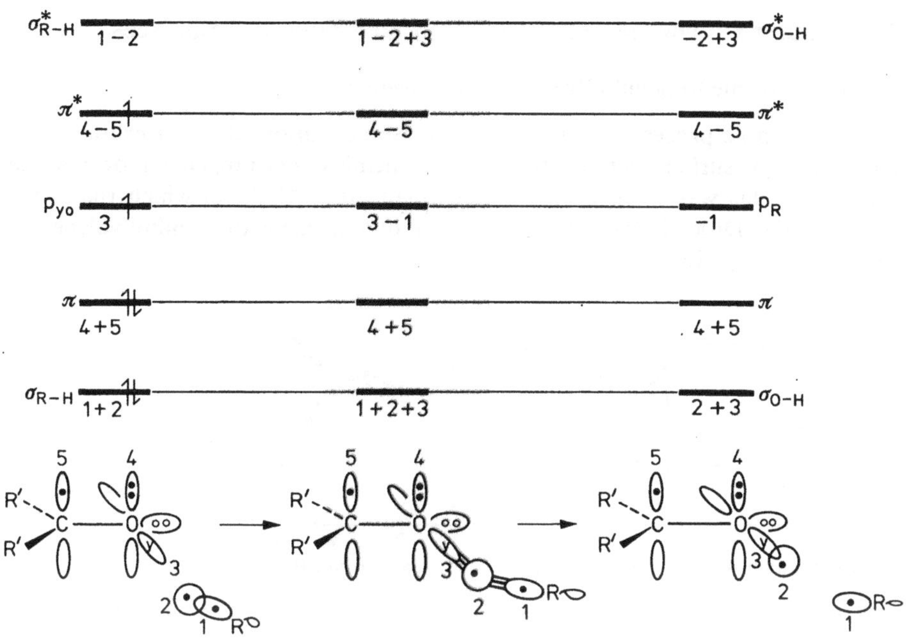

Fig. 11. Correlation Diagram for n-π* Hydrogen Abstraction

It is seen with such correlation diagrams that the n-π* excited state in reactions such as hydrogen abstraction, Type II fission, etc. proceeds directly to the ground state of product (note the heavy line in Fig. 12). This paraphrases the Zimmerman 1961–63 n-π* theory in slightly different language but arrives at this same conclusion.

A novel addition to this picture was added by Salem in 1974 [43] in which he noted that the ground state of the starting materials (bottom left in Fig. 12) correlate with the excited state of products; this is depicted in Fig. 12 with a dashed correlation line. It is not suprising that this reaction does not occur. Also, as noted by Salem [43] the two correlation surfaces differ in symmetry and do not avoid one another.

The two types of correlations, the first deriving from the work of Zimmerman [1, 3–5], i.e. the excited reactant state to product ground state, and the second deriving from the hypothetical reactant ground state to product excited state — have been termed "Salem diagrams" [e.g. [42]].

The type of crossing here is quite different from the avoided type we discuss below and we shall see that its relevance to photochemical mechanisms is minimal due to this difference (vide infra).

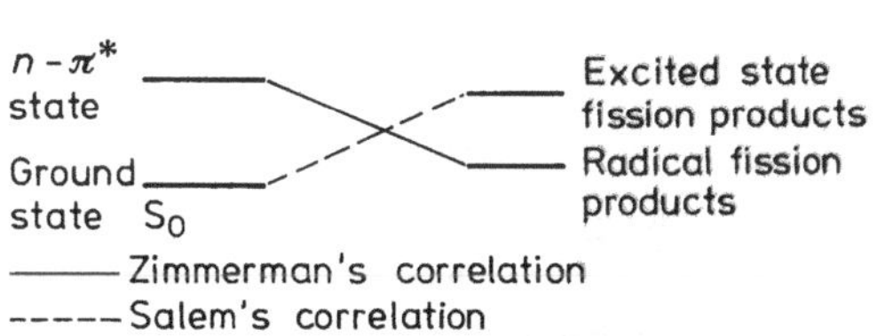

Fig. 12. Direct Formation of Ground State Product in Some n-π* Reactions

4.2 Computer Calculation of Correlation Diagrams and Surfaces

4.2.1 The Barrelene to Semibullvalene Rearrangement

Aside from simple processes such as cis-trans isomerization, the first example of a potential energy surface derived for a photochemical rearrangement or reaction was in 1967. This was work by the Zimmerman group [44–47] in which the hypersurface for the Di-π-Methane rearrangement of barrelene to semibullvalene was obtained; note Fig. 13.

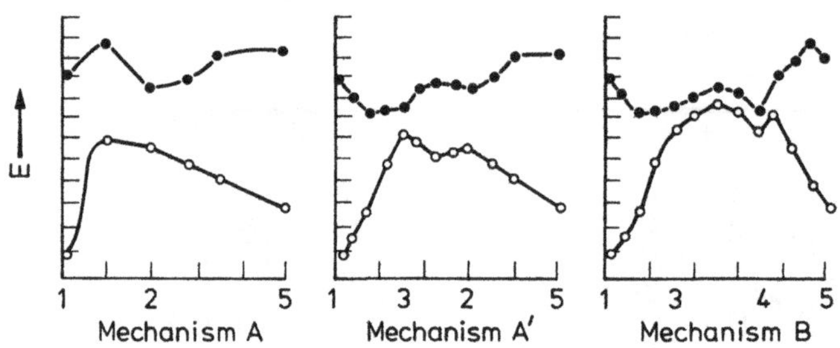

Fig. 13. Potential Energy vs. reaction coordinate for three mechanisms: ○, ground-state curve, ●, excited-state curve

The three corresponding mechanisms are given in Scheme 4.

Although only simple three-dimensional Hückel theory (i.e. similar to extended Hückel) was employed in this early study, the results are qualitatively correct. Inclusion of electron-electron interaction is unlikely to change the qualitative situation in this case.

What is noted is that in Mechanism A, the vertical excited state is born in an energy well and unlikely to react. Indeed, it was Mechanism A which is the simplest and was initially favored on the basis of Occam's Razor type arguments. In retrospect it may be shown that this mechanism involves a Möbius array of 8 orbitals with 8 electrons, and is excited state forbidden. A variation in this mechanism (i.e. A′) in which the first two bridging steps are separated in time, clearly has improved matters so that

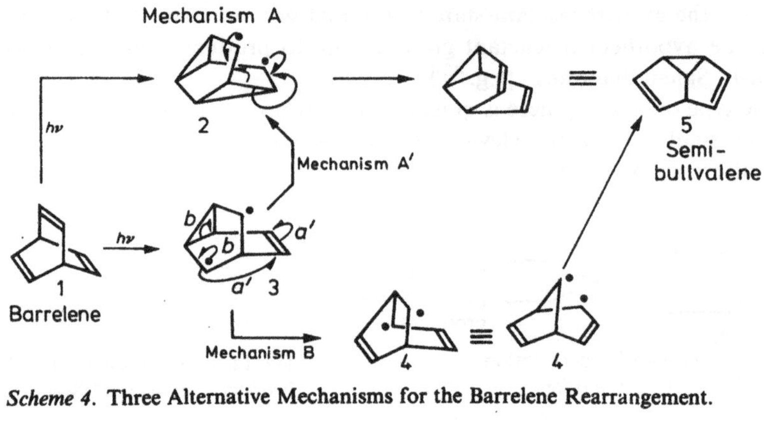

Scheme 4. Three Alternative Mechanisms for the Barrelene Rearrangement.

the vertical excited state may now reach an energy minimum positioned near a ground state maximum. However, the approach of ground and excited states, a "bifunnel", is not far along the reaction coordinate and is canted with decay being most facile back to ground state reactant. Note below for more about bifunnels and canted bifunnels.

The third mechanism, B, on vertical excitation leads to an excited state born with excess vibrational energy (i.e. on the slope of an energy well). This energy is seen to be enough to allow the molecule to reach the next excited state minimum which corresponds to the penultimate species of the barrelene to semibullvalene rearrangement (an allylic biradical). Close approach of ground and excited state surfaces allows radiationless decay with the preferred pathway on the ground state surface then leading to product semibullvalene.

4.2.2 Application of Computer Generated Surfaces and Calculations to the Di-π-Methane and Bicycle Rearrangements

Since that time there has been ample literature discussion of the different possible kinds of combinations of ground and excited state surfaces.

One way to generate surfaces is by explicit QM calculation of species as they are followed through some mechanism. SCF-CI calculations have proven of considerable value in the author's research. The philosophy here has been to include as basis orbitals only those atomic and hybrid orbitals which are part of chromophores or make up bonds which are altered, broken, formed or modified, during the photochemical transformation. Additionally, basis orbitals aimed along the directions of bonds are used, since then the SCF wavefunctions are linear combinations of recognizable orbitals of bonds rather then arbitrary vertically and horizontally oriented atomic orbitals.

This approach has been applied to a number of reactions of interest to the author [48-53]: the Di-π-Methane Rearrangement [48-53], the Bicycle Rearrangement [49-53], and the reverse Di-π-Methane Rearrangement [48-53].

One example [53] of these will suffice. Thus, the photochemical reactions interconverting 1,1-dicarbomethoxy-3,3,5,5-tetraphenyl-1,4-pentadiene, 1,1,2,2-tetraphenyl-3-(2′,2′-dicarbomethoxyvinyl)cyclopropane, and 1,1-dicarbomethoxy-2,2-diphenyl-3-(2′,2′-diphenylvinyl)cyclopropane are of interest in several ways. The singlet processes provide an example typifying the treatment of many of the cases referenced above.

Thus the singlet photochemistry is summarized in Scheme 5.

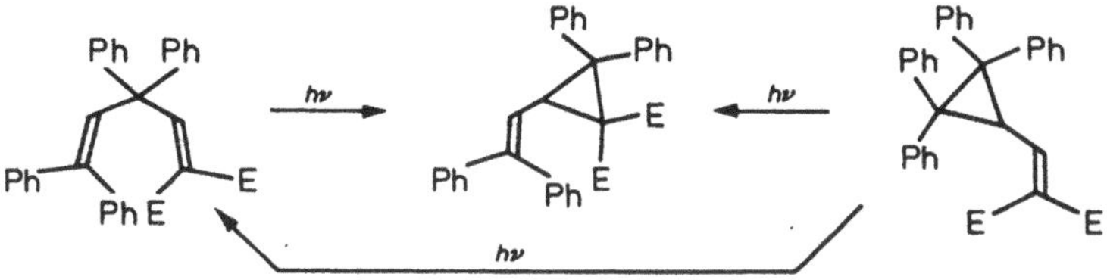

Scheme 5. Singlet Photochemistry of the Dicarbomethoxy Tetraphenyl Pentadiene and Vinylcyclopropanes.

It is noted that there are three isomeric compounds which interconvert photochemically and that there is an intermediate structure cyclopropyldicarbinyl diradical species common to the mechanisms involving the three. Hence it is convenient to plot the SCF-dervied correlation diagram as a triptych following the example of our earlier work [52]. This is given in Fig. 14.

For this set of calculations, as an approximation, each pair of phenyls was

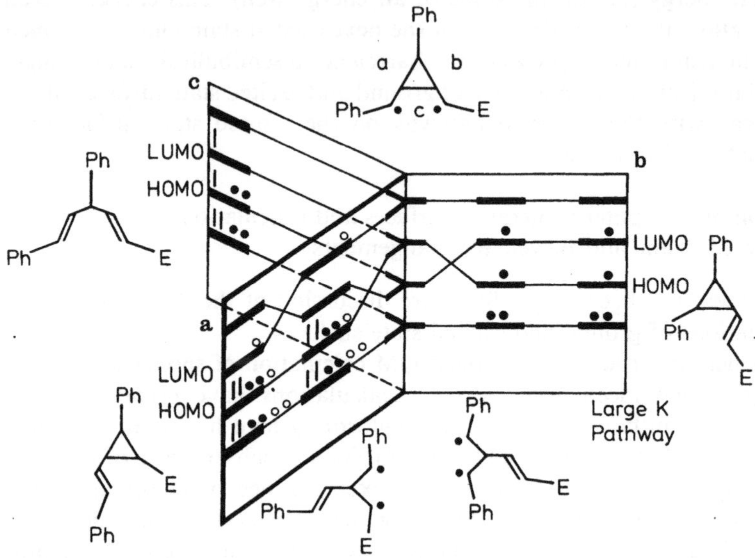

Fig. 14. MO Triptych for the Di-π-Methane and Acyclic Bicycle Rearrangements of the Dicarbomethoxy Compounds

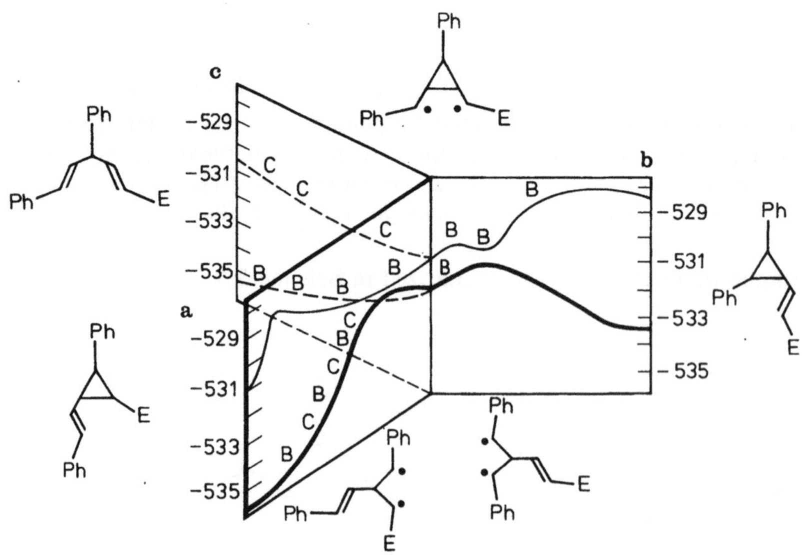

Fig. 15. Surfaces Derived for the Rearrangement of the Dicarbomethoxy Compounds

simulated by one such group as was each pair of carbomethoxyls simulated by one ester function.

In addition to the SCF calculations which afford molecular orbitals, CI was also employed. The results of the configuration interaction calculations are given in a parallel triptych in Fig. 15. Here surfaces, or states, are obtained rather than MO's.

Inspection of the triptych of Fig. 15 shows that the S_1 state of the cyclopropyldicarbinyl diradical is necessarily formed when starting with the pentadiene, that is, in the Di-π-Methane rearrangement. This is seen in two ways. In the MO triptych of Fig. 14, we see no HOMO-LUMO crossing in this wing of the triptych. In the state triptych of Fig. 15, the surfaces then do not approach one another in this wing. Hence there is no mechanism for interconversion to ground state enroute to the diradical from the S_1 diene.

Having arrived at the central triptych axis electronically excited, the diradical a priori might select either front branch of the triptych to follow. The MO version of Fig. 14 affords no indication of prejudice. However, the SCF-CI triptych of Fig. 15 shows a lower energy surface leading off to the left branch and there an approach of the excited and ground state surfaces occurs, thus allowing internal conversion, via this "bifunnel", to ground state. The bifunnel is seen to occur at the point along the reaction coordinate where in the MO triptych a HOMO-LUMO crossing is present. This is in accord with our 1966 reasoning about internal conversions and HOMO-LUMO crossings [39] as well as our reasoning about bifunnels [50,51,53]..

Starting with the diphenylvinylcyclopropane on the left-front branch of the triptych, known not to react photochemically, we see (note Fig. 15) the vertical excited state formed in a well with a sizeable barrier; the lack of reactivity is understandable. The source of this barrier is seen on reference to the MO triptych of Fig. 14. Here LUMO and LUMO + 1 cross very quickly in the front-left branch of the triptych and an upper excited configuration would be formed adiabatically (i.e. without change in electron assignment).

If we now consider the reaction of the dicarbomethoxyvinylcyclopropane, we note that the reacting molecules B in the state triptych may arrive at the central axis of the triptych either still electronically excited (i.e. as S_1) adiabatically, or may arrive at the axis as ground state cyclopropyldicarbinyl diradicals by use of the bifunnel in the right-front triptych wing of Fig. 15; the bifunnel is a poor one with imperfect approach and both possibilities seem to occur. Thus, we see that there are at least two cyclopropyldicarbinyl diradical species involved in organic reactions. The S_0 diradical can be seen to lead most readily via the back wing of the triptych to the reverse Di-π-Methane rearrangement product, the diene. This is expected by every expert on ground state diradicals. The S_1 diradical cannot get to diene ground state but must utilize the front left wing of the triptych where it finds a relatively efficient bifunnel. Internal conversion via the bifunnel leads to the diphenylvinylcyclopropane (i.e. the bicycle product) as observed experimentally. The same conclusions can be reached for the most part by use of the MO diagram in Fig. 14.

4.2.3 Bifunnels, Canted Bifunnels, Inefficient vs. Efficient Bifunnels

The preceding dealt with computer calculations which permit detailed discussion of photochemical reactions. The concept of HOMO-LUMO crossings controlling conversion to ground state [39] has been discussed earlier and its relation to the occurrence of bifunnels has been made clear.

It should be noted that very elegant and early detailed discussions of funnels (our bifunnels) were presented by Michl [54]. Still earlier, the discussion of Longuet-Higgins and Abrahamson [55] on the surfaces expected corresponding to a HOMO-LUMO crossing are basic to all these discussions. Note also research by the author relating molecular twisting of 1-phenylcycloalkenes to the rates of S_1 and T_1 radiationless decay [56].

One interesting point is that the bifunnels encountered are not invariably positioned vertically. Thus the bifunnel may be canted with the expectation then that decay will preferentially occur to one side of the ground state maximum. One example of this was found in one study [50] of the bicycle rearrangement. This is depicted in Fig. 16. In this instance the canting derives from a change in the HOMO and LUMO energies prior to and following the crossing with the product MO's both being of higher energies than their counterparts prior to the crossing (note Fig. 16).

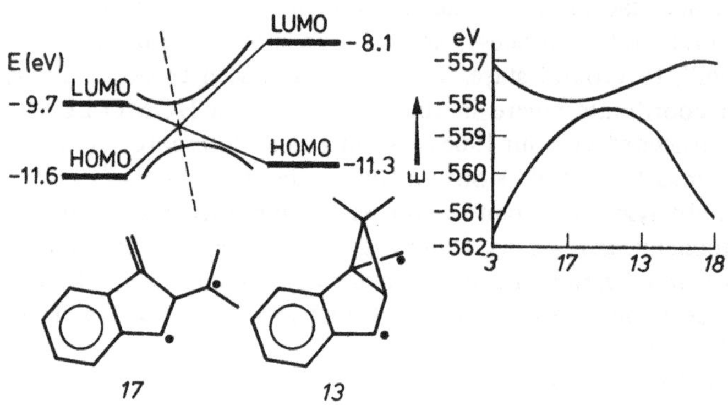

a Canting towards *17*; from SCF level calculations

b Displacement towards *3* from CI level calculations

Fig. 16a and b. Energetics and Canting of a Bifunnel

Beyond this factor affecting the nature of the decay in bifunnels, there is the tendency for a molecule to proceed along the same direction in entering and leaving the bifunnel. This momentum effect, originally described by Teller [57], would not apply to a molecule which had time to equilibrate in the upper funnel.

Additionally, with increased separation between the upper and lower surfaces of the bifunnel, the efficiency of decay to S_0 would decrease. Such an effect has been observed [50] in our studies on the bicycle rearrangement. With appreciable surface separation, then, the momentum effect should be inhibited [50]. However the main consequence of an inefficient bifunnel is a diminished tendency toward internal conversion.

4.2.4 The Role of S_0-S_2-S_1 Mixing

An intriguing point has been made by Oosterhoff [58] in the photochemical butadiene to cyclobutene closure. It was noted that the usual MO correlation diagrams reveal a HOMO-LUMO crossing. In parallel fashion the state diagrams of the Longuet-Higgins and Abrahamson type reveal an avoided crossing of S_0 and S_2 [55]. However, it is S_1 which is generally of photochemical interest.

Additionally, if only S_0 and S_1 are considered, these configurations cannot mix as a consequence of Brillouin's theorem. Nevertheless, the situation is not as serious as it appears. Thus it is seen that in an MO approximation the S_0 and S_1 configurations do become degenerate where HOMO-LUMO crossing occurs, and this signifies that where Möbius-Hückel theory predicts a degeneracy, surfaces will at least approach one another.

An intriguing point is to be seen, however. Thus the Oosterhoff state correlation, in which S_1 did not interact with the S_0 and S_2 configurations, results not only from Brillouin's theorem precluding interaction of S_1 with S_0 but also from the exact symmetry of the cyclobutadiene — cyclobutene reacting system which imparts different symmetries to the S_1 and S_2 configurations. With different symmetries, S_1 and S_2 also cannot interact. More generally, a reacting system will not have perfect symmetry. Even in the butadiene — cyclobutene case, one can expect molecular deformations to break up this symmetry.

In such rearrangements as the bicycle process it has been observed that the S_1 state becomes progressively more heavily weighted in S_0 and S_2 configurations as the bifunnel is approached [53]. Furthermore, an interesting and general analysis is possible. Thus, in absence of symmetry effects the S_2 configuration will admix with S_0 and S_1 but S_0 and S_1 will not mix. This is analogous to the MO problem of the allyl species in that there are three basis orbitals interacting in a "linear" fashion, except that here the basis orbitals are configurational wavefunctions rather than atomic orbitals. Also here the matrix elements leading to admixture are positive (energy raising, deriving from repulsions) whereas in the allyl problem the off-diagonal elements are negative (bonding). The fact that the basis orbitals are of unequal energy in the surface problem is of lesser consequence. Hence we have a S_0--S_2--S_1 linear array and an "inverted allyl-like" resultant set of states (surfaces). The highest energy surface takes S_0, S_2, and S_1 in positive combination. The lowest state, S_0, not to be confused with the configuration S_0 before configuration interaction, should be weighted as $S_0 - S_2 + S_1$. The S_1 state should have a single change in sign along the string of three configurations with the center configuration vanishing only if the energies were identical which they are not. S_1 is found to have the form $S_0 + S_2 - S_1$ [53].

The important conclusion is that the S_1 state becomes progressively more heavily weighted in S_0 and S_1 configurations as a bifunnel is approached. Thus Möbius-Hückel considerations are relevant.

5 Two Useful MO and QM Methods

5.1 Delta-P and Delta-E Matrices: Prediction of Molecular Geometric Relaxation, and Excitation Energy Distribution

A challenging problem is determination of the distribution of electronic excitation energy in a molecule. A priori, one might inspect the excited state wavefunction and try to compare it with the ground state wavefunction in different portions of the molecule. This is rather tedious and difficult. Easier is the approach we described some years ago [50,52,53,59-63]. This involves obtaining an overlap population matrix or bond order matrix P^* for the excited state of interest; any off-diagonal element P_{rt} gives the bond order between orbitals r and t. Then one subtracts the ground state matrix, P_0, as in Equation 9 to obtain a delta-P matrix.

$$\Delta P = P^* - P_0 \tag{9}$$

Where the bond orders do not really differ in the excited state from those of the ground state, the corresponding ΔP_{rt} element will be near zero. Alkyl substituents included in calculations will tend to contribute such small ΔP_{rt} elements as will other portions of the molecule not appreciably excited. Where the ΔP_{rt} term is negative, this overlap has become more antibonding in the excited state than in the ground state; this corresponds to a more energy-rich part of the molecule where excitation energy is concentrated. Such overlaps tend to diminish by stretching of bonds or twisting of π type bonding. Such diminution of antibonding drives the molecule towards a bifunnel by diminishing the $S_0 - S_1$ or $S_0 - T_1$ energy gap. For ΔP_{rt} elements which are positive, that locale of the molecule is more bonding than in the ground state and energy poor. Such overlaps tend to increase with bond compression, also leading the molecule towards a bifunnel.

Where hybrid orbitals are involved it is more precise to use corresponding ΔE_{rt} values as has been noted in these studies; note Equation 10.

Each term can be dissected into one- and two-center components which then are individually summed to give elements of a ΔE matrix. Thus energetic and unperturbed portions of an excited species can be described.

With either variation, the method nicely indicates the fate of excitation energy and

$$\Delta E\,(k \rightarrow l) = \underbrace{H_{ll} - H_{kk}}_{\substack{\text{One-electron}\\ \text{promotion}\\ \text{energy}}} + \underbrace{2\sum_{w}^{occ} G_{wlwl} - 2\sum_{w}^{occ} G_{wkwk}}_{\substack{\text{Change in electron-electron}\\ \text{repulsion with one e each in}\\ \text{MO k and l compared with GS}}} \tag{10}$$

$$\underbrace{+ \sum_{w}^{occ} G_{wkkw} - \sum_{w}^{occ} G_{wllw} + 2G_{kllk} - G_{klkl}}_{\text{Change in exchange energy stabilization on excitation}}$$

helps predict reactivity. Since P matrices are available for all levels of sophistication ranging from simple Hückel calculations to SCF-CI, the method is of broad applicability.

Often qualitative MO's can be used to predict reactivity [59]. Thus reference to Fig. 17 provides one example of the qualitative approach. Here alpha-expulsion is considered. The four orbital sequence is butadienoid and hence allows writing the wavefunction for the π system in qualitative form as shown. The negative ΔP_{34} indicates that this bond will relax by stretching.

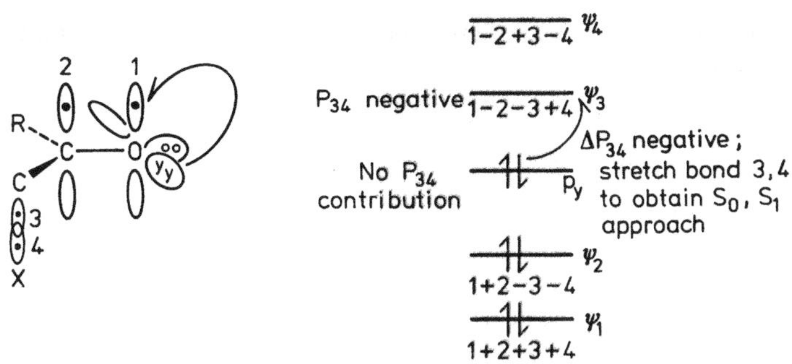

Fig. 17. Alpha-Expulsion and the Delta-P Method

5.2 The Large K — Small K Concept; Control of Reaction Course by Multiplicity

Occasionally in photochemistry it is observed that the singlet reacts differently than the triplet. A useful generalization has been developed in our research which deals with this phenomenon [49, 53, 64]. This is seen most readily in terms of the chemistry of 1,1,3,3-tetraphenyl-5,5-dicarbomethoxy-1,4-pentadiene in Scheme 6.

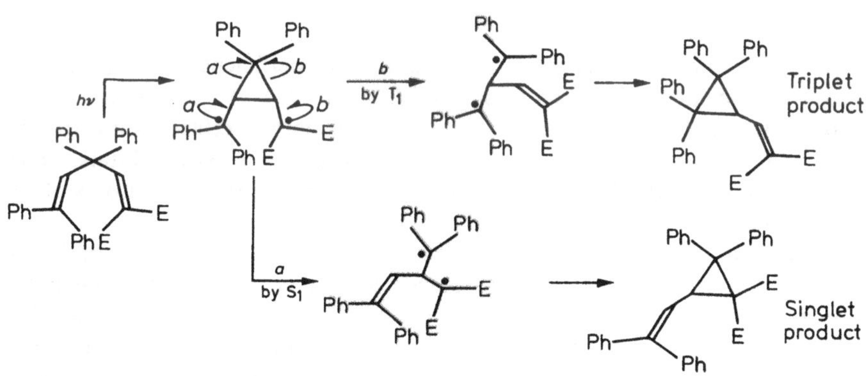

Scheme 6. Multiplicity Control of the Di-π-Methane Rearrangement

Here which bond of the cyclopropyldicarbinyl diradical opens, and hence which product is formed, is determined by the spins of the odd electrons. Fig. 18 schematically shows the energies of the two alternative pathways available to the diradical (i.e. energies taken from the calculated potential energy surfaces). It is seen that for situations such as that in Fig. 18, the lower energy singlet pathway is the one designated the "Small K" process, and the lower energy triplet pathway is the "Large K" route. K here is the exchange integral, which before configuration interaction is just half the singlet-triplet (i.e. $S_1 - T_1$) energy splitting.

In general we formulate the rule that singlets prefer "small K pathways" while triplets prefer "large K pathways".

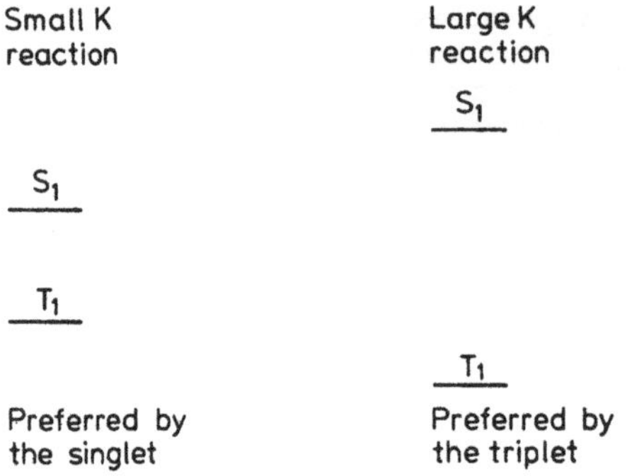

Fig. 18. Large K Versus Small K Processes

Then the question is how one determines independently whether a reaction is of the small K or large K. type. One can obtain the singlet-triplet splittings from poly-electron calculation as has been done in the above cases in order to understand the source of the effect.

Secondly, one can obtain the value of K approximately by use of Hückel or SCF LCAO MO coefficients (Equation 11) [65]. We note that

$$K_{kl} = \sum_{r,t} C_{rk}C_{rl}C_{tk}C_{tl}G_{rt} \tag{11}$$

this involves the summation of pairs of products of coefficients (as $C_{rk}C_{rl}$) with one Hückel coefficient being taken from HOMO (i.e. MO k) and the other from LUMO (i.e. MO l); but the largest G_{rt} occurs when both coefficients are for the same atom r (or t). The summation is over all pairs of atoms or basis orbitals. Thus, where HOMO and LUMO do not appear heavily localized at the same atoms of the molecule, K tends to be be small.

While Equation 11 can be used to estimate the relative magnitudes of K for different excited state transition states or species, a number of generalizations prove more convenient. For example: 1. Pericyclic reactions tend to have small K's (HOMO and LUMO tend not to match). 2. Double bond twisting tends to give a large K. 3. Diradical species with electron withdrawing groups on diradical centers have diminished K's. 4. Processes separating odd-electron centers increase K. And, there are further rules.

6 The Utility of Resonance and Lewis Structure Reasoning in Organic Photochemistry

A final comment is required about our mechanistic approach to organic photochemistry. It is clear that this is bifurcated. On one hand one uses MO calculations to obtain predictions and excited state descriptions. On the other hand, he writes

Lewis type structures to represent excited states and then uses arrow notation reminiscent of ground state chemistry to predict reactivity. The question is whether the latter is too naive. The answer is no! Throughout our studies we have found that excited state reactivity follows the guidelines one uses for ground state behavior, except that the structures with which one is dealing are now electronically excited. With reasonable approximations for excited state structures, organic intuition based on precedent and the requirement for continuous electron redistribution leads one, if not to the actual reaction product, at least to a potential product. The requirement for continuous electron redistribution was one we postulated 21 years ago and it has proven useful. Electron pushing is rapid and convenient. QM methods such as MO Following, surface calculations, etc. tend to provide detail. The two approaches are complementary and one makes maximum progress by using both.

7 References

1. H. E. Zimmerman: Abstracts 17th Nat. Organic Symp., Bloomington, IN, 1961, page 31.
2. H. E. Zimmerman, D. I. Schuster: J. Amer. Chem. Soc. *83*, 4486 (1961).
3. H. E. Zimmerman, D. I. Schuster: J. Amer. Chem. Soc. *84*, 4527 (1962).
4. H. E. Zimmerman: "Advances in Photochemistry", 1, 193–208 (1963) (A. Noyes, Jr., G. S. Hammond, J. N. Pitts, Jr. Eds.) Interscience, New York 1963.
5. H. E. Zimmerman: Tetrahedron *19*, Suppl. 2, 393 (1962).
6. H. E. Zimmerman: Pure and Appl. Chem. *9*, 493 (1964).
7. This was also presented in some detail in a 1959 NSF Proposal and numerous lectures by the author.
8. E. Havinga, R. O. DeJongh, W. Dorst: Rec. Trav. Chim. *75*, 378 (1956); note also ref. 9.
9. For a useful review to recent publications note E. Havinga and J. Cornelisse: Pure Appl. Chem. *47* (1976).
10. H. E. Zimmerman, S. Somasekhara: J. Amer. Chem. Soc. *95*, 922 (1963).
11. H. E. Zimmerman, V. R. Sandel: J. Amer. Chem. Soc. *95*, 915 (1963).
12. H. E. Zimmerman: Science *153*, 837 (1966).
13. (a) R. Daudel, R. Lefebvre, C. Moser: Quantum Chemistry, Interscience, New York 1955; (b) Ng. Ph. Buu-Hoi, P. Daudel, R. Daudel, P. Jacquignon, G. Morin, R. Muxart, and C. Sandorfy: Bull Soc. Chim. France C. 132 (1951).
14. H. E. Zimmerman, R. W. Binkley, R. S. Givens, and M. Sherwin: J. Amer. Chem. Soc. *89*, 3932–3933 (1967).
15. H. E. Zimmerman, P. S. Mariano: J. Amer. Chem. Soc. *91*, 1718–1727 (1969).
16. H. E. Zimmerman: "Rearrangements in Ground and Excited States", *3*, 131—166 (P. DeMayo (ed.)) Academic Press, NY 1981.
17. H. E. Zimmerman, J. S. Swenton: J. Amer. Chem. Soc. *89*, 1436 (1964).
18. H. E. Zimmerman, J. S.: Swenton: J. Amer. Chem. Soc. *89*, 906 (1967).
19. H. E. Zimmerman, R. W. Binkley, J. J. McCullough, and G. A. Zimmerman: J. Amer. Chem. Soc. *89*, 6589 (1967).
20. H. E. Zimmerman: Angewandte Chemie, Intern. Edit., *8*, 1 (1969).
21. (a) H. E. Zimmerman, G. E. Keck: J. Amer. Chem. Soc. *97*, 3527 (1975); (b) H. E. Zimmerman, G. E. Keck, J. L. Pflederer: J. Amer. Chem. Soc. *97*, 5574 (1975).
22. (a) H. E. Zimmerman, R. J. Pasteris: J. Org. Chem. *45*, 4864 (1980).
23. H. E. Zimmerman, R. J. Pasteris: J. Org. Chem. *45*, 4876 (1980).
24. H. E. Zimmerman, D. Döpp, P. S. Huyffer: J. Amer. Chem. Soc. *88*, 5352 (1966).
25. H. E. Zimmerman, D. S. Crumrine: J. Amer. Chem. Soc. *90*, 5612 (1968).
26. H. E. Zimmerman, D. S. Crumrine, D. Döpp, and P. S. Huyffer: J. Amer. Chem. Soc. *91*, 434 (1969).

27. M. Kasha: "Comparative Effects of Radiation" (ed. M. Burton, J. S. Kirby-Smith, and J. L. Magee) Wiley, New York 1960, p. 72–97, note esp. pages 87–89; (b) The presentation in Ref. 27a needs to be modified slightly in the present discussion to conform to current organic mechanisms. For example, hydrogen abstraction is from a carbinol carbon rather than from the oxygen.
28. M. Kasha: "Light and Life" (W. D. McElroy, B. Glass (eds.) Johns Hopkins Univ. Press, Baltimore, 1961, note esp. pg. 54 for mention of radical-like reactivity.
29. A. Padwa: Tetrahedron Letts 3465 (1964).
30. C. Walling, M. J. Gibian: J. Amer. Chem. Soc. *87*, 3361 (1965).
31. This is a modification of notation used by G. Wheland, "Resonance in Organic Chemistry", Wiley, New York 1955, pg. 282, which carried only two dimensional connotation.
32. (a) R. Hoffmann: J. Chem. Phys. *39*, 1397 (1963); (b) idem, ibid., 2480 (1964).
33. Note N. C. Yang, D. H. Yang: J. Amer. Chem. Soc. *80*, 2913 (1958) for an early report of this reaction.
34. H. E. Zimmerman, B. R. Cowley, C-Y. Tseng, and J. W. Wilson: J. Amer. Chem. Soc. *86*, 947 (1964).
35. H. E. Zimmerman: J. Amer. Chem. Soc. *88*, 1564 (1966).
36. (a) M. J. S. Dewar: Tetrahedron, Suppl., *8*, 75 (1966).
This paper, in referencing the earlier Zimmerman Möbius-Hückel publication, agrees in conclusions and philosophy; (b) Similarly, a subsequent publication, M. J. S. Dewar: "Aromaticity", The Chemical Society, London 1967, esp. pages 212—213, concurred in philosophy and Möbius-Hückel nomenclature but referenced only 36a.
37. A. Frost, B. Musulin: J. Chem. Phys. *21*, 572 (1953).
38. H. E. Zimmerman: Accounts Chem. Res. *4*, 272 (1971).
39. H. E. Zimmerman: J. Amer. Chem. Soc. *88*, 1566 (1966).
40. (a) R. B. Woodward, R. Hoffmann: "The Conservation of Orbital Symmetry, Verlag Chemie, Weinheim .1970; (b) This publication has convenient generalized Selection Rules for Pericyclic Reactions not available in the earlier publications and which nicely show the interrelationship between this approach and the Möbius-Hückel one.
41. H. E. Zimmerman: Accounts Chem. Res. *5*, 393 (1972).
42. (a) B. Bigot, A. Devaquet, N. J. Turro: J. Chem. Soc. *103*, 6 (1981); (b) The derived diagrams were said to be "natural correlations" showing intended crossings. However, just as MO's 1 and 2 of allyl show no intention of being degenerate, there appears no reason for crossing here.
43. L. Salem: J. Amer. Chem. Soc. *96*, 3486 (1974).
44. H. E. Zimmerman, R. W. Binkley, R. S. Givens, and M. A. Sherwin: J. Amer. Chem. Soc. *89*, 3932 (1967).
45. H. E. Zimmerman: Photochem. and Photobiol., *7*, 519 (1968).
46. H. E. Zimmerman, R. W. Binkley, R. S. Givens, G. L. Grunwald, and M. A. Sherwin: J. Amer. Chem. Soc. *91*, 3316 (1969).
47. H. E. Zimmerman, R. J. Boettcher, N. Buehler, G. E. Keck, and M. G. Steinmetz: J. Amer. Chem. Soc. *98*, 7680 (1976).
48. H. E. Zimmerman, D. R. Diehl: J. Amer. Chem. Soc. *101*, 1841 (1979).
49. H. E. Zimmerman, D. Armesto, M. G. Amezua, T. P. Gannett, and R. P. Johnson: J. Amer. Chem. Soc. *101*, 6367 (1979).
50. H. E. Zimmerman, R. E. Factor: J. Amer. Chem. Soc. *102*, 3538 (1980).
51. H. E. Zimmerman, T. P. Cutler: Chem. Communic. 232 (1978).
52. H. E. Zimmerman, T. P. Cutler: J. Org. Chem. *43*, 3283 (1978).
53. H. E. Zimmerman, R. E. Factor: Tetrahedron *37*, Suppl. 1, 125 (1981).
54. J. Michl: Molec. Photochem. *4*, 243 (1972).
55. H. C. Longuet-Higgins, E. W. Abrahamson: J. Amer. Chem. Soc. *87*, 2045 (1965).
56. H. E. Zimmerman, K. S. Kamm, D. P. Werthemann: J. Amer. Chem. Soc. *97*, 3718 (1975).
57. E. Teller: J. Phys. Chem. *41*, 109 (1937).
58. W. Th. A. M. van der Lugt, L. Oosterhoff: J. Amer. Chem. Soc. *91*, 6042 (1969).
59. H. E. Zimmerman, M. G. Steinmetz: Chem. Communic. 230 (1978).
60. H. E. Zimmerman, W. T. Gruenbaum, R. T. Klun, M. G. Steinmetz, and T. R. Welter: Chem. Communic. 228 (1978).

61. H. E. Zimmerman, R. T. Klun: Tetrahedron *34*, 1775 (1978).
62. H. E. Zimmerman, M. G. Steinmetz, C. L. Kreil: J. Amer. Chem. Soc. *100*, 4146 (1978).
63. H. E. Zimmerman, T. R. Welter: J. Amer. Chem. Soc. *100*, 4131 (1978).
64. H. E. Zimmerman, J. H. Penn: Proceedings Natl. Sci. U.S.A. *78*, 2021 (1981).
65. H. E. Zimmerman: Quantum Mechanics For Organic Chemists, Academic Press, New York, 1975.

New Developments in Polymer Synthesis

Takeo Saegusa

Department of Synthetic Chemistry, Faculty of Engineering, Kyoto University, Kyoto, Japan

Table of Contents

Chemistry of polymer synthesis is still making a steady progress and becoming diversified. During the past few years, many interesting discoveries have been made and various new materials of novel functions have been reported. The fundamentals of polymerization chemistry are

1. the design of polymerization catalysts,
2. the molecular design of monomers,
3. the design of polymerization reactions, and
4. the molecular engineering of polymers.

The present article describes new developments in polymer synthesis and illustrates some examples of the above four principles. It is not a comprehensive review, and the topics have been picked out according to the author's interests of research.

1 Photoinitiators for Cationic Polymerization [1)]

As an example of the design of polymerization catalysts, a group of new polymerization catalysts are described here, which are activated by UV irradiation to initiate cationic polymerization.

Photoinitiation of radical polymerization has long been known. Recently, a group of photoinitiators for cationic polymerization hase been discovered and developed by Crivello et al. [1)]. They include diaryliodonium (*1*), [2)] triarylsulfonium (*2*), [3-5)] dialkylphenacylsulfonium (*3*), [6)] and dialkyl-4-hydroxyphenylsulfonium salts (*4*) [7)].

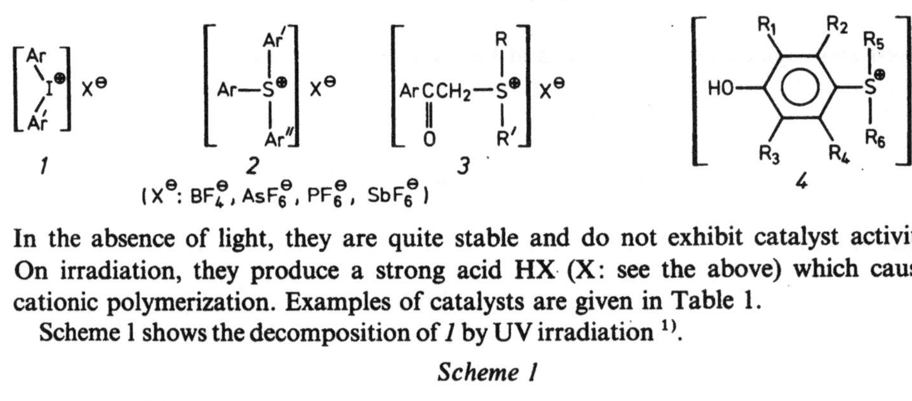

1 2 3 4

($X^{\ominus}$: $BF_4^{\ominus}$, $AsF_6^{\ominus}$, $PF_6^{\ominus}$, $SbF_6^{\ominus}$)

In the absence of light, they are quite stable and do not exhibit catalyst activity. On irradiation, they produce a strong acid HX (X: see the above) which causes cationic polymerization. Examples of catalysts are given in Table 1.

Scheme 1 shows the decomposition of *1* by UV irradiation [1)].

Scheme 1

Major

$$Ar_2I^+X^- \xrightarrow{h\nu} [Ar_2I^+X^-]^* \quad (1)$$

$$[Ar_2I^+X^-]^* \longrightarrow Ar{-}I^{+\cdot} + Ar\cdot + X^- \quad (2)$$

$$Ar{-}I^{+\cdot} + Y{-}H \longrightarrow Ar{-}I^+{-}H + Y\cdot \quad (3)$$

$$Ar{-}I^+{-}H \longrightarrow Ar{-}I + H^+ \quad (4)$$

Minor

$$[Ar_2I^+X^-]^* + Y{-}H \longrightarrow [Ar{-}Y{-}H]^+ + ArI + X^- \quad (5)$$

$$[Ar{-}Y{-}H]^+ \longrightarrow ArY + H^+ \quad (6)$$

(Y—H: solvent, monomer)

Table 1. Iodonium and sulfonium salts as photoinitiators for cationic polymerization

Diaryliodonium salts [2)]

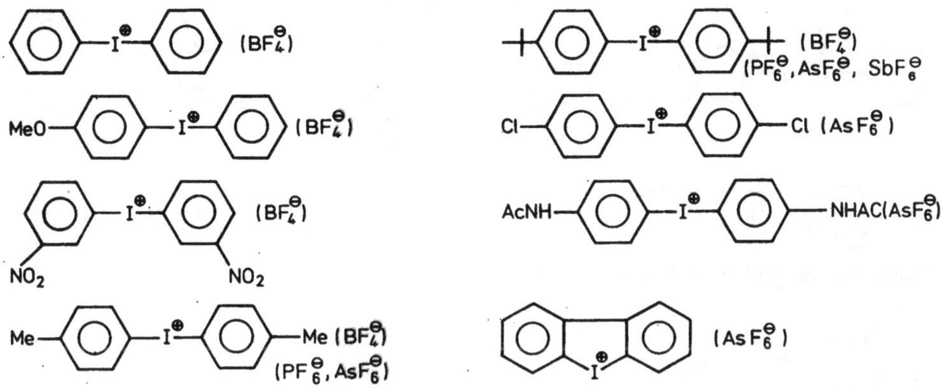

Triarylsulfonium salts [3,4)]

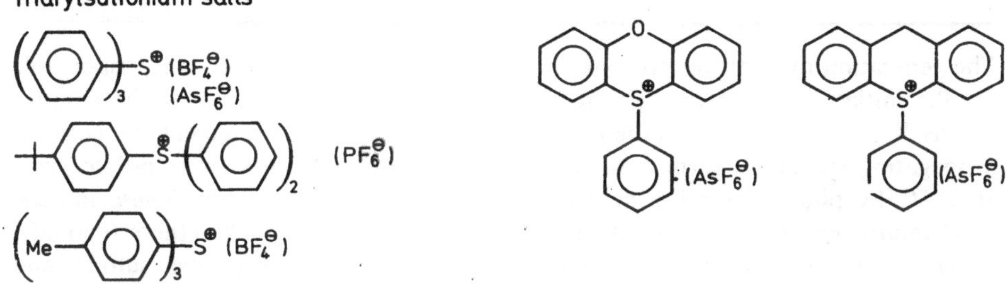

Dialkylphenacylsulfonium salts [5)]

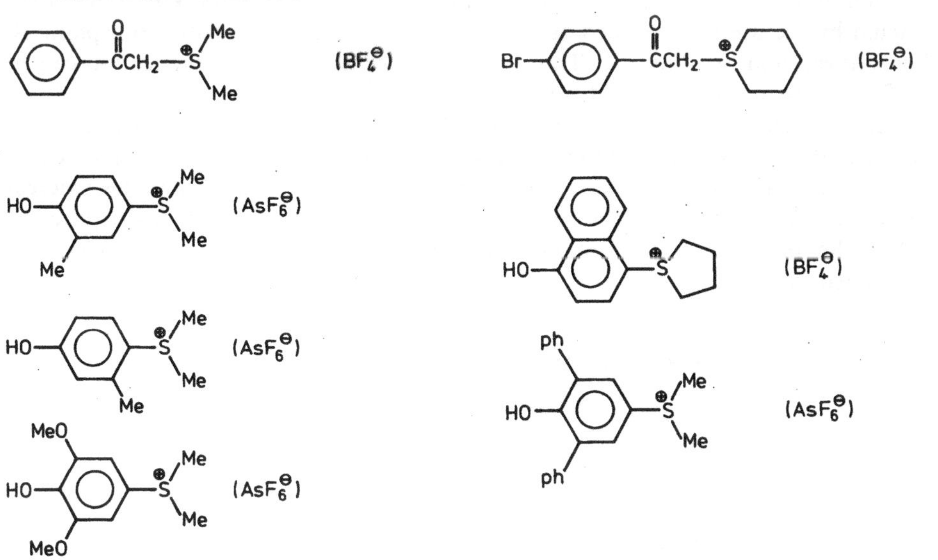

Table 1. (Continued)

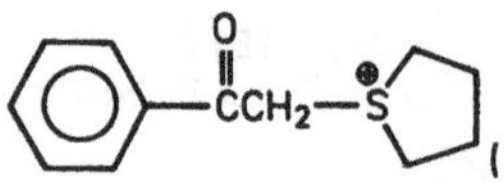

$(BF_4^{\ominus})$ $(PF_6^{\ominus}, AsF_6^{\ominus}, SbF_6^{\ominus})$

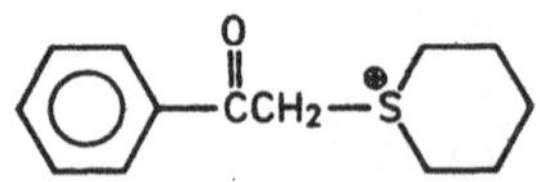

$(AsF_6^{\ominus})$

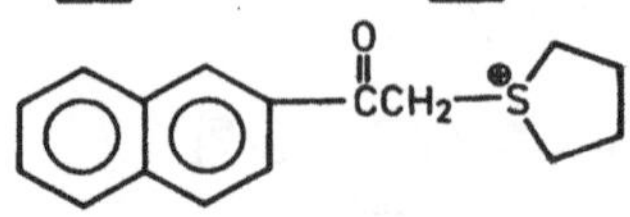

$(BF_4^{\ominus})$

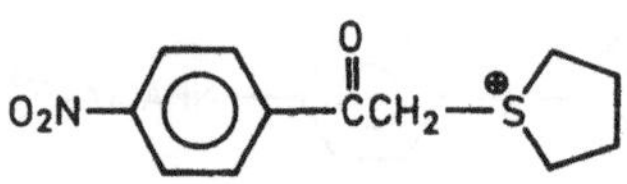

$(BF_4^{\ominus})$

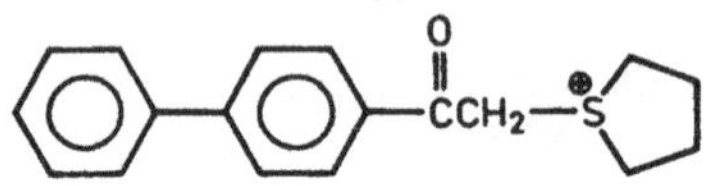

$(BF_4^{\ominus})$

Dialkyl-4-hydroxyphenylsulfonium salts [e]

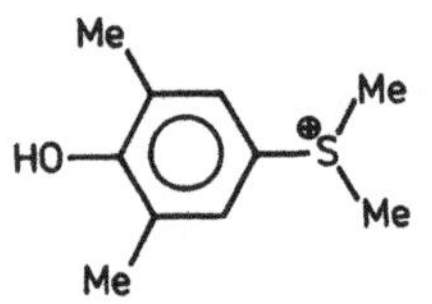

$(BF_4^{\ominus})$ $(AsF_6^{\ominus})$

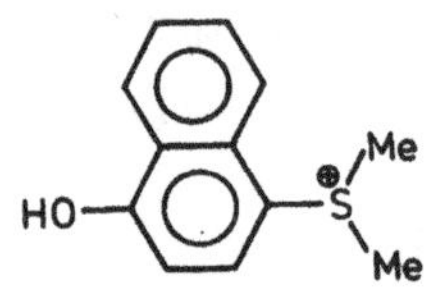

$(BF_4^{\ominus})$

The generation of a strong protonic acid HX in the above photolysis is responsible for the initiation of cationic polymerization. It should be noted that the counter anions X^- are very weak (stable) nucleophiles which do not intercept the cationic propagating species to form covalent bonds. Various monomers were polymerized at 25 °C by photoirradiation at wavelengths shorter than 360 nm. These include vinyl monomers (styrene, α-methylstyrene and vinyl ether), cyclic ethers (epoxide, oxetane, tetrahydrofuran and trioxane), cyclic sulfides (propylene sulfide and thietane), lactones (ε-caprolactone) and spiro bicyclic orthoester. The photodecomposition of these diaryldiazonium salts *1* can be sensitized at wavelengths longer than 360 nm by the use of dyes such as Acridine orange, Acridine yellow, Phosphine R, Benzoflavin, and Setoflavin T. Thus, photoinitiated cationic polymerization can be performed by incadescent light sources or even ambient sunlight [1].

The polymerization rate depends on both the reactivity of monomers and the nature of the counter anion of the initiator salt. In the fastest case (3-vinylcyclohexene oxide), a quantitative conversion was attained at 25 °C within 1.5 minutes whereas in the slowest case (ε-caprolactone), irradiation at 60 °C for 60 min was required.

For the photodecomposition of triarylsulfonium salts *2*, Scheme 2 has been postulated [3].

Scheme 2

$$Ar_3S^+X^- \xrightarrow{h\nu} [Ar_3S^+X^-]^* \tag{7}$$

$$[Ar_3S^+X^-]^* \longrightarrow Ar_2S^{\dot{+}} + Ar\cdot + X^- \tag{8}$$

$$Ar_2S^{\dot{+}} + YH \longrightarrow Ar_2S^+{-}H + Y\cdot \tag{9}$$

$$Ar_2S^+{-}H \longrightarrow Ar_2S + H^+ \tag{10}$$

$$2\,Ar\cdot \longrightarrow Ar{-}Ar \tag{11}$$

$$Ar\cdot + YH \longrightarrow ArH + Y\cdot \tag{12}$$

The generation of a protonic acid (HX) is responsible for the initiation of cationic polymerization. Various monomers are polymerized by sulfonium photoinitiators. Especially this system has been shown to be potentially useful for UV curable coatings of metal and plastics with epoxy resins.

As is seen in the above scheme (Eqs. (7) to (12)), some free radical species (Ar· and Y·) are also produced as transient intermediates. Therefore, the photolysis of these sulfonium salts also initiates free-radical polymerization[4]. The amphi-functional character of sulfonium salts was demonstrated by the following series of experiments. Irradiation of an equimolar mixture of 1,4-cyclohexene oxide (7-oxabicyclo[4.1.0]heptane) and methyl methacrylate including $Ph_3S^+ \cdot SbF_6^-$ as the photoinitiator gave a mixture of two homopolymers. Thus, both cationic (cyclohexene oxide) and free-radical (methyl methacrylate) polymerizations took place independently. The same system containing 2,6-di-*t*-butyl-4-methylphenol (radical inhibitor) gave only poly(cyclohexene oxide). Alternatively, the system with triethylamine (poison for cationic species) yielded only poly(methyl methacrylate). Monomers such as glycidyl acrylate and glycidyl methacrylate which contain functional groups capable of cationic and free-radical polymerizations are converted into a cross-linked insoluble polymer.

The photolysis of dialkylphenacylsulfonium salts *3* generates an ylid *5* and a strong acid HX (Eq. (13)), the latter being responsible for the initiation of cationic polymerization[6].

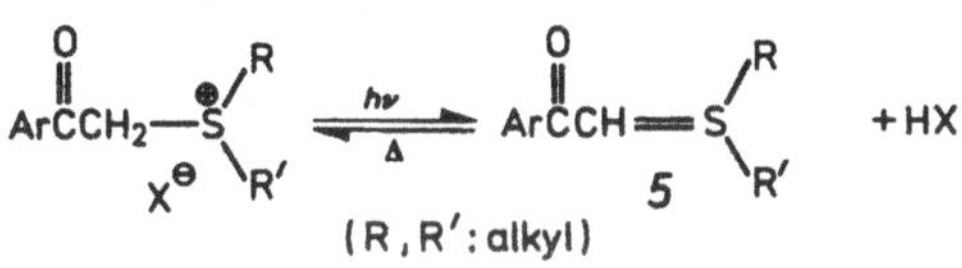

(13)

The above reaction is reversible, and the monomer competes with ylid *5* for the reaction with HX. Those monomers which are more nucleophilic than *5* can be polymerized, i.e. they are epoxides, vinyl ethers and cyclic acetals.

The fourth type of photoirradiated cationic initiator is dialkyl-4-hydroxyphenyl-sulfonium salt *4*[7] (Table 1). Photoexcitation of *4* gives rise to the formation of a resonance-stabilized ylid *6* and an acid HX.

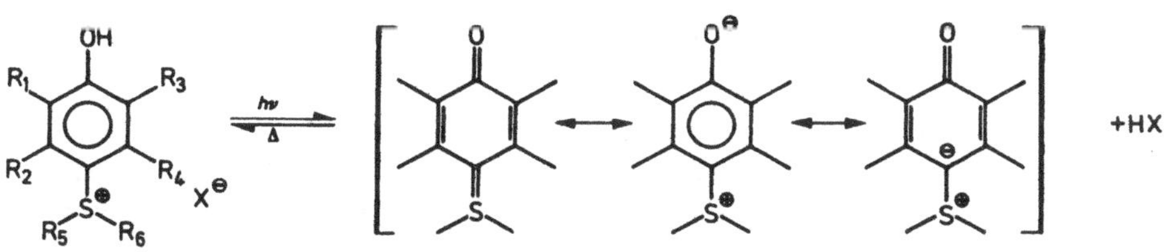

4 *6*

The monomers of styrene oxide, 1,4-cyclohexene oxide, trioxane, and vinyl ether were polymerized at satisfactory rates. However, tetrahydrofuran, ε-caprolactone, and α-methylstyrene could not be polymerized[7].

Photocurable coatings are widely used for metal, plastics wood and paper. Photoinitiated free-radical polymerization, however, can only be applied to vinyl monomers. The studies of Crivello have broadened the scope of monomers. In addition, photoinitiated cationic polymerization is not sensitive toward oxygen (air). Photoinitiated free-radical polymerization sometimes requires working in inert atmosphere in order to avoid the inhibition through oxygen [1].

2 Free-Radical Ring-Opening Polymerization

Ring-opening polymerization is an important field of research in the chemistry of polymer synthesis. Usually, it proceeds by ionic mechanisms, i.e. cationic, anionic and coordinate anionic mechanisms. Research on ring-opening polymerization proceeding via free-radical propagating species in which the so-called "molecular design of monomer" plays an important role has recently been reported.

Several examples of ring-opening polymerization proceeding via free-radical mechanism have been reported, e.g. the free-radical polymerization of vinylcyclopropane *7* and its derivatives having alkoxycarbonyl substituents *9* [8,9]:

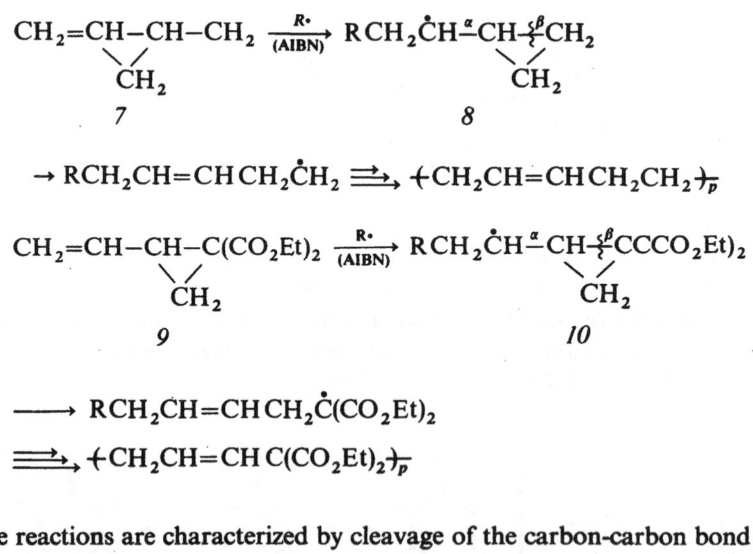

The above reactions are characterized by cleavage of the carbon-carbon bond of the cyclopropane ring at C_β with respect to the carbon radical in the transient species of *8* and *10*. In other words, the opening of the cyclopropane ring takes place according to the so-called "β-scission rule". The driving force for the ring-opening is the relief of the strain in cyclopropane ring.

Another example is the polymerization of the dimer of o-xylylene (*11*), which is also characterized by β-scission of the carbon-carbon bond in a key intermediate species *12*. Acquisition of the resonance-stabilization energy of the benzene ring is the driving force of the reaction.

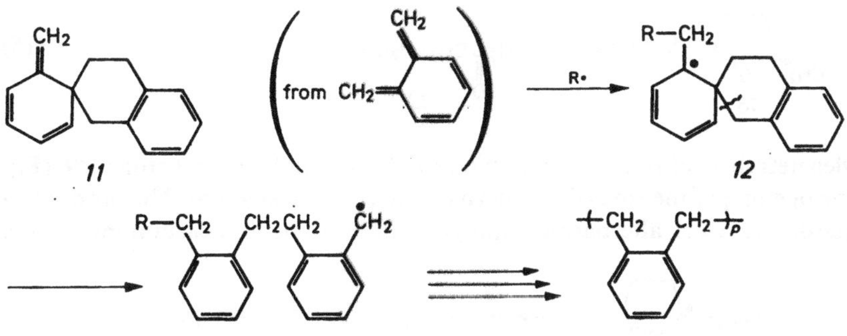

Polymerizations of 1,2-dithiolane *13* [11] and tetrafluorothiirane [12] *14* are also known to proceed via free-radical propagating species.

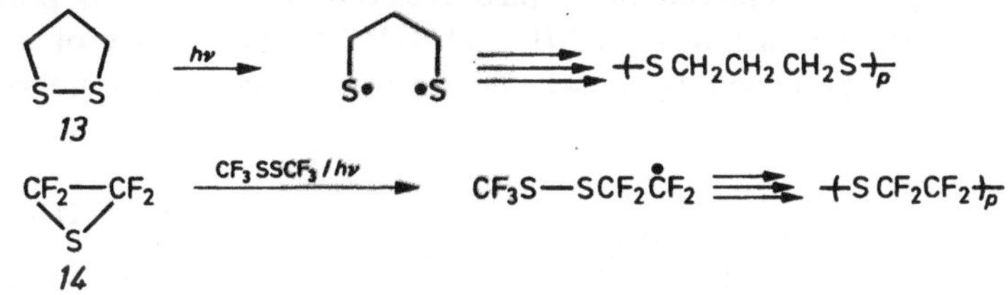

Very recently, Bailey and Endo have enlarged the scope of the free-radical ring-opening polymerization. [13–15] Several examples of their studies are described below. All the polymerizations are represented by the following general pattern of fundamental reaction. It is seen that the free-radical ring-opening by a β-scission mechanism is coupled with the addition of a free radical to the carbon-carbon double bond [13].

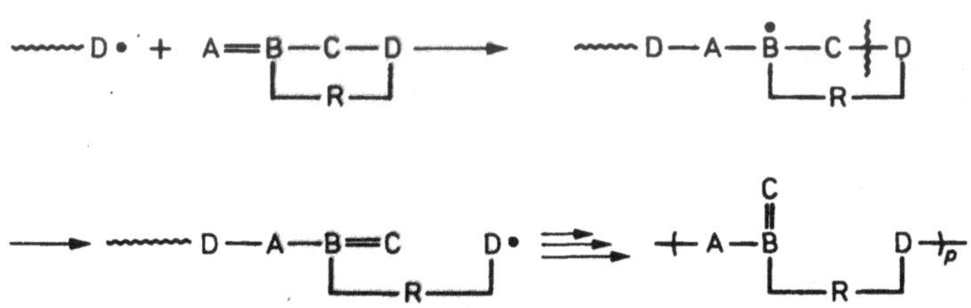

2-Methylene-1,3-dioxolane (*15*) is polymerized by peroxide initiator to produce a polyester *16* which is regarded as the product of the hypothetical ring-opening polymerization of the non-polymerizable heterocycle of γ-butyrolactone (Eq. (14)). [13]. The opening of the cyclic species bearing a free radical *17* is the key step.

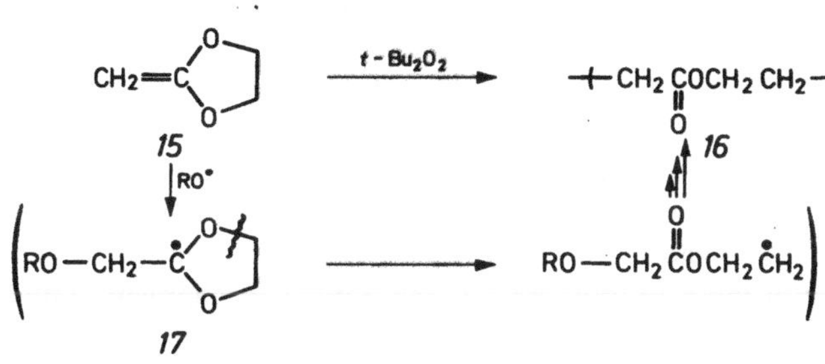

(14)

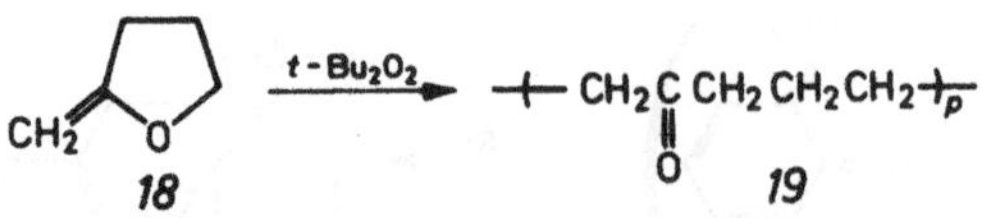

(15)

2-Methylenetetrahydrofuran *18* is polymerized to a polymeric ketone *19* (Eq. (15))[13]. The opening of the ring of an intermediate *20* is the key step. The product *19* may be regarded as a 2:1 alternating copolymer of ethylene and carbon monoxide.

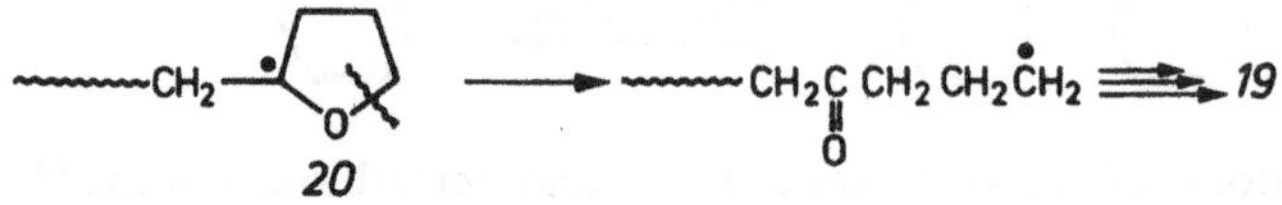

A bis-methylene compound of a spiro skeleton structure *21* is polymerized to produce a poly(ether carbonate) *22* (Eq. (16))[13-14]. The course of the polymerization has been explained by a scheme involving free-radical intermediates *23* and *24*.

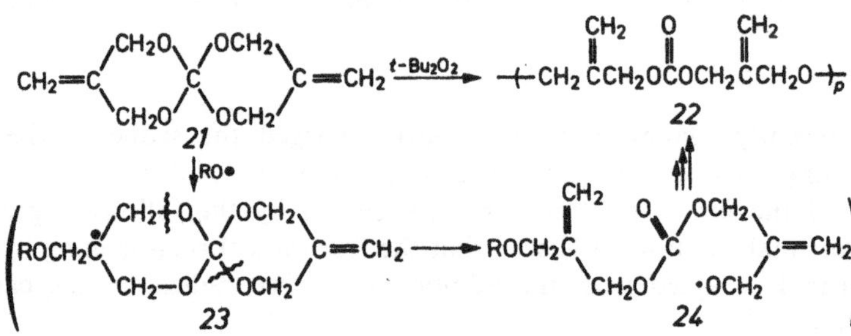

(16)

The same polymer was also obtained by cationic polymerization of *21*. The *21*→*22* polymerization is characterized by volume expansion.

The following cyclic compounds *25* and *26* were also polymerized by a free radical initiator[13,15].

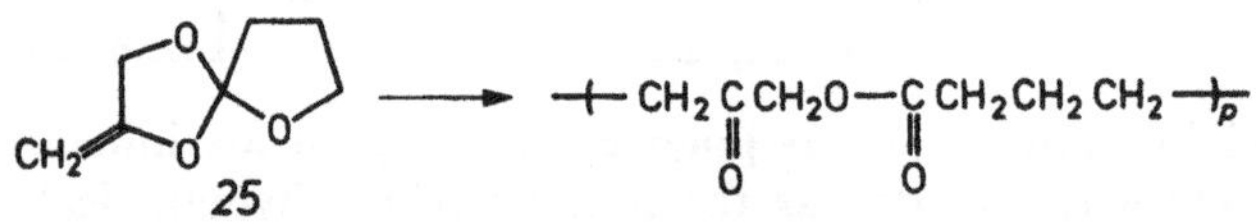

(17)

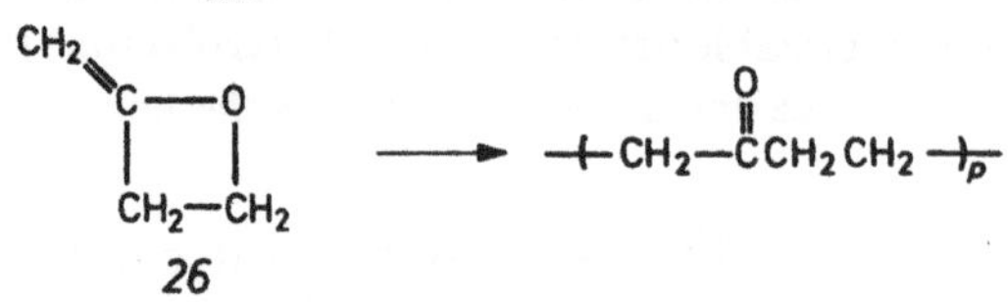

(18)

The above examples of free-radical ring-opening polymerization, which have been explored by Bailey and Endo, produce polymers containing ketonic carbonyl and/or ester groups in the main chain. In addition, these cyclic monomers can be copolymerized with vinyl monomers by free-radical mechanism. Thus, the variety of the polymers produced by radical polymerization has been enlarged.

3 No Catalyst Copolymerization via Zwitterion Intermediates

The above subject has been selected as an example of the design of new polymerization reactions. It is concerned with a concept of copolymerization by spontaneous initiation and subsequent propagation via zwitterion intermediates. This copolymerization, which was invented and has been developed by the author, is called "no catalyst copolymerization". It is very characteristic since usual polymerization reactions require either an initiator or a catalyst.

The above concept is based on the fact that in organic chemistry the reaction between a nucleophile and an electrophile proceeds without any catalyst. The new copolymerization consists of the combination between a monomer (M_N) having nucleophilic reactivity and a monomer (M_E) having electrophilic reactivity. The interaction between these two monomers generates a zwitterion *27* called "genetic zwitterion".

$$M_N + M_E \rightarrow \underset{27}{{}^{+}M_N - M_E^{-}} \qquad (19)$$

Genetic zwitterion *27* is the key intermediate of the copolymerization.

Two moles of *27* react with each other to produce the propagating species *28* (Eq. (20)) which grows by successive addition of *27* (Eq. (21)).

$$27 + 27 \rightarrow \underset{28}{{}^{+}M_N - M_E M_N - M_E^{-}} \qquad (20)$$

$${}^{+}M_N - M_E M_N - M_E^{-} + 27 \times n \rightarrow \underset{29'}{{}^{+}M_N (M_E M_N)_{n+1} M_E^{-}} \qquad (21)$$

The reaction between propagating zwitterions *28* and *29* takes place when their concentration becomes high as the polymerization proceeds (Eq. (22)), whereby the molecular weight of the zwitterion sharply increases.

$$\begin{aligned} &{}^{+}M_N (M_E M_N)_m M_E^{-} + {}^{+}M_N (M_E M_N)_n M_E^{-} \\ &\qquad \rightarrow {}^{+}M_N (M_E M_N)_{m+n+1} M_E^{-} \end{aligned} \qquad (22)$$

A series of the above reactions (Eq. (19) to (22)) gives rise to the formation of alternating copolymers. Thus, the above new copolymerization has two characteristics, the one is the spontaneous initiation without any catalyst and the other is the production of a 1:1 alternating copolymer.

A typical and illustrative example is the copolymerization of 2-oxazoline *30* with β-propiolactone *31* which yields a 1:1 alternating copolymer *32* at room temperature. A genetic zwitterion *33* is produced by ring opening of *31* upon attack of the nucleophile *30* (Eq. (23)).

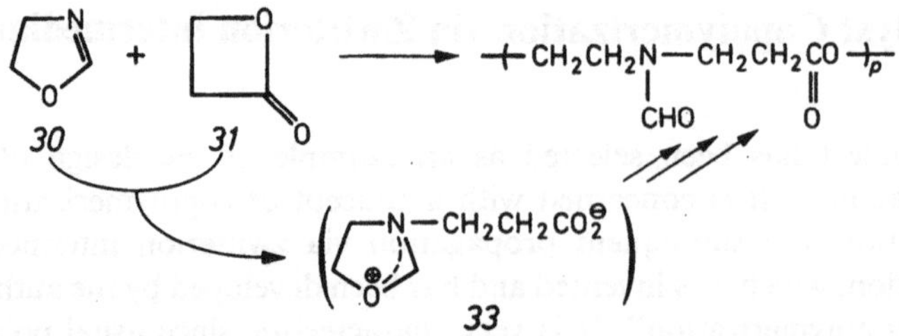

(23)

The growth of propagating zwitterions is due to the opening of the oxazolinium ring of one zwitterion by nucleophilic attack of the carboxylate group of another zwitterion (Eq. (24)).

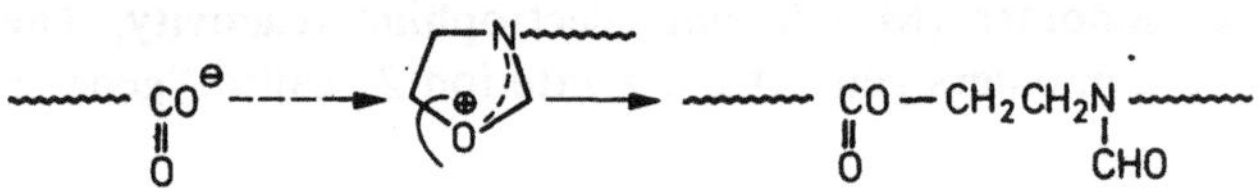

(24)

On the basis of the above concept, many new copolymerizations have been explored using combinations of various M_N and M_E monomers. Table 2 shows some typical examples [13c]. The number of combinations between M_N and M_E monomers is $6 \times 7 = 42$. Among 42 combinations, copolymerizations of over 20 significant combinations were examined and shown to occur spontaneously.

Table 2. Typical M_N and M_E monomers

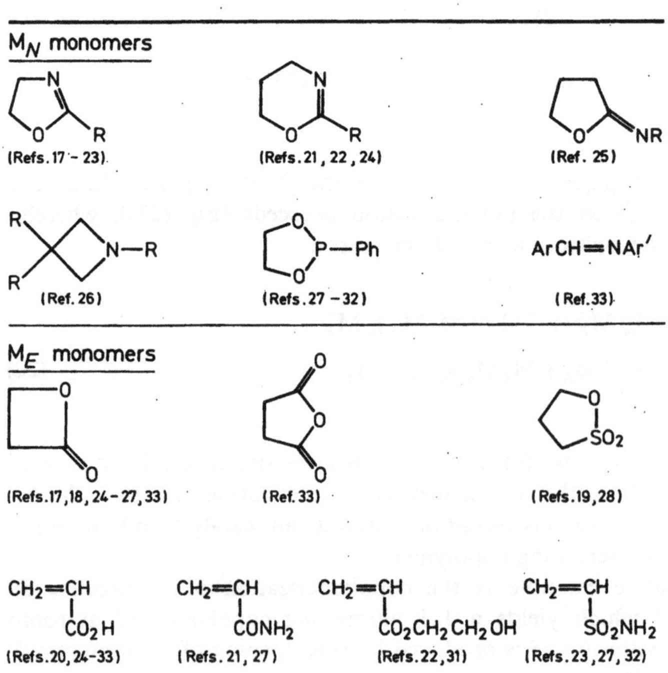

The concept of "no catalyst copolymerization" has demonstrated its usefulness in the exploration of phosphorus-containing polymers. For example, the cyclic phosphonite *34* was successfully copolymerized both with β-propiolactone (*31*) and acrylic acid [35]. These copolymerizations proceed via a common zwitterion *36* and hence produce the same copolymer *37*.

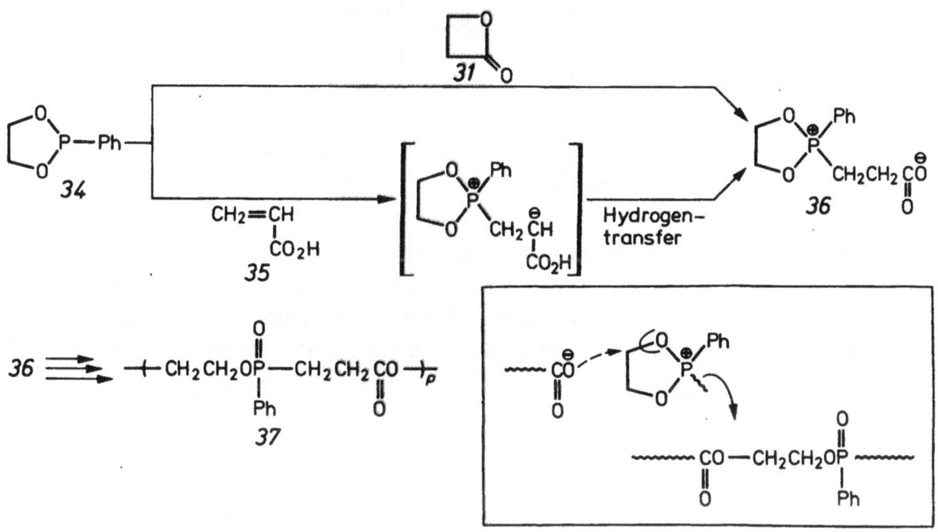

Propagation proceeds through opening of the phosphonium ring, which is located at the end of one zwitterion, by nucleophilic attack of the carboxylate group of another zwitterion. This reaction pattern belongs to the family of the Arbusov reaction (in the rectangle).

In addition to the copolymerizations involving *34* and M_E monomers (Table 2), several new copolymerizations of P(III) compounds have been discovered. One prototype is seen in the combination of a cyclic phosphite *38* and an α-keto acid *39* (Eq. (25)) [34, 35].

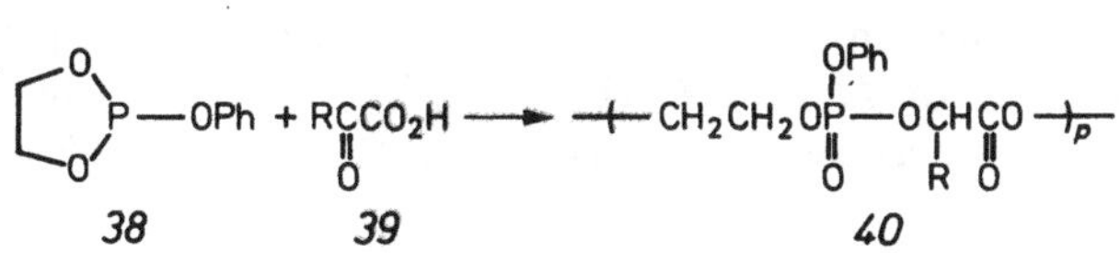
(25)

According to Eq. (25), a cyclic phosphite monomer (M_N) *38* is oxidized to a phosphate unit yielding copolymer *40* whereas the α-keto acid monomer (M_E) *39* is reduced to the corresponding α-hydroxy acid ester. Thus, the term "redox copolymerization" has been proposed to designate this type of copolymerization in which one monomer is reduced and the other monomer oxidized. The redox copolymerization clearly differs from the so-called "redox polymerization" in classical polymer chemistry where the redox reaction between the two catalyst components (oxidant and reductant) is responsible for the production of free radicals.

The key intermediate of the above redox copolymerization is a zwitterion *41*. The propagation step involves opening of the phosphonium ring of a zwitterion by nucleophilic attack of the carboxylate group of another zwitterion (Eq. (26)). Thus, the scheme below also represents the pattern of the Arbusov reaction.

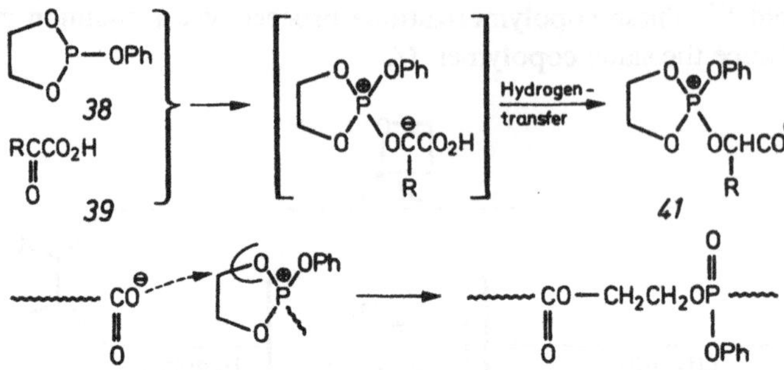

(26)

p-Benzoquinone (*42*) is a reactive and highly polarizable oxidant which is readily involved in redox copolymerizations with various P(III) monomers [36–38]. The copolymerization of salicyl phenyl phosphite *43* with *42* proceeds at room temperature. The opening of the phosphonium ring in zwitterion *45* occurs readily due to of the highly reactive linkage of — $\overset{+}{P}$—OC(O)— in the ring.

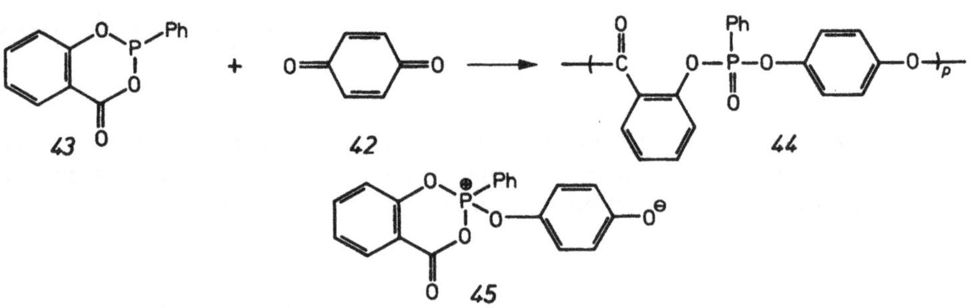

The scope of the spontaneous copolymerization of P(III) monomers has been extended to copolymerizations with more sophisticated regulations of the arrangements of monomeric units in copolymers. They include a 2:1 sequence-ordered binary copolymerization of *43* with *46* (Eq. (27)) [30] and 1:1:1 sequence-ordered terpolymerizations of *34*/acrylate *47*/CO_2 (Eq. (28)) [39] and *48*/*49*/*39* (Eq. (29)) [40].

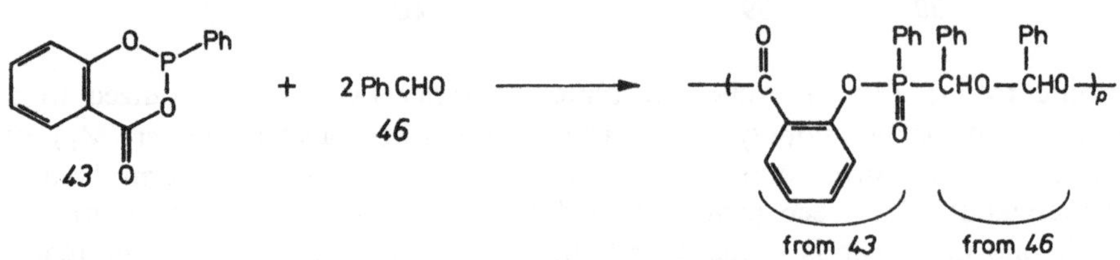

(27)

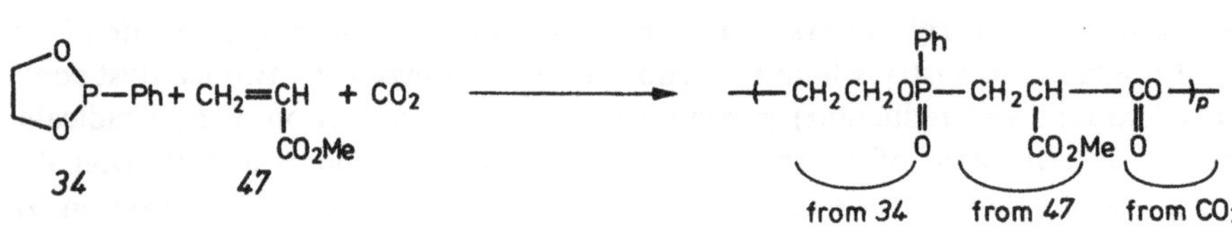

(28)

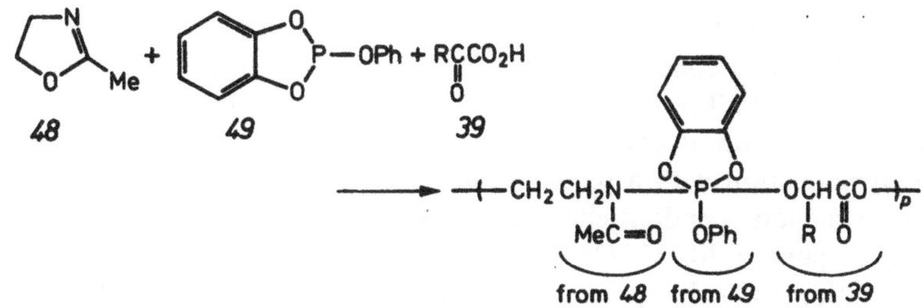

(29)

4 New Developments in the Synthesis of End-Reactive Oligomers

The term "end-reactive oligomers" means oligomers which have a polymerizable group at one or both ends of the molecule. The end-reactive oligomer is the building component of graft and block copolymers. As the industrial importance of graft and block copolymers increases, the chemistry of the synthesis of end-reactive oligomers has been developed. In this chapter, recent developments of the synthesis of end-reactive oligomers are described under the topic of molecular engineering of polymers. The end-reactive oligomers are classified into two groups, i.e. one-end reactive oligomers and two-end reactive oligomers.

4.1 One-end Reactive Oligomers

Table 3. Typical Polymerizable Groups of MACROMERS®

Olefin	$—CH{=}CH_2$
	$—C(CH_3){=}CH_2$
Vinyl ether	$—OCH{=}CH_2$
p-Styryl	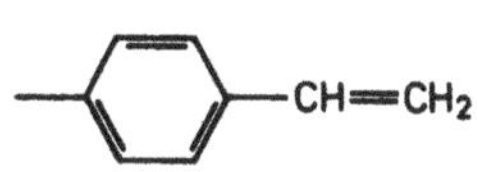
Epoxy	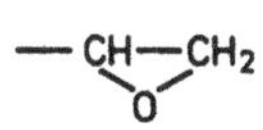
Methacrylate	$—OC({=}O)C(CH_3){=}CH_2$
Meleate and fumarate (semi ester)	$—OC({=}O)CH{=}CHC({=}O)OH$
Vinyl ester	$—CH_2C({=}O)OCH{=}CH_2$
Glycol	$—CH(OH)—CH_2(OH)$

Among the one-end reactive oligomers, the so-called "MACROMER®" is the most popular. The concept of MACROMER® was first described in patents by Milkovich et al.[41]. The term MACROMER® means oligomeric materials having a polymerizable group at one end of the molecule, which are conveniently subjected to copolymerization with a second reactive monomer to produce graft copolymers possessing the long pendant chain of the MACROMER®.

At the beginning, MACROMER® was basically polystyrene prepared by anionic polymerization and subsequent reaction of the living end to produce a polymerizable group selected from the groups shown in Table 3[42].

The scope of MACROMER® has been extended to generally designate the end-reactive oligomers of various monomers. A variety of MACROMERS® have been reported. Hydroxy end-reactive oligomers were esterified with methacroyl chloride to yield methacrylate end-reactive MACROMERS®[41,43]:

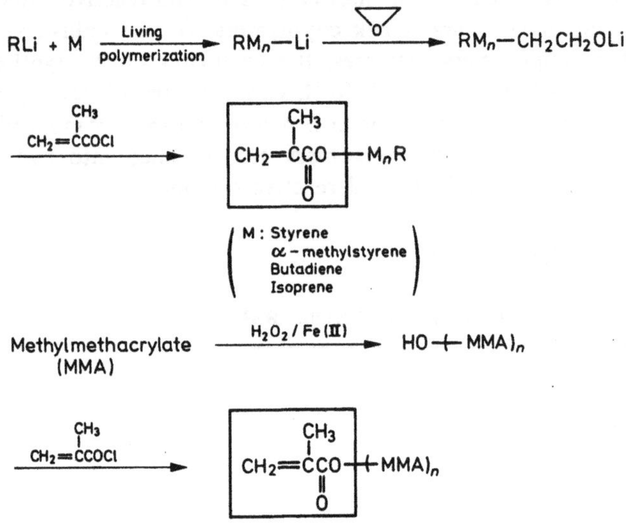

Kennedy[44] reported the synthesis of a MACROMER® on the basis of the cationic polymerization of isobutylene.

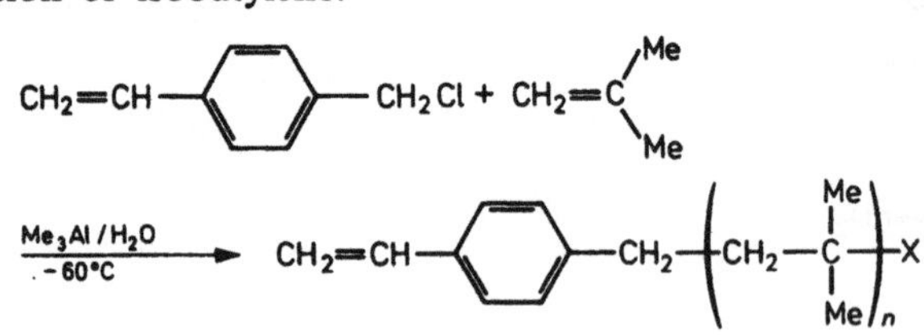

A polyaddition reaction has also been employed in the MACROMER® synthesis[45–47].

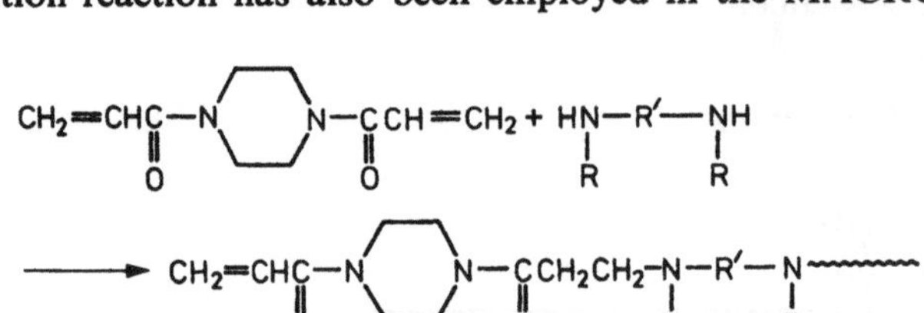

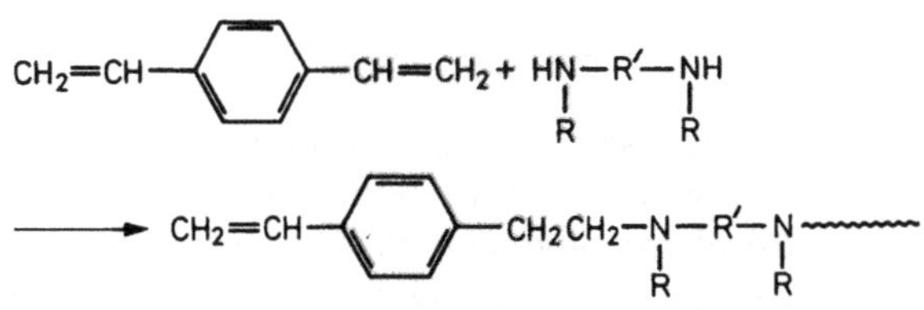

Cationic living-opening polymerization of *tert*-butylaziridine served to prepare a polyamine MACROMER® [48].

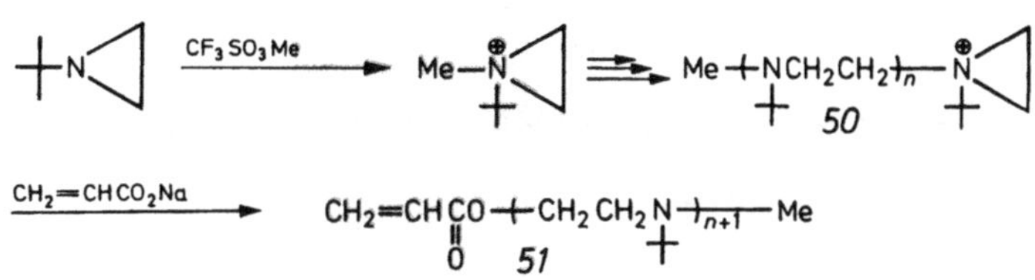

In the above example, both *50* and *51* are MACROMERS®. The cyclic ammonium group of *50* can also be transformed into several other functional groups, for example [48]:

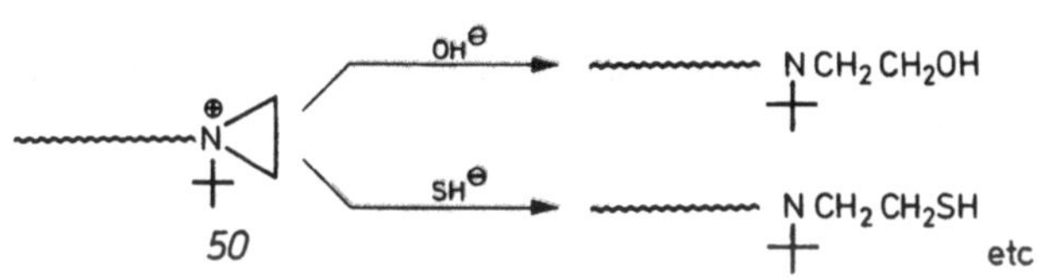

Another type of one-end reactive oligomers are oligomeric initiators which have an end group capable of initiating the polymerization of other monomers. Polymeric azo compounds of the type *52* and *53* were prepared by reaction of anionic living polymers with azo-bis-isobutyronitrile [49]:

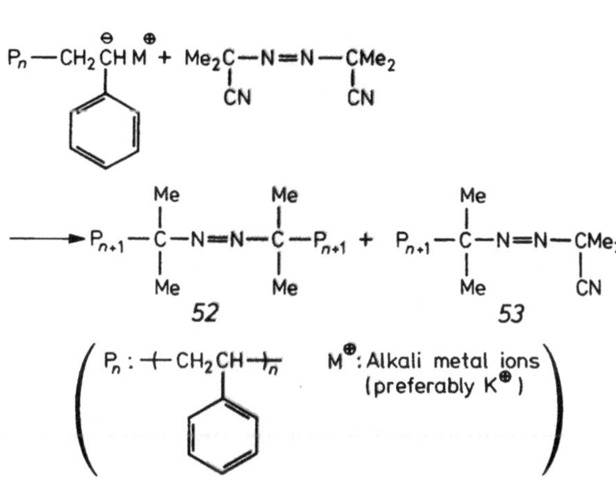

The polymeric initiator was heated with vinyl monomers to produce block copolymers[49].

The end-hydroxy oligomer may serve as an end-initiator oligomer when it is transformed into an initiating catalyst of ring-opening polymerization of other monomers. For example, end-hydroxy oligomers were converted into metal aldoxides *54* which functioned as a catalyst in the living polymerization of ε-caprolactone (CL)[50].

$$(RO)_2AlOZnOAl(OR)_2 + P\sim\sim OH$$

$$\xrightarrow{-ROH} (P\sim\sim O)_2AlOZnOAl(O\sim\sim P)_2 \quad \mathit{54}$$

$$\xrightarrow{\varepsilon-CL} [P\sim\sim O-(CL)_n]_2AlOZnOAl[(CL)_n-O\sim\sim P]_2$$

$$\xrightarrow{H_2O} P\sim\sim O-(CL)_n-H \quad \mathit{55}$$

A block copolymer *55* consisting of a polystyrene block and a poly (CL) block was found to be an effective blending agent for the combination between polystyrene and poly(vinyl chloride). A polybutadiene/poly(CL) block copolymer was an efficient blending agent for a mixture of polybutadiene and polyacrylonitrile. In addition, a block copolymer of poly(hydrogenated 1,4-butadiene) with poly(CL) allows a three-component mixture of polystyrene, polyethylene and poly(vinyl chloride) to be blended.

4.2 Two-end Reactive Oligomers

Linear oligomers having functional groups at both ends of the molecule have long been known as the building component of block copolymers. A typical example are polymeric glycols formed in the production of polyurethane.

$$HO-R-OH + OCN-R'-NCO$$

$$\rightarrow OCN-R'\!\left(NH\underset{\underset{O}{\|}}{C}O-R-O\underset{\underset{O}{\|}}{C}NH-R'\right)_{n}NCO$$

$$\Longrightarrow \text{polyurethane}$$

In the polyurethane industry, the polymeric glycols are prepared by anionic polymerization of epoxides such as ethylene oxide and propylene oxide. Poly(tetramethylene glycol), which was prepared by polymerization of tetrahydrofuran, was subjected to chain extension by reaction with diisocyanate (polyurethane formation) and with dimethyl terephthalate (polyester by alcoholysis).

Kennedy prepared poly(isobutylene) glycol by the following scheme of reactions[44,51]:

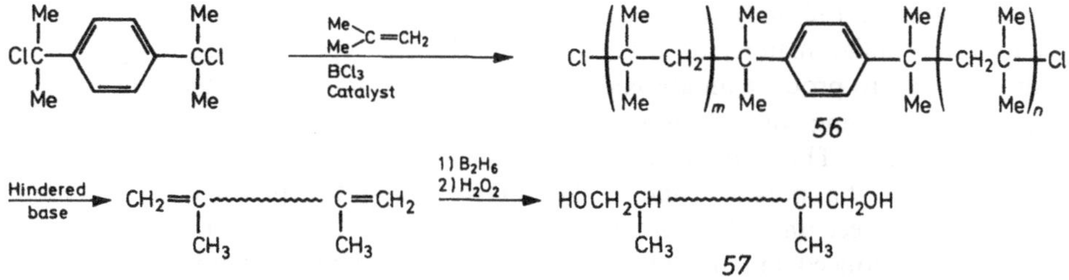

Poly(isobutylene) glycol is a suitable rubbery segment in thermoplastic elastomers. An oligomer such as *56* with a tertiary chloro group at both ends could be employed also as the initiator of cationic polymerization of α-methylstyrene (α-MeSt) in the synthesis of a three-block copolymer of poly(α:MeSt)-polyisobutylene-poly(α-MeSt)[52].

Poly(isobutylene) dicarboxylic acid was prepared by oxidation of the copolymer of isobutylene with a diene[53,54]. The most efficient oxidizing agent was the system $KMnO_4$-periodic acid. Oxidation of a copolymer of isobutylene and 2,3-dimethylbutadiene afforded a polymeric bis-ketone[54].

5 Polymers Containing Cyclic Ether Units in the Main Chain

The reaction of high polymers is another research field of molecular engineering of polymers. In classical polymer chemistry, the synthesis of poly(vinyl alcohol) by hydrolysis of poly(vinyl acetate) may be quoted as a typical example. The chemistry of polymer reactions is still advancing, and many interesting studies are being carried out. In this chapter, a study of Smith et al.[55] is described, which illustrates the characteristics of polymer reactions and the production of a functional polymer.

A group of polymers *58–61*, whose main chains consist of cyclic ether units, were prepared starting from polymers having carbon-carbon double bond-containing

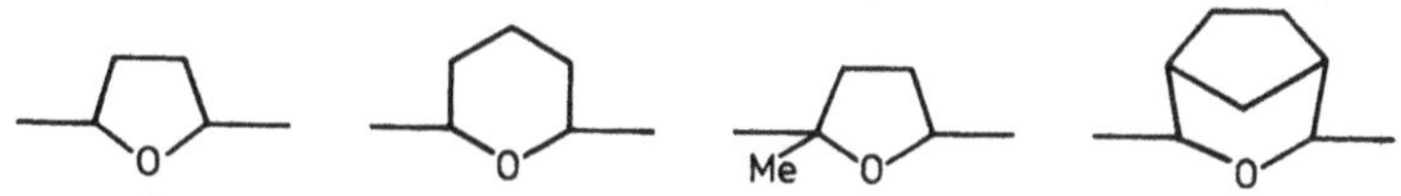

58 59 60 61

units. A series of reactions for the preparation of poly(2,5-tetrahydrofuranediyl) *58* in which cis poly-1,4-butadiene *62* (MW 100,000; *cis* 98%) is the starting material is given below.

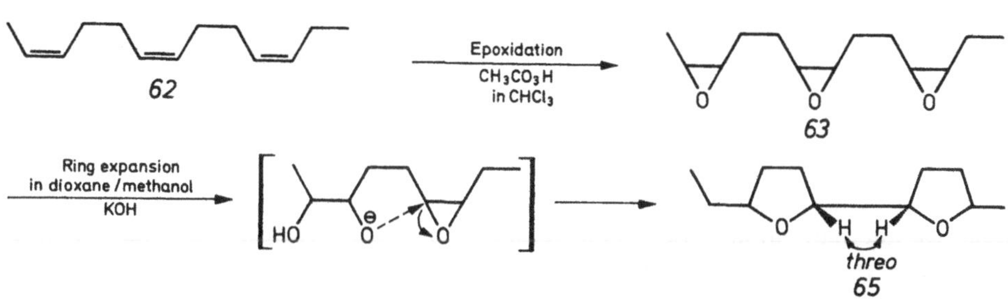

The epoxidation (*62→63*) proceeds quantitatively. The ring expansion reaction occurs in a similar manner as the chain reaction. An epoxide ring is opened by hydroxide ions to produce an alkoxide which, in turn, attacks the adjoining epoxide rings. The opening of an epoxide ring by the OH^- catalyst is the rate-determining step. The subsequent ring expansion takes place rapidly along the main chain of the polymer and is terminated when the propagating alkoxide encounters a defect unit of a foreign structure. The 2,5-tetrahydrofuranediyl units (THF diyl) produced are linked in sequences. On the other hand, the original, unreacted epoxide units remain also in sequences. At the stage of intermediate conversion, a block copolymer consisting of sequences of epoxide units and THF diyl units is formed. It was shown that a polymer containing 66% THF diyl units and 27% epoxide units exhibits the crystallinity of the original epoxide polymer *63*.

The above type of reaction is characteristic of the so-called "polymer reactions". The rate itself and the activation parameters may be quite different from those of the conventional anionic ring-opening reaction of epoxides. In the above reaction of a polymeric epoxide, a polymer consisting of 77% THF diyl units and 16% epoxide units was isolated in a soluble form. If the opening of the epoxide ring by an alkoxide group had occurred in a random fashion, cross-linking would have taken place to produce an insoluble polymer.

Polymers containing over 60 mol-% THF diyl units and prepared from *cis*-1,4-polybutadiene showed a strong coordinating tendency toward metal ions whereas the polymer with 68 mol-% THF diyl units prepared from *trans*-1,4-polybutadiene does not form metal complexes. The difference in the coordination properties between these two types of polymers has been ascribed to the difference in the steric structure between the two polymers.

The previously described reaction scheme for the preparation of these polymers involves an S_N2 inversion during the opening of the epoxide ring. Consequently, the two hydrogen atoms at the ring junctions (which are *cis* to each other in the preceeding epoxide ring) are "*threo*" to each other in the poly(THF diyl) unit of *cis*-1,4-polybutadiene, regardless of the steric relation of the adjacent epoxide units, whereas the two hydrogen atoms are "*erythro*" in the poly(THF diyl) unit from *trans*-1,4-polybutadiene.

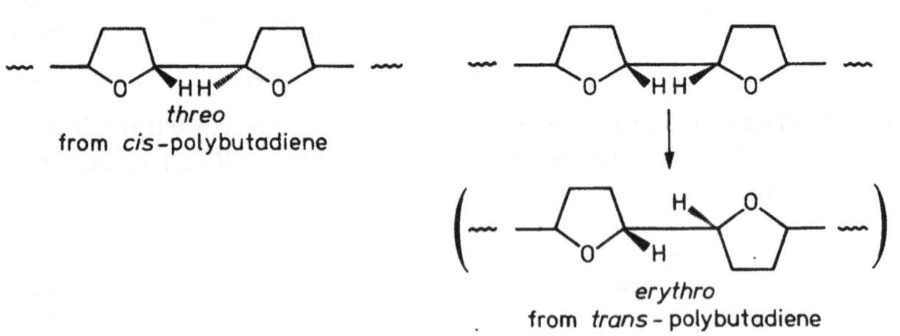

The CPK space-filling molecular model of the *threo* structure represents a helical loop in which all oxygen atoms point toward the center (Fig. 1) whereas that of the *erythro* structure tends to form an extended rigid chain in which oxygen atoms tend to alternate along the chain because of steric crowding of the methine hydrogens (Fig. 2). As to the *threo* polymer, it is reasonable to assume by

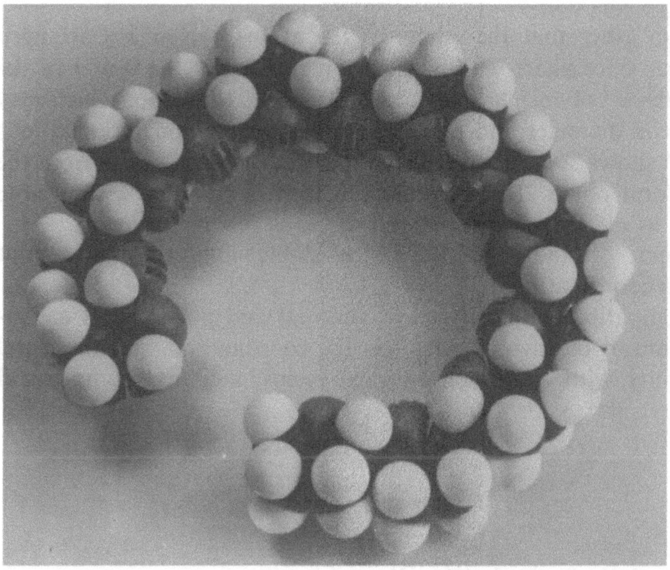

Fig. 1. Molecular model of poly(THF diyl) (*threo* form in *cis*-1,4-polybutadiene) [55]

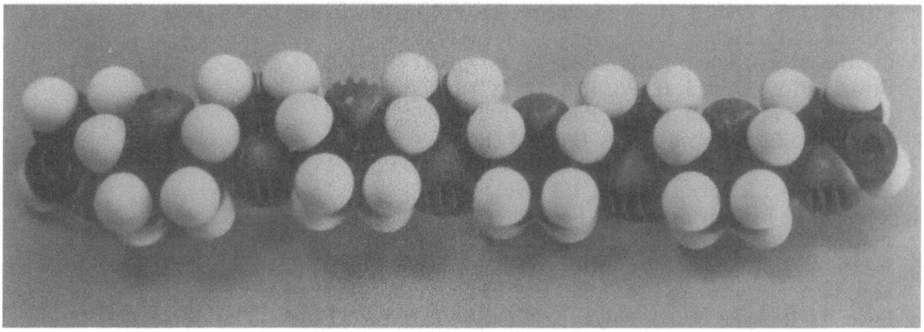

Fig. 2. Molecular model of poly(THF diyl) (*erythro* form in *trans*-1,4-polybutadiene) [55]

Table 4. Phase transfer of picrate salts (25 °C)[a] [55]

Cation	Complexing agents	Salt transferred (%)
Li^+	18-crown-6	63
	poly(THF diyl 77%) (*threo*)	37
K^+	18-crown-6	74
	poly(THF diyl 58%) (*threo*)	30
	poly(THF diyl 77%) (*threo*)	53
	poly(THF diyl 65%) (*erythro*)	0
Ba^{++}	poly(THF diyl 77%) (*threo*)	67

[a] [picrate] in water 0.025 g/l, [complexing agent] in $CHCl_3$ 2.5 g/l

analogy with the crown ether that the ether oxygen atoms, which are arranged inside the helix loop, can coordinate with a metal ion situated at the center of the loop. In addition, the *threo* polymer has some conformational freedom which may cause changes in the helix diameter and pitch-features to accommodate metal ions of various size. Table 4 shows some results of the transfer of picrate salts from the aqueous to the chloroform phase. This polymer can form a complex with a large delocalized cation such as methylene blue cation.
For the *erythro* polymer, on the other hand, the stable comformation cannot provide a multidentate a coordination.

Poly(THF-diyl) is compatible with poly(vinyl chloride) and poly(methyl methacrylate). The combination of the two features, i.e. the coordination with the metal ion and the compatibility with some conventional resins, will find some useful applications.

6 References

1. A review article; Crivello, J. V.: Chemtech. *1980*, 625
2. Crivello, J. V., Lam, J. H. W.: Macromolecules *10*, 1307 (1977); Fourth Internat. Symp. Cationic Polymerization, Akron, USA, June 1976; J. Polym. Sci., Symp. *56*, 383 (1976) Crivello, J. V. et al.: J. Radical. Curing, *4*, 2 (1977)
3. Crivello, J. V., Lam, J. H. W.: J. Polym. Sci., Polym. Chem. Ed. *17*, 977 (1979)
4. Crivello, J. V., Lam, J. H. W.: ibid. *17*, 759 (1979)
5. Crivello, J. V., Lam, J. H. W.: ibid. *18*, 2677, 2697 (1980)
6. Crivello, J. V., Lam, J. H. W.: ibid. *17*, 2877 (1979)
7. Crivello, J. V., Lam, J. H. W.: ibid. *18*, 1021 (1980)
8. Takahashi, T.: J. Polym. Sci. A-1, *6*, 403 (1968)
9. Chow, L., Ahn, K. D.: J. Polym. Sci., Polym. Lett. Ed. *15*, 751 (1977)
10. Errede, L. A.: J. Polym. Sci. *49*, 253 (1961)
11. Whitney, R. B., Calvin, M.: J. Chem. Phys. *23*, 1750 (1955)
12. Brasen, W. R. et al.: J. Org. Chem. *30*, 4188 (1965)
13. Bailey, W. J.: Internat. Symp. Ring-Opening Polymerizations, Karlovy Vary (Czechoslovakia), Sept. 1980
14. Endo, T., Bailey, J. W.: J. Polym. Sci., Polym. Lett. Ed. *13*, 193 (1975)
15. Endo, T., Bailey, J. W.: ibid. *18*, 25 (1980)
16. Review articles of "No Catalyst Copolymerizations" (a) Saegusa, T.: Chem. Technology *5*, 295 (1975); (b) Saegusa, T., Kobayashi, S., Kimura, Y.: Pure Appl. Chem. *48*, 3071 (1976); (c) Saegusa, T.: Angew. Chem. *89*, 867 (1977), Angew. Chem., Internat. Ed. (English) *16*, 826 (1977); (d) Saegusa, T., Kobayashi, S.: J. Polym. Sci., Polym. Symp. *62*, 79 (1978); (e) Saegusa, T., Kobayashi, S.: Pure Appl. Chem. *50*, 281 (1978); (f) Saegusa, T.: Makromol. Chem., Suppl. *3*, 157 (1979)
17. Saegusa, T., Ikeda, H., Fujii, H.: Macromolecules *5*, 354 (1972)
18. Saegusa, T., Kobayashi, S., Kimura, Y.: ibid. *7*, 1 (1974)
19. Saegusa, T., et al.: ibid. *8*, 259 (1975)
20. Saegusa, T., Kobayashi, S., Kimura, Y.: ibid. *7*, 139 (1974)
21. Saegusa, T., Kobayashi, S., Kimura, Y.: ibid. *8*, 374 (1975)
22. Saegusa, T., Kimura, Y., Kobayashi, S.: Macromolecules *10*, 239 (1977)
23. Saegusa, T., Kobayashi, S., Furukawa, J.: ibid. *9*, 728 (1976)
24. Saegusa, T., Kimura, Y., Kobayashi, S.: ibid. *10*, 236 (1977)
25. Saegusa, T., et al.: ibid. *7*, 546 (1974)
26. Saegusa, T., et al.: Macromolecules *7*, 956 (1974)
27. Saegusa, T., et al.: ibid. *9*, 724 (1976)
28. Saegusa, T., Kobayashi, S., Furukawa, J.: ibid. *10*, 73 (1977)

29. Saegusa, T., et al.: Polym. Bull. *1*, 91 (1978)
30. Saegusa, T., Kobayashi, T., Kobayashi, S.: Polym. Bull. *1*, 259 (1979)
31. Saegusa, T., Niwano, M., Kobayashi, S.: Polym. Bull. *2*, 249 (1980)
32. Saegusa, T., Kobayashi, S., Furukawa, J.: Macromolecules *11*, 1027 (1978)
33. Saegusa, T., Kobayashi, S., Furukawa, J.: ibid. *8*, 703 (1975)
34. Saegusa, T., Yokoyama, T., Kobayashi, S.: Polym. Bull. *1*, 55 (1978)
35. Saegusa, T., et al.: Macromolecules *10*, 791 (1977)
36. Saegusa, T., et al.: ibid. *12*, 533 (1979)
37. Kobayashi, S., Kobayashi, T., Saegusa, T.: Polym. Preprints Japan *28*, 762 (1979)
38. Kobayashi, S., Kobayashi, T., Saegusa, T.: Macromolecules, May/June (1981)
39. Saegusa, T., Kobayashi, S., Kimura, Y.: Macromolecules, *10*, 68 (1977)
40. Saegusa, T., Kobayashi, T., Kobayashi, S.: Polym. Preprints Japan, *27*, 44 and 832 (1978)
41. Milkovich, R., Chiang, M. T.: US Pat. 3,842,050 (1974); 3,842,057, 3,842,059 (1974); 3,846,393 (1974); 3,862,098 (1975); 3,862,101 (1975)
42. Milkovich, R.: Polym. Preprints (Amer. Chem. Soc., Div. Polym. Chem.) *1980*, 40
43. Palit, S. R., Mandal, B. M.: J. Macromol. Sci. *C2*, 225 (1972)
44. Kennedy, J. P.: Fifth Internat. Symp. Cationic and Other Ionic Polymerization, Kyoto, April 1980 (Preprints, p. 6), Polym. J. *12*, 609 (1980)
45. Ferruti, P., et al.: Polymer *18*, 387 (1977); J. Polym. Sci., Polym. Chem. Ed. *15*, 2151 (1977)
46. Nitadori, Y., Tsuruta, T.: Polym. Preprints *20*, 539 (1979); Makromol. Chem. *180*, 1877 (1979)
47. Tsuruta, T.: Internat. Symp. Polymeric Amines and Ammonium Salts Ghent, Belgium, Sept. 1979; Proc. p. 163 Pergamon Press 1980
48. Goethals, E. J.: Internat. Symp. Ring-Opening Polymerization, Karlovy Vary, Sept. 1980; Pure Appl. Chem., in press.
49. Riess, G., Reeb, R.: Polym. Preprints (Amer. Chem. Soc., Div. Polym. Chem.), *1980*, 55
50. Teyssié, P.: Internat. Symp. Polymerization Mechanism, Liverpool, Sept. 1980
51. Kennedy, J. P., Smith, R. A.: J. Polym. Sci., Polym. Chem. Ed. *18*, 1523 (1980)
52. Kennedy, J. P., Smith, R. A.: ibid. *18*, 1539 (1980)
53. Guizard, C., Cheradame, H.: Europ. Polym. J. *15*, 689 (1979)
54. Cheradame, H., Gandini, A.: Fifth Internat. Symp. Cationic and Other Ionic Polymerizations, Kyoto, April 1980 (Preprints p. 102)
55. Smith, S.: IUPAC Internat. Symp. Macromolecules, Florence, Sept. 1980; Schultz, U. J., et al.: J. Amer. Chem. Soc. *102*, 798 (1980)

Biochemical Engineering

Hossein Janshekar and Armin Fiechter

Swiss Federal Institute of Technology Hönggerberg, CH-8093 Zürich, Switzerland

Table of Contents

Hossein Janshekar and Armin Fiechter

1 Introduction

For a long time, microbes have been used as catalysts of biogenic processes. After the discovery in the 19th century that living cells can act as biocatalysts, bioprocesses were adopted by the industry for the production of useful products. In most cases, only spontaneous reactions with complex media (natural substrates) were applied. Aseptic conditions were not maintained. The production of ethanol including alcoholic beverages (wine, beer, brandy) is such an example. It is noteworthy that alcohol production still comprises the major part of the bioindustry. But also other products of the classical fermentation industry are based on spontaneous reactions occurring in natural feedstocks, in spite of the great progress made in process development during the past 50 to 80 years. It should not be overlooked, however, that many improvements were the result of an empirical approach and the cell was considered as a black box only. The biological aspects were at best limited to the selection of high-performance strains, whereby classical genetics contributed decisively to the improvement of strains.

The first steps towards the actual integration of technical and biological measures (i.e. biotechnology) were made after the development of the operon theory in the 50's. The efforts of genetic engineering in the 60's resulted in a more systematic development of process improvement including a growing knowledge of metabolic control (gene expression) and engineering aspects. On this basis, present-day biotechnology clearly differs from the possibilities of the classical "fermentations".

The changing situation which took place in the 70's confronts science, industry and the authorities responsible for R + D with many problems.

In a far-sighted way, sound guidelines were established in Germany for example, in order to systematically promote modern biotechnology. The results of these efforts are reflected in a study by the Ministry of Science and Technology (BMFT) who published a second edition in January 1974. In this document not only the field of biotechnology was defined but also the objectives for R + D were developed including both the biological and technical aspects. Submerged cultures of microbial (including pathogens), plant and animal cells are considered as a unit operation. Many known and possible new products have been listed for the demonstration of the high potentials of living cells. Since that time, the development has shown that Germany's conclusions to support biotechnology have caused other countries to do the same. Thus, all industrialized countries now stress the importance of the promotion of these new technologies.

This review deals with the most relevant categories of products prepared by biotechnological methods.

The products mentioned therein were not selected on the basis of their economical importance but rather on the observable changes occuring in their production and application. It is hoped that a realistic picture will emerge of current biotechnology characterizing the rapid changes that took place in the past years and which we expect to continue during the years of the oncoming decade.

2 Biological Products

2.1 Single-Cell Protein (SCP)

Man has traditionally obtained the major components of his diet, e.g. proteins, carbohydrates and fats from animal and vegetable sources. However, these sources cannot meet the increasing demand of the world population for edible proteins. Therefore, new sources of edible proteins must be sought to counteract the forecasted world shortage of proteins.

The term single-cell protein (SCP) was coined at MIT in 1966. It includes the mass production of different microorganisms (bacteria, yeast, fungi, and algae). The advantage of microorganisms for protein production is their high protein content and rapid growth. For this purpose, a wide variety of carbon substrates, including n-paraffins, methane, methanol, ethanol, acetate, CO_2, cellulosic materials from different by-products and waste sources are used. Western Europe [2,3], Russia [4] and Japan [5] have designed plans for 100000 to 1000000 tons per year feed protein plants. Countries in Asia, Africa and South America, because of their geographical, political and economic situations, are more interested in village-level technology and systems producing 1000 to 10000 tons per year of feed SCP.

Specific growth rates of microorganisms, yields, pH and temperature tolerance, aeration requirements, protein content and amino acid profiles, genetic stability, and non pathogenicity to plants, animals or humans are the criteria for selecting the organism for SCP production. The bacteria have high growth rates (generation times of some minutes) as compared to 2–3 hours for yeasts and 16 hours or more for fungi and algae. The typical yields of microorganisms grown on hydrocarbons are double those obtianed from carbohydrates. Increasingly higher costs of hydrocarbons have made SCP production on the base of these raw materials less attractive in western countries than renewable resources such as agricultural wastes of by-products. Efforts are being made to improve the yield of protein from such carbon sources as methane, methanol, n-paraffin and gas oil which will reduce the production costs of SCP.

Cultivations are usually carried out in submerged cultures in bioreactors. Fungi can be cultivated on solid wastes such as straw. In those countries which have enough land and·sun, and in regions where naturally alkaline lakes are present, SCP from photoautotrophically grown algae could be economically feasible.

Since the biomass production is an aerobic process and air should be supplied to cultures, efforts are being made to develop bioreactors having higher oxygen transfer rates and lower power requirements.

Imperical Chemical Industries have developed a pressure-cycle air lift bioreactor for the production of SCP from methanol. The design of the bioreactor takes advantage of the hydrostatic property of the broth to increase the oxygen transfer rate and reduce power requirements for aeration and agitation [2]. The application of continuous [6,7] and incremental feeding [8], and cell-recycle techniques [9] have increased the biomass production rates. Microbial regulatory phenomena like the glucose and the Pasteur effect [10] which influence the biomass production rate have been investigated.

Biomass production is an exothermic process and heat is liberated during growth.

Most of the microorganisms have highest specific growth rates in the range 30–35 °C. For the maintenance of constant temperature cooling is necessary, especially at high production rates. Efforts are being made to select thermotolerant and thermophilic strains which lead to a considerable reduction in the cooling costs [11,12]. For the separation of biomass from the broth, different techniques are used. These include flocculation, centrifugation, filtration, decantation, ion-exchange, phase separation with solvents, washing with surfactants, solvent extraction, evaporation or combinations of these techniques. Factors influencing the separation costs are the microbial cell sizes and cell densities as well as the growth media. Bacterial cells have a smaller size (1—2 µm) than other microorganisms. The costs for the separation of bacterial cells by continuous centrifugation are higher than those for yeasts and fungi. Research to find other alternative methods like two-zone froth-flotation, electrochemical coagulation [3], and food- or feed-grade bioflocculants [13] in order to reduce the cost and duration of cell recovery is progressing.

The quality of produced biomass is judged by its high protein and low nucleic acid content and the absence of harmful compounds. The nutritional value of SCP depends on its amino acid pattern. Techniques for the reduction of nucleis acids [14] and the correction of amino acid patterns are being developed.

Investigations on nutritional and toxicological aspects were immense before SCP could be utilized for the generation of feed and food additives [15].

2.2 Alcohols and Polyols

Some yeasts and bacteria are able to produce different alcohols like ethanol and butanol as well as polyols like glycerin and 2,3-butandiol. These compounds are used in drinks such as beer and wines, and also may be used in or as solvents, drugs, chemicals, oils, waxes, lacquers, antifreezing and antifoaming agents, precipitants, dyestuff, pomades, raw materials for chemical syntheses, motor fuels, and carbon sources for SCP production. These products are mainly synthesized from petroleum — derived materials like ethylene and acetaldehyde. However, because of the insufficient availability and high prices of the raw materials, the microbial production of alcohols has become an interesting area for many researchers.

Ethanol is formed by the anaerobic metabolism of yeasts like *Saccharomyces* and many other species. In the presence of sulfite salts or in alkaline solutions, the alcohol formation can be changed to glycerin formation. *Clostridium* and *Bacillus* species participate in the production of butanol-acetone-butyric acid. Besides n-butanol, acetone and butyric acid, other organic compounds like propionic and lactic acids, 2-propanol, ethanol, and acetyl methylcarbinol (3-oxo-2-butanol) as well as CO_2 and H_2 are produced as by-products. Some bacteria generate 2-propanol from acetone and others form acetone from ethanol.

Molasses, fruit juice, corns, bagasse, Jerusalem artichockes, cassava, whey, sulfite liquor, saw dust and other wood by-products are used as substrates for alcohol and glycerin production. Starch-based substrates should be first saccharified by amylases prepared from barley, fungi or bacteria. Cellulosic materials must also be chemically or enzymatically hydrolyzed before being used as substrates for alcohol production. *Clostridium* species contain amylases and are able to convert starch and cellulose directly [16].

The yeast growth is diauxic [17]. Under the conditions of glucose repression, ethanol formation takes place even in the presence of oxygen. Yeasts require a small but finite oxygen supply for synthesis of unsaturated fatty acids, sterols, and nicotinic acid. These compounds which are essential to membrane functions are synthesized only aerobically [18].

Alcohol is distilled up to a content of 96% in one or more stages. About 1% of ethanol consists of fusel oils (degradation products of amino acids) which can be used as solvents for lacquers and resins. Solids from the processed liquor containing proteins, carbohydrates, mineral salts, riboflavin and other vitamins are used in poultry, swine and cattle feeds. CO_2 and H_2 produced in butanol-acetone-butyric acid production can be used for the chemical synthesis of methanol and ammonia, or are burned.

All of the microbial alcohol production processes are confronted with two basic problems, namely product toxicity and energy-consuming product separation. However, recent progress made in distillation techniques allows ethanol production at economical feasible costs.

Efforts for the isolation or genetic manipulation of ethanol-resistant strains, which are less sensitive to alcohol inhibition [19], are being made. The selection of thermotolerant and thermophilic strains is feasible [20].

The application of continuous processes helps to find the optimum conditions for maximum productivity and the best solvent ratios. Vacuum processes, cell recycling and immobilized yeast cells [21] have been used to increase productivity and the yield of alcohol. Cysewski et al. [22] showed an increase in the alcohol production rate from 7 to 80 $g\,l^{-1}\,h^{-1}$ by the application of continuous methods with recycling at 0.066 bar oxygen pressure.

2.3 Antibiotics and other Pharmaceuticals

Antibiotics are microbial metabolites like amino sugars, poly- and oligosaccharides, poly- and olygopeptides etc. with different pharmacological activities. Since the historical discovery of penicillin by Alexander Fleming in 1929, tons of thousands of antibiotics of microbial origin have been isolated and every year new antibiotics are discovered. Some antibiotics produced on a commercial scale have been listed by Perlman [23]. The world production of antibiotics has now reached 100000 tons per year.

Antibiotics are used as cardiofonic, hypocholesterolemic, antiflammatory, anti-hypertensive, diabetogenic and neuromuscular blocking agents. They are anti-microbials and some display antitumoric properties. Their addition to feed promotes the growth of domestic animals and poultry. Agricultural antibiotics can control bacterial and fungal plant diseases. Furthermore, they can be used as insecticides, herbicides and plant regulators. They are biodegradable and their use on the land instead of chemical products can reduce the possibility of environmental pollution. Therefore, antibiotics can be considered as the most significant products used for the modification of microbial activity to meet human needs [24].

Many antibiotics are produced by molds. Some including streptomycins and polypeptide antibiotics are synthesized by bacteria. Actinomycetes, lichens and plants are other sources of antibiotics. Among the important commercially produced antibiotics are penicillins, streptomycins and tetracyclines. The most prominent penicillin currently produced is penicillin G, the acyl moiety being phenylacetyl. By the addition of different compounds to a growing culture, different groups can be incorporated into the acyl portion of the penicillin molecule and new antibiotics are produced. Acylation may be also be achieved by chemical means instead of by the action of fungi. In this way, new types of semisynthetic penicillins are prepared [23].

Surface or submerged cultures are used for the production of antibiotics from molds. In the production three phases are involved. First, mycelium formation takes place. Then, antibiotic (penicillin) is formed while growth and respiration are retarded. In the final phase, mycelium autolysis occurs and penicillin production ceases. Biomass is separated from the broth in a drum filter and penicillin is usually extracted with amyl acetate, dewatered and precipitated. The process is carried out batchwise or semicontinuously. A corn-steep solution is mostly used as a carbon source. All the antibiotics are produced under high levels of aeration and vigorous agitation. Strict attention is paid to sterility. Areas of investigation include selection and mutation to obtain good yields and stable strains in submerged cultures, the reduction of manufacturing costs, [23] exploitation of genetic methods in the search for new antibiotics [25], biosynthesis and mechanism of action [26], antibiotic resistance of the microorganism [27], microbial transformation [28], and development of cytotoxic and antitumor antibiotics [29].

Pharmacologically active compounds of microbial origin are not limited to antibiotics. Viable or dead microorganisms are used as antigens and toxin-formers for the production of antibodies and antitoxins. Pathogenic bacteria are mass-produced in surface and submerged cultures for vaccine production. Special bacteria are used in the form of oral medication. For example, viable cells of *Lactobacillus* and *Escherichia coli* are employed in the treatment of maladjusted intestinal microflora [30].

Viruses and double-stranded amino acids from phage-infected *E. coli* cells are used as inductors in the synthesis of interferons [31]. Interferons are just being synthesized on a large scale by a new technology [31].

2.4 Biotransformation Products

Microorganisms are able to carry out different reactions in which a compound is converted into a structurally related product. These processes, which are catalyzed by one or several enzymes of cells, are stereospecific. Reaction conditions are mild, therefore convenient, and in many cases are preferable to chemical routes. The concentration of the desired products may be increased by the selection of properly blocked mutants. In Table 1 are listed some of many types of reactions that have been carried out with microorganisms. The product yields are usually high (Table 2).

The production of vinegar from ethanol, gluconic acid from glucose and many steroids are examples of currently used industrial-scale bioconversions. The production of acetic acid from ethanol is characteristic of *Acetobacter* or *Gluconobacter* species. One gram of ethanol theoretically produces 1.304 g acetic acid [34]. *Aspergillus*,

Table 1. Types of bioconversion reactions [32]

Acylation	Demethylation	Isomerization
Amidation	Epimerization	Methylation
Amination	Epoxidation	Oxidation
Cleavage of C—C bonds	Esterification	Phosphorylation
Condensation	Halogenation	Racemization
Deamination	Hydration	Reduction
Decarboxylation	Hydrolysis	Transglycosidation
Dehydration	Hydroxylation	

Table 2. Yields of some bioconversions [33]

Substrate	Product	Microorganism	yield (%)
Sorbitol	Sorbose	*Gluconobacter suboxidans*	98
Mannitol	Fructose	*Gluconobacter suboxidans*	95
Glycerol	Dihydroxyacetone	*Gluconobacter suboxidans*	95
Glucose	5-Ketogluconic acid	*Gluconobacter suboxidans*	90
Glucose	2-Ketogluconic acid	*Pseudomonas mildenbergii*	100
Glucose	Gluconic acid	*Aspergillus niger*	97
L-Tyrosine	L-DOPA	*Aspergillus oryzae*	100
Progesterone	11α-Hydroxy-progesterone	*Rhizopus nigricans*	90

Penicillium and other fungal species convert glucose to gluconic acid in cases of calcium deficiency. The utilization of microbial transformations allows new derivatives of antibiotics to be produced [28]. Ketogenic processes involving conversion of polyalcohols into ketoses are another example of microbial transformation. The most important processes of this type are the oxidation of D-sorbitol to L-sorbose exploited for the manufacture of vitamin C and the oxidation of glycerol to 1,3-dihydroxyacetone. Both processes are carried out on an industrial scale using *Acetobacter* strains [35]. Steroids are complex polycyclic lipids including a wide variety of compounds such as sterols, bile acids and hormones. Reactions involved in the microbial transformation of steroids include oxidation, reduction, hydrolysis and esterification. Microbial modifications also have a significant effect on the cost of the desired hormones. In 1949 α-hydroxylation of progesterone at C-11 caused a reduction in the price of cortisone from $ 200 to $ 6 per gram within a very brief period and, due to further improvements, the price is below $ 1 per gram at present [36].

2.5 Enzymes

Enzymes are used as catalysts in chemical reactions taking place in the food, pharmaceutical and biochemical industries. The reactions carried out under the catabolic action of enzymes do not require extreme conditions (temperature, pH etc.). The advantages are: high turnover rates, stereospecificity, production of less by-products and lower reprocessing losses. Industrial enzymes are obtained from microorganisms and, more recently, also from animal and plant tissues. Mostly, enzymes obtained from plant and animal sources are more difficult to isolate than enzymes from bacterial sources. For both technical and economical reasons, microbial enzymes are becoming increasingly important. In contrast to plant and animal enzymes, microbial enzymes are indefinitely available and independent of climatic seasonal and weather variations.

Microbial cells contain or produce at least 2500 different enzymes which can catalyze biochemical reactions leading to growth, respiration and product formation. Most of these enzymes can readily be separated from cells and can display their catalytic activities independently of the cells. Although microbial enzyme activities have been observed for many centuries, only recently have microbial enzymes been commercially utilized.

The industrially produced enzymes include amylases, protease, pectinase, invertase, catalase, penicillinase, glucose-oxidase, streptokinase-streptodornase, cellulase, and others. Amylase and protease are used in the textile, baking and leather industries. Amylase can also be used instead of malt in starch hydrolysis. Pectinase is employed in the manufacture of fruit juices. Invertase can increase the saccharose solubility by hydrolyzing it to the corresponding reducing sugars. It is used as sweetener and plasticizer. Catalase decomposes hydrogen peroxide to water and oxygen. It is used in bleaching and milk sterilization. Penicillinase can inactivate penicillin. Glucose oxidase removes glucose from egg albumen in powdered egg production and is also used as an analytical reagent. Streptokinase-streptodornase is applied as an anti-inflammatory agent. Cellulase is used in the conversion of wood and wood by-products and as a pharmaceutical digestive acid. Cytochrome P-450, a complex mixture of enzymes in the liver which is responsible for the metabolism of drugs and other xenobiotics, has been subjected to large-scale purification. It may have applications in medicine, act as an environmental cleaner for the transformation of many chemicals including pesticides and be involved in the oxidation of a variety of amines and hydrazines in the near future [37].

Bacteria, fungi and yeasts are used in enzyme production. The pathogenic and biological stability of the strains are important factors governing culture selection. The most frequently employed bacteria are *Bacillus*, *Streptomyces*, *Clostridium*, *Cellvibrio*, *Cellulomonas*, and *Actinomycete* species. The fungi generally involved include *Trichoderma*, *Aspergillus* and *Penicillin* species. Among the yeasts, *Saccharomyces cerevisiae* and *Candida utilis* may be mentioned. Increasing use is being made of mutagenic agents in the production of mutant strains displaying increased enzyme-synthesizing capabilities.

Bacterial enzymes are mostly produced in liquid-surface cultures. For fungi, solid-surface or submerged culture methods are employed. Submerged cultures are used in reactors of 10000–100000 l or greater volume [38]. Mashes are usually composed of mixtures of carbohydrates. The processes are usually performed batchwise.

Most enzymes of industrial importance like amylase, protease, cellulase, etc. are extracellular. By the addition of surface-active agents, enzyme excretion through cell membranes and consequently the yield of these enzymes can be increased [38]. Extracellular enzymes are separated from microbial cells by filtration and, if necessary, in addition by enrichment. Intracellular enzymes are released by disruption of the cells which is usually accomplished by autolysis or mechanical disruption. Enzymes for medical and analytical purposes must be pure. For purification, different techniques like precipitation, differential adsorption and elution, electrophoresis, dialysis lyophylization, and cristalization are applied [32]. Great care is taken throughout the recovery and precipitation process to maintain the enzyme activity.

Because of their low concentration in living tissues, enzymes can be produced and isolated only in small quantities. These expensive compounds are soluble and, after

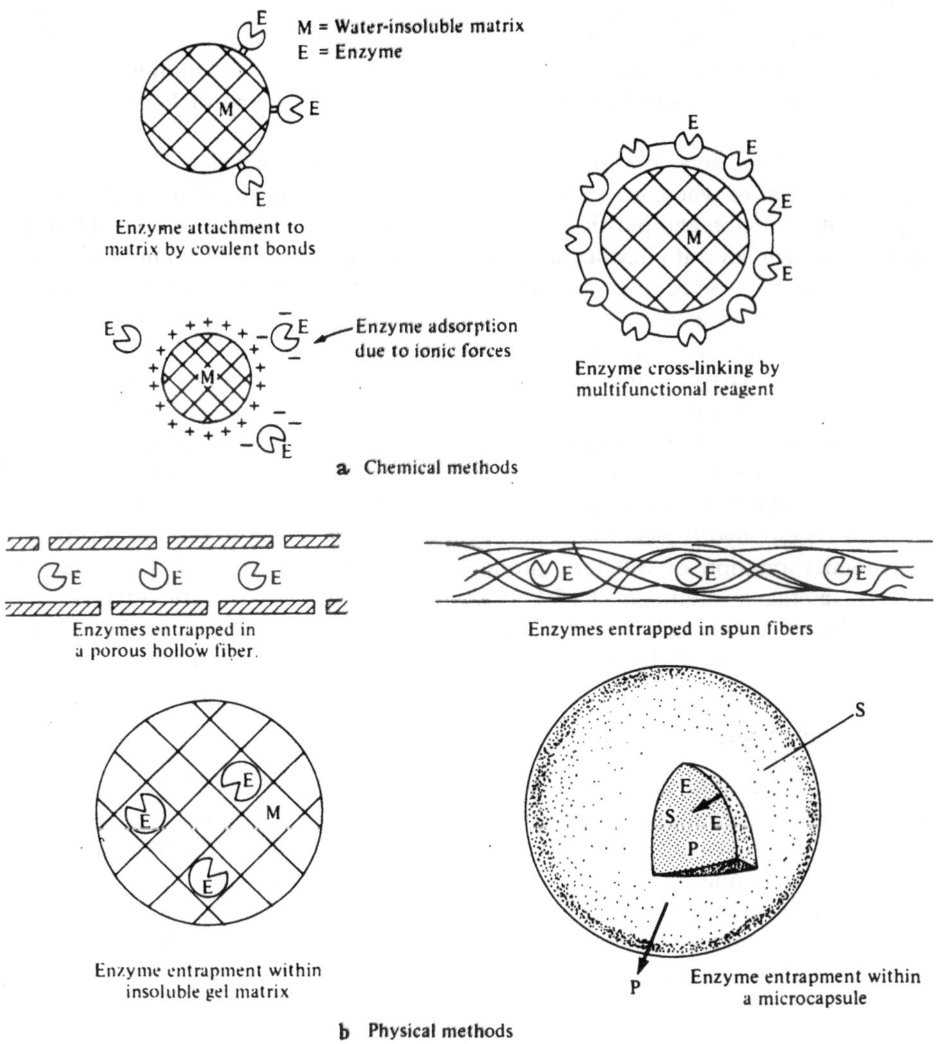

Fig. 1 a and b. Summary of techniques for enzyme immobilization. An enzyme can be immobilized by fixing it on the surface of a macroscopic material or by trapping it inside a matrix which is permeable to the enzyme's substrate. (Reprinted from Bailey, J. E., Ollis, D. F.: Biochemical Engineering Fundamentals, p. 184, New York-London-Tokyo, McGraw-Hill 1977)

transformation, are partially lost. In addition, enzymes contained in the products may be undesirable impurities which must be removed. The latter affect the economical efficiency of enzyme-involving processes. The search for a better and more efficient utilization of enzymes has led to the development of a relatively new enzymatic technique, namely enzyme immobilization [39]. Immobilization of an enzyme means to confine or localize it so that it can be reused continuously. Immobilization of more than 200 enzymes has already been reported.

Enzymes are immobilized by attachment to or confinement in water-insoluble materials (Fig. 1). Enzymes can be immobilized by adsorption on biologically inert carriers like organic polymers, glass, mineral salts, metal oxides, and different silicates. Since enzymes retain their activity for a longer time in an undissolved form, many reactions catalyzed by enzymes can be carried out in continuous systems. Immobilized enzymes can be used in agitated vessels, fluidized or fixed bed tower reactors [40].

The immobilization concept was later extended and applied to living cells [41]. Immobilization of whole cells rather than purified enzymes reduced the expense of separation, isolation and purification of the enzyme. Furthermore, in multistep reactions, in which several enzymes are involved, the application of immobilized cells is advantageous. Since the enzymes are in their native state their stability is enhanced. Such systems may widely be applied, which is not possible with isolated pure enzymes, and are less expensive than processes based on free intact cells [42].

Immobilized enzyme and cell techniques are now expanding into different branches of biochemical engineering like the transformation of carbohydrates, e.g. isomerization of glucose to fructose, production of carboxylic acids and amino acids, waste water treatment, separations of D,L-amino acids, biotransformation of steroid hormones and antibiotics, hydrolysis of proteins to peptides and amino acids, treatment of cheese, hydrolysis of milk sugar, hydrolysis of starch to glucose syrup, and microbial formation of different products like hydrogen, ethanol, Coenzyme A, NADP, methane, aflatoxin etc. [43]. Immobilized enzymes are very often used in biomedicine [44]. This technique has already significantly contributed to the development of new enzymatic electrodes [45] and chromatographic methods.

2.6 Amino Acids

Amino acids are monomeric units of polypeptides and proteins. They are widely used in the food and chemical industries as flavor enhancers, seasonings and sweeteners e.g. for the improvement of bread quality, also in the production of drugs, cosmetics, synthetic leather and surfactants, in medicine for infusions and as therapeutic agents. Amino acids are produced by chemical synthesis or extraction from protein hydrolyzate. They may be also produced by microbiological methods.

Microbial amino acids are mostly produced in Japan [46]. The annual production of amino acids in Japan had reached a level of 300 million dollars in 1977 [47]. Microbial amino acids can be produced directly from intermediates or by enzymatic methods. A variety of substrates are used for microbial growth. These include molasses (especially beet), hydrolyzate, glucose, xylose, acetic acid, methanol, ethanol, benzoic acid, and n-paraffin. Investigations are being made in the search for inexpensive and easily available carbon sources [48].

The strains used are either wild types or mutants. Wild types from the genera *Arthrobacter*, *Bacillus*, *Brevibacterium*, *Corynebacterium* and *Microbacterium* are mostly employed in glutamic acid and alanine production [48]. The yields, depending on the carbon source and bacterial species, are between 10–80%. Other amino acids are also accumulated in wild types; however, yields are lower.

The terminal amino acids are under strict metabolic control. Some act as feedback inhibitors or repressors. Their synthesis is in equilibrium with metabolic requirement. This equilibrium position prevents their accumulation and hence the yield of these compounds is low. By changing the growth requirement (environmental stimulus) or by genetic manipulation, mutants could be found with limited or removed feedback inhibitors and repressors, e.g. auxotrophic and regulatory mutants [49]. This needed a better understanding of biosynthesis and regulation of amino acid production. By selection of these mutants it became possible to alter microbial metabolism which led to the accumulation of the desired amino acids.

The biotechnologists have also clarified the regulation mechanism of glutamic acid excretion in bacterial cell membranes and discovered the effect of medium composition, osmotic pressure, biotin, antibiotic, detergent, saturated fatty acid and oleic acid addition on the excretion of amino acids [49].

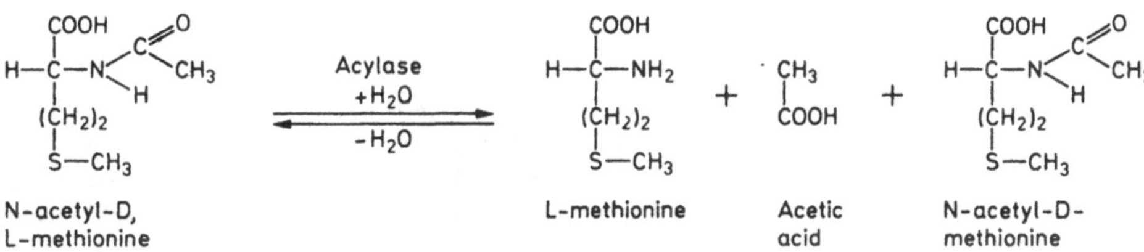

Fig. 2. Reaction System [53]

Table 3. Operation conditions of the extended time test for measuring the deactivation constants [53]

Racemate concentration	(m mol l^{-1})	600
Temperature	(°C)	37
pH	(—)	7.0
Effector, Co^{2+}	(m mol l^{-1})	0.5
Reactor volume	(l)	1.07
Membrane area	(m^2)	1
Membrane cut-off	(MG)	70,000
Flow rate	(ml)	9.6
Residence time	(min)	111
Initial amount of enzyme	(g)	0.835
Flushed amount of enzyme	(g)	~0.350
Residual amount of enzyme	(g)	0.485
Enzyme concentration	(g l^{-1})	0.453
Conversion	(%)	80.2
Measured reaction rate	(m mol l^{-1} min^{-1})	2.16
Calculated reaction rate	(m mol l^{-1} min^{-1})	3.75
Effectivity	(%)	57.6
Deactivation constant k_{de}	(d^{-1})	0.0344

For amino acids like tryptophan and serine, because of the difficulty in avoiding metabolic control, another technique was developed, i.e. intermediate addition of the corresponding amino acids [50]. Amino acids are also produced by intact cells, cell-free enzyme preparations or immobilized enzymes [51, 52].

A recent example of a racemate separation process (D,L-methionine) by immobilized acylase has been described [53] indicating the upcoming trend of enzyme technology in amino acid production (Fig. 2, Table 3).

2.7 Organic Acids

Various microorganisms under certain conditions are able to excrete intermediate products (organic acids) from or closely related to the tricarboxylic acid cycle (Fig. 3, Table 4). For example, *Clostridium* produces acetic acid and butyric acid, *Lactobacillus* and *Streptococcus* species produce lactic acid, *Acetobacter* species acetic, gluconic, and ketogluconic acids, and *Pseudomonas* species 2-ketogluconic and α-ketoglutaric acids.

Acid accumulation may also proceed via other biopathways. In the case of acute nutrient deficiency like shortage of trace metals, the extent of acid accumulation in several instances may be increased. Organic-acid forming microorganisms may accumulate other organic acids in addition to the acid of interest. Careful selection and mutation of strains as well as optimization of media and culture conditions could lead to processes that yield high concentrations of particular organic acids. There are about 60 known kinds of organic acids, derived from microorganisms, some of which are being produced on a commercial scale. For example, the citric acid production in the United States and Europe amounts to about 105000 tons anually. In addition, Japan produces about 5000 tons and about 10000 tons are

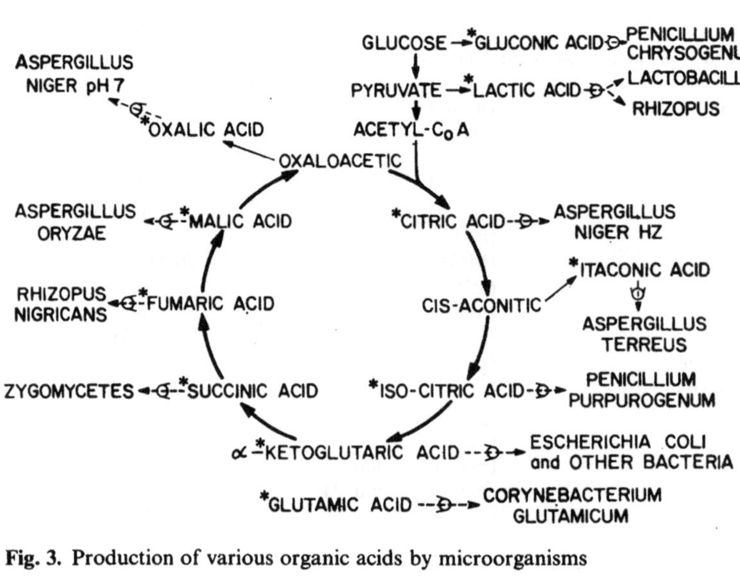

Fig. 3. Production of various organic acids by microorganisms

Table 4. Main Organic Acids Produced by Microorganisms [48]

Acid	Microorganism	Yield % (C-source)
Acetic acid	*Acetobacter aceti*	95 (ethanol)
Propionic acid	*Propionibact. shermanii*	60 (glucose)
Pyruvic acid	*Ps. aeruginosa*	50 (glucose)
Lactic acid[a]	*L. delbrückii*	90 (glucose)
Succinic acid	*Bacterium succinicum*	57 (malic acid)
Tartaric acid	*Gluconobact. suboxydans*	27 (glucose)
Fumaric acid	*Rhizopus delemar*	58 (glucose)
Malic acid	*L. brevis*	100 (glucose)
Itaconic acid[a]	*Asp. terreus*	60 (glucose)
α-Ketoglutaric acid	*C. hydrocarbofumarica*	84 (n-paraffin)
Citric acid[a]	*Asp. niger*	85 (sucrose)
	C. lipolytica	140 (n-paraffin)
L(+)Isocitric acid	*C. brumptii*	28 (glucose)
L(+)Alloisocitric acid	*Pen. purpurogenum*	40 (glucose)
Gluconic acid[a]	*Asp. niger*	95 (glucose)
2-Ketogluconic acid[a]	*Ps. fluorescens*	90 (glucose)
5-Ketogluconic acid	*Gluconobact. suboxidans*	90 (glucose)
D-Araboascorbic acid	*Pen. notatum*	45 (glucose)
Kojic acid	*Asp. oryzae*	50 (glucose)

[a] Industrialized in Japan

produced in Brazil, Israel, Mexico, Colombia, and Argentina [54]. Organic acids are mostly used in the food, pharmaceutical and cosmetic industries.

Starch wastes, wheat and rice bran, molasses, ethanol, wine, apple juice as well as n-paraffins are used as raw materials. Starch materials are first saccharified by amylase. The acid production may be conducted by either surface or submerged culture techniques. The application of microbial film reactions has also been suggested for lactic acid production [55]. Continuous culture techniques are not usually considered to be suitable for use in organic acid production. The maximum rate of acid production occurs during a period in which the rate of growth of mycellium is insignificant. There is no requirement for efficient aeration of the culture during acid formation. The pH should be controlled carefully since otherwise this could lead to the formation of other simultaneously produced acids and to a decrease in yield. For example, acetic acid production is performed in a strong acidic range (pH about 2) to repress oxalic acid formation. Microbial organic acids may further be processed to other compounds. For example, oxalic, itaconic and aconitic acids used as plasticizers are readily produced by oxidation, heat treatment and dehydration of citric acid, respectively.

2.8 Vitamins, Pigments and Coenzymes

Several pigments, vitamins and coenzymes like carotenoids, riboflavin (B_2), cobalamine (B_{12}), ascorbic acid (C), ergosterol (D_2), biotin (H), gibberellin, etc. are produced during the normal metabolism of microorganisms. Microbial production of some of these compounds like riboflavin and cobalamin is commercially applied. The microbial production of other substances such as biotin has not yet been commercially utilized due to the competition from chemical synthesis. These compounds are used in feed and food as growth stimulants, coloring [56] and flavor [57] enhancers.

Carotenoids, the most important family of natural pigments, are employed as food additives for intensifying or modifying the color in fats, oils, cheese, and beverages. Carotenoids are produced by many species of algae and fungi. β-Carotine, a precursor of vitamin A, is generated by some members of *Coanephoraceae* (*Phycomycetes*). By mixed cultivation of two sexual forms of this species, the mycelium carotenoid content can be increased by 15–20 times [58].

Vitamin B_{12} has been extracted from activated sludge or as a by-product of the acetone-butanol process. The vitamin is also synthesized by microorganisms in intestinal habitats. For commercial purposes, *Bacillus*, *Propionibacterium* or *Pseudomonas* species and more recently methanogenic bacteria are utilized. Yields of up to 50 mg l^{-1} can be achieved. The reaction mixture is evaporated and used as a feed supplement for various domestic animals or further purified and crystallized for medical use [59].

Riboflavin is produced by *Clostridium*, *Ascomycetes* and *Candida* species. The yield can be as high as 5 g l^{-1} after 7 days [60]. *Gibberella fujikuroi* is utilized for the synthesis of gibberellins, a group of plant hormones used for plant growth promotion. Glucose, molasses, lipid (corn oil) are usually used as carbon sources. Vitamin B_{12} may also be synthesized from alcohols and hydrocarbons.

In industrial-scale productions, submerged and surface-cultivation methods are routinely applied whereas an immobilized-cell technique for the production of vitamin C has only recently been reported.

Extensive studies in this field including genetic studies, strain improvement by mutation, medium economization and improvement of extraction and purification techniques have led to substantial improvements in process development generally yielding an efficient and economic production.

2.9 Bioinsecticides

The control of scale insects, whiteflies, plant hoppers, aphids, chinch bugs, phytophagous beetles, flies, caterpillars, masquitoes, and stored product pests costs more than one-half billion dollars annually in the United States alone. In the last decade great progress was made in the commercial production of microbial insecticides which can partially replace synthetically prepared insecticides.

Bacteria, viruses, fungi and protozoa are used for the production of microbial insecticides. The most useful bacteria are spore formers. Among entomopathogenic bacteria, *Bacillus moritai*, *B. popillae*, and *B. thuringiensis* are being developed to commercially applicable microbial insecticides [62]. The most useful viruses are those protected by a proteinaceous inclusion body. *Baculovirus*, *Entomopoxvirus* and *Iridovirus* are prominent viral insecticides and each year new insect viruses are discovered [63]. Fungi can be effective pathogens if field conditions are optimal for spore germination, invasion and growth. They are generally less specific than other groups of entomopathogens. Most of the promising fungi are *Phycomycete* and *Deuteromycete* species [64]. Protozoa of *Nosema*, *Mattesia*, and *Vairimorpha* species are currently investigated in great detail for the production of microbial insecticides [64].

Entomopathogenic bacteria are mass-produced in submerged cultures, fungi being usually cultivated by surface growth. Viruses and protozoa are produced in in vivo processes, that is in living insects. Insect viruses, because of their specificity, i.e. they are obligate parasites, are more difficult to produce on an industrial scale than bacteria or fungi. However, more than a dozen commercial preparations of viral insecticides have been produced. Commercial fungal insecticides are produced in the Soviet Union. Molasses, dextrose, grain mash and by-products are used as carbon sources. Entomopathogenic protozoa are difficult to cultivate and have never been industrially utilized. Production feasibility, safety, and effectiveness against important pests are major criteria involved in the evaluation of mass production of microbial insecticides.

Research in this field will not only lead to the production of effective, cheap and environmentally friendly insecticides but will also help to acquire a better understanding and control of infectious diseases of useful insects such as the honey bee and silk worm.

2.10 Other Biological Products

The above mentioned compounds are not all that organisms can produce. The present article is too short to elucidate in detail all the possible compounds which can be produced by microbial action. There are other useful products which could substitute for chemical synthesis. Some of these include: nucleosides and nucleotides used as flavor-enhancing materials [65], extracellular polysaccharides like xanthan, dextran, pullulan and other gums [66], fatty acids and fats [67], biosurfactants [68], rennets [69], hydrogen [70], radioactive compounds [71], and compounds exhibiting electrical properties [72].

Table 5 gives examples of neutral and anionic polysaccharides. Of the polymers shown, dextran, pullulan, xanthan, and alginates are of most commercial interest. About 8000 tonnes per year of xanthan are currently produced for use in food industry, in oil-well drilling and numerous other areas [73].

Not all of the above mentioned microbial processes have reached the stage of large-scale production. The reason being either that there are other routes which for the present time can produce the same compounds in large quantities and at less cost or the microbial systems involved are not yet fully understood. However, this does not interfere with the fact that microorganisms are the largest and most powerful manufacturing companies on our planet.

3 Biological Conversion and Degradation

3.1 Biological Waste Water Treatment

Removal and purification of waste waters, especially in densely populated and industrial areas, has become a serious problem. Waste waters are very different. They generally contain a complex mixture of solids and dissolved components of different

Table 5. Some neutral and anionic microbial polysaccharides [73)]

Trivial name	Organism(s)	Special solution properties and (potential) uses
Dextran	*Acetobacter spp.* *Leuconostoc mesenteroides*	Plastic flow; as source for utilizing dextran derivatives for pharmaceutical purposes
Pullulan	*Pullularia pullulans*	Gels on heating; O_2-impermeable films formed; readily biodegradable
Curdlan	*Agrobacterium*	Gels on heating or acidification; gelled foods; stable salts and acids
Scleroglucan	*Sclerotium glucanicum*	Highly viscous and pseudoplastic, drilling muds, latex points
Cellulose	*Acetobacter spp.*	Water-insoluble
Levan	*Bacillus spp.* *Leuconostoc mesenteroides*	—
—	*Seratia marescens*	—
—	*Pseudomonas* NCLB 11264	Highly viscous and pseudoplastic
Alginate	*Pseudomonas aeruginosa* *Azotobacter vinelandii*	Range of viscosity types; gels with Ca^{2+}; textile printing, food applications
Xanthan	*Xanthomonas campestris*	Highly viscous and pseudoplastic; gels with galactomannes; resistant to acid, alkali and biodegradation; oil-well drilling, food industry
Phosphomannane	*Hansenula capsulata* *Hansenula holstii*	Thixotrophic at high contration, gels with borax
—	*Physarum polycephalum*	—
—	*Rhizobium meliloti*	—

nature and concentration. Domestic waste (sewage) consists of garbage, laundry water and excrement. Furthermore the sewage contains a varied population of soil and intestinal microorganisms including pathogens. The composition of industrial wastes strongly depends on the source. Waste waters from sugar and starch factories, creameries, breweries and yeast fabrication contain large quantities of carbohydrates, fats and protein. The waste water from chemical and metallurgical industries may contain considerable amounts of toxic, corrosive and even explosive substances. A very serious form of pollution is the release of heavy metal compounds like iron, copper, lead, metalloids, etc. into the aqueous environment. The oil lost into the sea during transportation is estimated to be 10^6 metric tons per year.

The aim of water treatment is to remove its dissolved and undissolved components, to eliminate its pathogenic microorganisms and to detoxify it so that it is non-toxic to man and will not affect the microflora nor pollute the water into which it is discharged. Because of the importance of waste water treatment in maintaining our biocenosis, the biological treatment of waste water has become the most important task of biochemical engineers. Every year thousands of millions of cubic meters of waste water are handled and various contaminants are removed before being discharged into the streams and seas [74)].

Waste water treatment is usually performed in three stages. In the primary stage the most easily separated contaminants like solids, which can settle out, oil films, etc are removed. In the secondary stage, colloids and soluble compounds, mostly

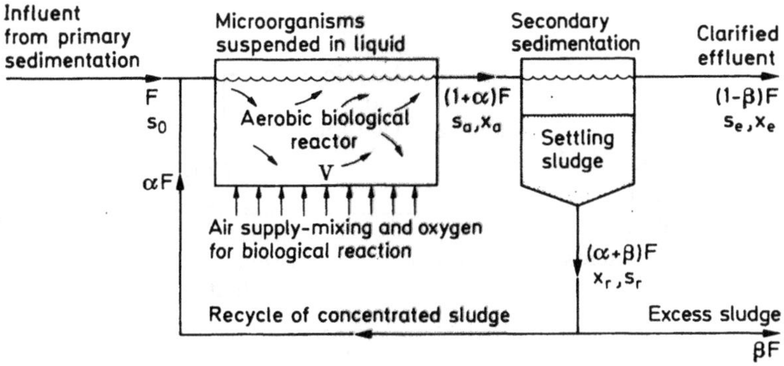

Fig. 4. Schematic representation of the activated sludge process (Reproduced from Bailey, J. E., Ollis, D. F.: Biochemical Engineering Fundamentals, p. 710, New York-London-Tokyo, McGraw-Hill 1977)

of organic origin, are biologically oxidized or adsorbed and cell sludge, CO_2, CO, NH_3, N_2, SH_2, H_2 etc. are formed. The remaining contaminants are removed in the tertiary stage by processes like electrodialysis, reverse osmosis, deep-bed filtration, adsorption etc.

The most common processes used in the secondary stage of waste water treatment include the activated sludge process, trickling biological filters, anaerobic digestion and the lagoon process. Most of these processes may be classified as a continuous-flow enrichment of microbial cultures. The predominant species are determined by the characteristics of the input wastes and environmental conditions. Fungi, algae and protozoa as well as higher animals like worms and fly larvae are found among the inhabitants of waste water treatment plants. A common bacterium in the activated sludge population is the *Pseudomonas* species *Zoogloea ramigera.*

In the activated sludge process, single- or multi-stage continuous-flow aerated reactors are used (Fig. 4). Effluent from the bioreactor is separated in a settling tank and a portion of sedimented sludge is usually recycled to the reactor providing a continuous sludge inoculation.

In the biological filter the effluent to be treated is trickled onto the top of the bed of porous materials such as Biolite colonized by microorganisms. This process is less subject to dynamic or shock loading of toxic substances than the activated sludge process. The activated sludge process often receives forced aeration. In the biological filter, air is circulated through the trickling filter by natural convection. Efficient transfer of oxygen is extremely important.

The sludge from the sewage-treatment process mainly contains organic materials. These are biologically decomposed in a large, closed container called an anaerobic digestor. A simplified scheme of the overall mechanism of anaerobic digestion is shown below.

Different groups of microorganisms participate in this anaerobic conversion. First, extracellular enzymes of hydrolytic bacteria liquefy the polymenic and insoluble substances. The produced monomers are then further degraded by acid-producing bacteria to acetic acid, H_2, CO_2 etc. These metabolites are the substrate for methane-producing bacteria. The end product, a gas mixture of 70% methane and

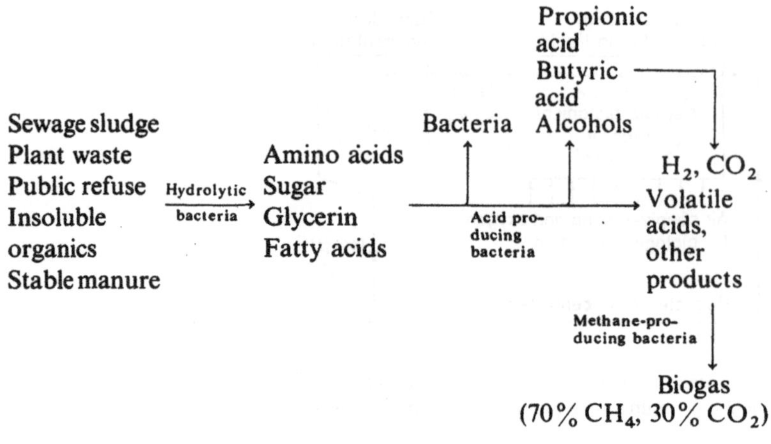

30% CO_2, is called biogas. This gas has a fuel value and can be used to reduce the energy costs of the waste-water treatment processes. From every kilogram of dried organic substances, 200–300 l biogas may be obtained. Substances other than sewage sludge, like public refuse, land and plant wastes, animal manure etc. can also be converted to biogas in an anaerobic digestor. In different countries efforts are being made to optimize the process in order to increase the production rate of biogas. An experimental plant in Florida has shown that the house refuse from a 400000 population can produce biogas which corresponds to 3×10^6 tons of oil. The sludge from the bottom of the anaerobic digestor is burned, composted or used as dung.

Effluents also contain nitrogenous compounds in the form of protein, amino acids, urea, and decomposition products as well as nitrogen usually as ammonium salts. Nitrogen can be removed by biological nitrification-denitrification processes. Biological nitrification converts organic and ammoniacal nitrogen (NH_3) to nitrates via the intermediate nitrites (NO_2^-). Nitrification is coupled with denitrification by facultative organisms in which nitrate nitrogen is converted to gaseous nitrogen [75]. Sulfates existing in many industrial effluents including excess sulfuric acid, gypsum, coal desulfurization by-products, acid mine water and metallurgical effluents can be disposed of by bacterial mutualism [76]. The biological removal of phosphorus, responsible for the eutrophication of lakes and streams, is still in the experimental stage [77].

3.2 Conversion of Agricultural and Forest Products

Petroleum and natural gas are still the most important sources of energy and almost anything based on the organic chemistry of today is likely to owe its origin to these materials. In the near future, these nonrenewable resources will not be available in sufficient quantities to fulfil the increasing demand of man. Impending shortages of natural gas and petroleum reserves have motivated intensive research efforts to find alternative renewable raw material and energy sources. Biochemical engineers paid their main attention to the development of commercial processes for the bio-

Table 6. Percent composition of dry agricultural residues and woods [79]

Material	Hexosans	Pentosans[a]	Lignin	Ash
Barley straw	40	19	14	11
Corn stover	36	16	15	4
Cotton gin trash	20	6	18	15
Rice hulls	36	15	19	20
Rice straw	39	17	10	12
Sorghum straw	33	18	15	10
Wheat straw	36	19	15	10
Corn stalks	35	15	19	5
Sycamore	46	14	22	1
Sweet gum	45	19	19	1
White fir	58	9	28	N.d.
Douglas fir	57	4	27	0.2
Redwood	44	6	34	0.4

[a] Poymerized xylose and arabinose; N.d.: not determined

conversion of plant biomass and agricultural wastes to produce liquid fuel and chemical feedstock. Plant biomass is the most abundant renewable resource on earth. It has been estimated that over 1.5×10^{11} tons of plant material is produced annually in the world by photosynthesis. Less than 1% of the forest resources is currently consumed, 30–50% of which end up as waste residues. For example, in the USA about 220 million tons of wood and bark residues like paper-pulp and wood sawdust are generated annually by the forest product industry. The annual production of crop residues (straw, stalks, etc.) amounts to 562 million tons [78].

The major organic components of land and plant biomass are carbohydrates (cellulose, hemicellulose, starch) and lignin. Table 6 shows the composition of some agricultural and wood residues [79]. Cellulose is a high molecular weight insoluble polymer of glucose. Hemicelluloses are alkali soluble, linear and branched heteropolymers of mainly pentose sugars. Lignin is a high molecular weight insoluble, three-dimensional random polymer of aromatic carbon skeletons.

As already mentioned, microorganisms such as *Clostridia* can produce acetate, lactate, acetone, butanol etc. from cellulosic wastes like sawdust, straw, bagasse, rice hulls etc. [80]. The utility of cellulose and hemicellulose increases if it is first hydroyzed to glucose and other soluble sugars. Hydrolysis of cellulosic materials can be accomplished by means of either acid or enzymatic treatment [81]. There are various bacteria and fungi which produce cellulose- and hemicellulose-degrading enzymes. Fig. 5 shows the saccharification of cellulose by the enzyme cellulase produced by the fungus *Trichoderma viride*. The soluble sugars produced can be separated from the culture or be converted to fuels, chemicals and proteins by other organisms like yeast [82]. *Clostridium thermocellum*, an anaerobic thermophile, hydrolyzes both cellulose and hemicellulose to sugars like glucose, cellobiose, xylose, xylobiose etc. which accumulate in the broth.

The ethanol from agricultural and plant wastes can be blended with petrol and used for running engines and vehicles. Although the biological production of ethanol is technically feasible, its economical feasibility has not yet been resolved.

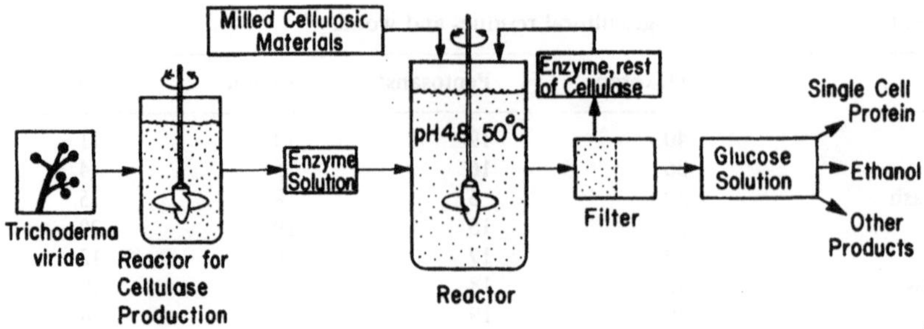

Fig. 5. Enzymatic conversion of cellulosic materials (Reproduced from Binder, H., Keune, H.: Forschung aktuell, Biotechnologie, p. 200, Frankfurt, Umschau 1978)

The lignin component of land and plant biomass, because of its structural complexity and its resistance to biodegradation, is much more difficult to deal with. The microbiology of lignin degradation is not yet well understood. Lignin-degrading microorganisms have been intensively studied in the past few years [83)]. Fungi are known to be active in the degradation of lignin. Bacteria are usually thought to degrade the smaller molecules arising from fungal attack or from industrial processes such as the manufacture of cellulose. When lignin is degraded by microorganisms, a variety of simple and complex aromatic compounds like phenols may be produced which are usually extracted from petroleum.

4 Genetic Engineering

Genetic engineering is the application of molecular genetics to induce predictable and hereditary changes in cells [84)]. The transposable genetic elements responsible for a specific interesting property of the cell like resistance to antibiotics, toxin production, etc. can be clipped out from their linked groups and placed into another organism (genetic recombination), an achievement that could not have been dreamed of 10 years ago. Recombination is an enormously efficient process for generating genetic variation which was practically impossible to achieve by natural evolution like transformation, transduction and conjugation.

Genetic recombination is mostly performed by transformation of a fragment of deoxyribonucleic acid, (DNA) in the cells. Once a gene has been isolated it can be altered in vitro. This can be accomplished by means of restriction enzymes, exonucleases, ligases and chemical mutagens to produce deletions, insertions, and point mutations. There are a number of in vitro techniques for testing changes in the function of altered genes. The altered genes can then be inserted into a host cell. The extra-chromosomal genetic elements used for this transportation, the cloning vectors, are plasmids or bacterial phages. By asexual reproduction, a group of cells having identical genetic characteristics are derived. Genetic transformation has been demonstrated in a number of species of the genera *Escherichia* and *Bacillus* [85)]. To date, many genes from a variety of prokaryotic and eukaryotic cells have been transferred into *E. coli*. Most of the new hybrid DNA's are stable in the host. However, there are some difficulties concerning eukaryotes. Much more about the structure and organization of eukaryotic genes and the factors that control their expression should be known [86]. Research in molecular biology is now focused on higher organisms, but *E. coli* remains an essential research tool.

Genetic recombination plays an important role in the production of new industrial microorganisms, especially prokaryotes. The powerful tool of molecular cloning may be utilized to introduce into *E. coli* genes specifying synthetic functions such as nitrogen fixation, antibiotic and enzyme production and synthesis of antibodies and hormones [87]. Gene cloning is used for strain improvement [88] to maximize the yield of a single metabolite, especially secondary products [89]. It is also used for the elucidation of metabolic pathways.

Resistance of organisms to bacterial phages and inhibitory chemical agents, such as penicillin, dyes etc., can be increased by gene cloning. The recombinant DNA technology has now provided appropriate tools for demonstrating genetic phenomena at the molecular level and for the study of biological evolution. Interferon, which may play several roles in response to tumors, could not be studied in sufficient detail because of the minute quantities previously available. This substance which, according to one estimate, could have world-wide sales of $ 3 billion a year by 1987, is now produced by recombinant DNA technology [90,91].

The participation of molecular biologists in industry is relatively new. It is not yet clear how soon industry will bring a product to market made on the basis of their techniques. However, there is no doubt that eventually various naturally occurring polypeptides such as insulin and interferon of which there is a shortage will be produced by this route.

5 Eucaryotic Cell Cultures

5.1 Plant Cell and Tissue Cultures

Higher plants are our most important source of food- and fodder stuffs. Many specific and valuable medical products such as alkaloids, steroids, and hormones are of plant origin. Some of these compounds are seldom found in microorganisms. Field cultivation, however, is limited by seasonal and climatic variations, area, pest, and field collection cost. Furthermore, the production of natural drugs by extraction from intact plants is wasteful, laboratory intensive, and subject to many limitations. These limitations could be overcome when the idea was conceived that the plant tissues and cells can be excised from the plant and grown in vitro on an appropriate medium. It took several years of intensive research till the methodological difficulties could be surmounted and the tissues were successfully dispersed and grown in agitated liquid media. Up to the present time, the growth of more than 200 different plants has been attempted in this way.

The new technique introduced new potential sources for studying the production of medicinally important compounds, their metabolic pathways and chemical regulation [92]. Activity in plant cell culture research has been considerably increased in the last decade [93]. This technique is used in studies on problems in plant propagation and crop improvement. Host-parasite relationship studies are facilitated by plant cell culture. Studies on enzyme regulation, plant pathogens, screening of substances toxic to plants, evaluation of herbicides, and growth regulators are other applications of plant cell culture [94]. Genetic engineering attempts to increase protein and amino acid content, create new organelles, and perform nuclear transformations using cell cultures. Gene transfer also is being attempted by the incorporation of microorganisms as new cytoplasmic organelles, an approach, that is especially interesting for the introduction of N_2-fixing genes into plants [95].

Plant tissue is obtained in a number of ways. Organs can be washed and desinfected chemically. The tissue is dissected from the interior of organs and cultured on solid media. Submerged cultures are normally initiated by transferring cell material from solid media. Single cells are obtained by filtering. Plant cells have been grown by batch cultivation for many years. The application of continuous culture methods is relatively new. Continuous culture methods enabled an understanding of the factors influencing plant cell secondary metabolism [96].

Examples of substrates released into the culture medium by plant tissue cultures are given in Table 7. The growth rate and yield can be improved by medium optimization. The products from plant tissue cultivations are either directly extracted from the cells or the medium or subjected to biotransformation and enzymatic synthesis.

Table 7. Substances released by plant tissue cultures

Acid phosphatase	Nucleotides
Alkaloids	Peroxidase
Allergens	Pigments
Amino acids	Polysaccharides
Amylase	Proteolytic enzymes
Antimicrobials	Saponins
Arginine-degrading enzymes	Scopoletin
Atropine	Steroids
Flavors	Vicinine
Nicotine	Volatile and vegetable oils

Biotransformations in tissue culture like microbial transformation are mostly focused on steroid biosynthesis [97]. By the addition of organic compounds to the medium, substrates of new structures with new chemical and biological properties can be synthesized.

Structure and regulatory aspects of plants are more complex than those of bacteria. The nutritive requirements of cell cultures are very different and cannot be generalized. They are very sensitive to mechanical effects like agitation and liquid motion, and produce high viscosities causing aeration difficulties. The cytological, morphological and physiological characteristics of cells as well as the stability of the culture vary greatly. Difficulties in the selection of high yielding strains, maintenance of their metabolic stability, high process costs and the technical problems of large-scale cultivation indicate that only expensive compounds can be produced in the near future [98,99].

5.2 Animal Cell and Tissue Cultures

The cells of living organisms in man and other mammals are specialized producers of a diversity of metabolites responsible for different phenomena occurring in the body. For example, hypophysis cells produce lipotrophin, a stimulator of the breakdown of fats, and somatotrophin, a protein hormone which promotes general body growth. Calcitonine and parathyroid hormones produced in the thyroid gland regulate the calcium and phosphate level in blood. Insulin and glucagon formed in the pancreas regulate the amount of glucose in blood. Some of the cellular products are defensive enzymes and play an important role in infectious diseases.

In the 19th century, the biologists realized that the death of an organism is not necessarily accompanied by the death of its individual parts. It was fascinating to observe that the cells and tissues in isolation and away from the controlling and modifying influences of other tissues in the body can retain their viability, continue to exhibit some of their normal functions, and produce products that they normally generate in the body. After the first successful transplant in vitro was performed by Wilhelm Roux in 1885, who kept alive a chick medullary plate in a physiological solution, the era of tissue culture began. Animal cell cultures developed into an active research area at the end of the 19th century and since then the field of tissue cultures has been rapidly expanding and developing. Now work with tissue cultures is within the scope of any laboratory and is used more and more by cytologists, geneticists, bacteriologists, virologists, parasitologists, entomologists, immunologists, pathologists, pharmacologists, radiobiologists, zoologists, physicians, and veterinarians.

Animal cell culture can be used for the production of viral vaccines, interferon, hormones, immunological reagents, and cellular biochemicals (Table 8). At the present time, vaccines are most widely used. Cancer research was one of the first areas in which tissue cultures have been extensively used. Plant cell cultures could also provide alternative possibilities of studying drugs and environmental pollutant metabolism which are traditionally investigated in live animals [101].

Cells are grown either in suspension in a free or immobilized form [102], or by adherence to a solid surface [100]. Materials used for promoting surface-dependent cell growth are glasses, metals, plastics, carbohydrate polymers etc. the media used contain substances such as blood plasma, amniotic fluids, tissue extracts, etc. [103]. Recent developments in animal cell culture are aimed at the improvement of strains and culture techniques, medium optimization, and scale-up. In contrast to plant cell culture, animal cell culture has already found its technical application. Large-scale

Table 8. Products produced from animal cells grown in monolayers [100]

1. Vaccines	2. Interferon
— polio	3. Hormones
— FMD	4. Immunological reagents
— mumps	5. Cellular biochemicals
— measles	— chalons
— rubella	— insulin
— Mareks	— enzymes
— yellow fever	— plasminogen
— rabies	— plasminogen activator
— other	

productions currently reach volumes of 2000 l [102]. One of the largest mammalian cell culture systems designed for the production of human interferon was recently completed by one of Japan's leading pharmaceutical companies.

6 Other Microbial Applications

6.1 Nitrogen Fixation

Plants need nitrogen for their growth. Theoretically, nitrogen from the air can cover the expected nutritional demand of a world population of 7 thousand million in the year 2000. However plants require ammoniacal nitrogen and cannot assimilate molecular nitrogen from the air. Therefore, farming has become limited because of the nitrogen deficiency of the soil. Such soils have to be supplied by artificial fertilizers. Nitrogen dungs are not only expensive to produce but can cause environmental pollution since only a small portion of dung is taken up by the plants. An alternative method for increasing the nitrogen content of the soil is biological nitrogen fixation. Biological nitrogen fixation is the conversion of molecular nitrogen to ammonium by the enzyme nitrogenase found in a few species of bacteria like *Azotobacter vinelandii* or *Klebsiella pneumoniae*.

Unfortunately, biological nitrogen fixation, like the synthesis of ammonia from nitrogen, requires a very large input of metabolic energy [104]. This is, however, derived from the inexhausible supply of the sun. The energy requirement is even higher under unfavorable conditions. Nitrogenase activity is controlled by a complex regulatory system. The end product, ammonium ion, acts as inhibitor of nitrogenase. Free-living N_2 fixing bacteria normally do not excrete ammonia from fixed N_2. They reduce only enough N_2 for their own growth.

In the last two decades intensive research has been undertaken into biological nitrogen fixation [105, 106, 107]. The first step was to isolate and mutate bacteria insensitive to repression of nitrogenase synthesis in order to fix high levels of N_2 even in the presence of excess ammonium. Mutant strains of *Rhizobium japonicum* with the ability to fix more nitrogen than their parent wild type were isolated [108]. It has been possible to construct mutant strains of *Klebsiella pneumoniae* which are blocked in NH_4^+ assimilation and derepressed for nitrogenase biosynthesis [104]. Genetic engineers transformed Nif-gene of *Klebsiella* to *Escherichia coli* to call forth Nif-gene to produce nitrogenase in an inferior environment [109].

The second step is to engineer N_2-fixing microorganisms to live in the roots of a cereal plant to donate fixed nitrogen to the plant. Many N_2-fixing bacteria do not enter into a symbiotic relationship. The already existing, naturally developed symbiotic systems between leguminous plants and *Rhizobia* have to be well understood and improved before being able to form N_2-fixing associations between other plants and N_2-fixing bacteria. Not more than 10% of several thousands of leguminous plants have been studied and there are many questions concerning root nodule formation and involvement in N_2-fixation which have yer to be answered. It is hoped that a better understanding of this complex biological system will enable man to apply biological N_2-fixation concepts to the improvement of soil, especially in the tropics [110].

6.2 Recovery of Metals

The demand of the world's industries for raw materials and environmental problems caused by metallurgical and mine industries are growing. The high cost of classical methods for the recovery of metals did not permit the economical extraction of metals from low-grade ores and recyling of metal wastes. When the uranium mine operations of the Elliot Lake area in Canada began to show that acid mine wastes contained significant amounts of soluble uranium, man discovered the possible utilization of microorganisms for the recovery of metals. This procedure was called microbial leaching and has been the subject of intensive research during the last two decades [111, 112].

Some microorganisms can catalyze certain oxido-reduction reactions like the oxidation of iron and manganese in water, the oxidation of sulfur compounds, and oxidation-reduction of nitrogen compounds. Aerobic autotrophic bacteria of the type *Thiobacillus* can release soluble iron, copper, and sulfuric acid as sulphates into water. These organisms can be found everywhere in nature wherever an acidic environment is maintained in the presence of sulfide-containing minerals.

Bacterial leaching of copper by *Thiobacillus ferrooxidans* has been practiced for years. Yearly, 250000 tons of copper are recovered by microbial leaching techniques in the USA. This process can be represented by the following simplified equations:

$$2\,FeSO_4 + H_2SO_4 + 1/2\,O_2 \xrightarrow{\text{bacterial action}} Fe_2(SO_4)_3 + H_2O$$

$$Fe_2(SO_4)_3 + \text{Cu-sulfide mineral} \longrightarrow CuSO_4 + FeSO_4$$

$$CuSO_4 + Fe \rightarrow Cu + FeSO_4$$

recycle

Uranium is also leached commercially by the same bacteria. Recently, bacterial leaching of gold in Africa has been patented by Pares and co-workers [113]. Also, chemical and microbially assisted leaching has been studied and is applied to remove vanadium from by-products of coke and coke ash refinery [112].

Because microbial leaching is an energy conserving and low cost method of recovering metals from ores, it is likely that it will find more large-scale application in the future than today and be applied to the recovery of other metals like nickel (Ni), zinc (Zn), tin (Sn), molybdenum (Mo), and others. Research and development in this field will also help to reduce the serious problem of microbial corrosion in cooling systems, fuel tanks, iron-containing pipelines etc. It has been estimated that biological deterioration of all materials excluding foodstoffs exceeds $ 5 billion a year.

Removal of heavy metals by algae [114] and volatilization of mercury by Estuarine bacteria [115] from effluents entering the natural environment have been studied.

7 Engineering Aspects

Biochemical engineering is a multidisciplinary subject which covers many fields. Efforts to make the industrial operation of a biological process effective and economically viable involves substantial interaction between microbiologists, biochemists, geneticists and engineers. The information concerning the characteristics of the bioprocesses such as internal regulation of metabolic systems is used by engineers to influence the extent and rate of the desired reactions. This is achieved by design and development of appropriate reactors to meet microbial requirements more effectively. Microbial experiments require complex and tedious operations. During all design stages general design factors such as the cleaning and sterilization of the systems as well as their aseptic demands have to be clearly kept in mind. For the control and accurate performance of a biosystem, measurement and monitoring of the involved parameters are necessary. Biosensors had to be developed to allow continuous readings of product and substrate concentrations, metabolic rates, and biomass concentration to be used in closed loop control. Devices for the measurement and control of physico- and biochemical parameters such as temperature, pH, dynamics and composition of culture, concentration of intra- and extra-cellular compounds, respiratory activity of cells etc. had to be constructed and adapted to biosystems. Some of the biological events can be translated into mathematical terms. Automation could be introduced into many well understood biosystems to facilate their analysis. The application of computers served to integrate some of the multitudinous events occurring in bioprocesses. The progress has now reached a stage which allows the process variables to be controlled by computers and, in some cases, the computer can be used as part of a feedback loop for the mathematical analysis of the system and optimization of the process. Engineers are confronted with new problems which arise once the system is to be scaled up and which require a solution. Attempts to apply modern engineering concepts and methods to processes like medium preparation, sterilization, product recovery etc. made a significant contribution to the economy of the systems and product improvement. In short, the contributions of engineers to the development and expansion of this field, especially in the last two decades, are phenomenal and going into further detail is beyond the scope of this article.

8 Outlook

Future trends in biotechnology point in many directions. A central problem facing biotechnology is increasing process efficiency. Enhancement of biopotentials and also of process concentrations are mandatory. A growth potential of more than $15\ g\ l^{-1}\ h^{-1}$ in productivity is still a futurist enterprise. Optimized high performance media and efficient equipment for the exploitation of high biopotentials are still lacking.

The prerequisites for high performances of this kind have been considerably improved during the past few years and substantial progress can be seen. However, a real breakthrough has not yet been realized and many processes still need substantial improvements in productivity.

As a second area of interest the research on immobilized biocatalysts has been greatly enforced.

In addition to purified and modified enzymes, cell preparations containing a set of enzymes catalyzing an entire sequence of reactions have been successfully immobilized. At present, a step ahead can be registered in the case of cofactor-dependent systems. Finally, immobilizing techniques have been introduced for whole cells. Useful applications are visible in industry and in analytical and clinical procedures.

A third trend is towards the search for cheap feed stock sources. This development started quite a while ago. Methanol used as a carbon source for microbial growth is of real interest at present. Cellulose and hemicellulose as components of wood are not yet an economic alternative, but recent progress is very impressing. It can easily be foreseen that wood will be utilized as a new resource for biotechnology within the next 10 years.

The efforts made in environmental protection can be observed as a fourth tendency. Improvements in sewage treatment and waste utilization are aimed at, as well as the recycling of valuable organic components. The improvement of methods will increase the possibilities of optimization and lead to economic integration of existing cycles.

This outlook can be terminated with an allusion to the consequences of genetic engineering, a quite efficient technique already today. Systematic implantation of genes from eucaryotes into procaryotic carriers is available for practical application with justifiable expenditure.

The catchword interferon sufficiently supports this statement, there is no clairvoyance needed to predict the rapid application of genetic engineering methods to other as yet unexplored areas.

The examples given may illustrate the fast development of biotechnology in most of its aspects and new trends are becoming strikingly visible.

9 References

1. DECHEMA: Biotechnology, study on R + D, January 1974
2. Gow, J. S. et al., in: Single-Cell Protein II (eds.), Tannenbaum, S. R., Wang, D. I. C., p. 370, Massachusetts, MIT Press 1975
3. Faust, U., Präve, P., Sukatsch, D. A.: J. Ferment. Technol. *55*, 609 (1977)
4. Chemical Week *120*, 9, 20 (1977)
5. Tung, T.: Food Eng. *47*, 24 (1975)
6. Dostalek, M., Haggström, L., Molin, N., in: Ferment. Technol. Today (ed.), Terui, G., p. 497, Japan Soc. Ferment. Technol. 1972
7. Janshekar, H., Fiechter, A., in: Proc. Problems with Molasses in the Yeast Industry Symp. (eds.), Sinda, E., Parkkinen, E., p. 81, Helsinki, Helsinki kauppakirjapaino 1980
8. Nyiri, L. K., in: Proc. Ist Int. Cong. IAMS (ed.), Hasegawa, T., Sci. Council of Japan *5*, 120 (1975)
9. Gold, D. S., Mohagheghi, A.: M. S. Thesis, MIT 1978
10. Fiechter, A., in: Proc. 4th Int. Symp. on Yeasts (eds.), Klaushofer, H., Sleytr, U. B., Part II, p. 17, Wien, Hochschülerschaft an der Hochschule für Bodenkultur 1974
11. Ljungdahl, L. G., Wiegel, J., in: 27th IUPAC Cong. Abstracts (eds.), Larinkari, J., Oksanen, J., p. 546, Helsinki, Alko offset 1979
12. Crawford, D. L., et al.: Biotechnol. Bioeng. *13*, 833 (1973)
13. Nakamura, J., Miyashiro, S., Hirose, Y.: Agr. Biol. Chem. *40*, 1565 (1976)
14. Litchfield, H. J.: Food Technol. *31*, 5, 175 (1977)
15. Protein Advisory Group., PAG Guidelines, FAO/WHO/UNICEF, United Nations, New York 1970–76
16. Weimer, P. J., Zeikus, J. G.: Appl. Environ. Microbiol. *33*, 289 (1977)
17. Fiechter, A., Fuhrmann, F. G., Käppeli, O.: Adv. Microbiol. Physiol. (in Press)

18. Schatzmann, H.: ETH thesis No. 5504, 1975
19. Gomez, R. F.: Proc. Colloque Cellulolyse Microbienne, Marseille Cedex 177 (1980)
20. Weigel, J., Ljungdahl, L. G.: Ann. Meeting of ASM, 93 (1978)
21. Criffith, W. L., Compare, A. L.: Develop. Ind. Microbiol. *17*, 241 (1976)
22. Cysewski, G. R., Wilke, C. R.: Biotechnol. Bioeng. *18*, 1297 (1976)
23. Perlman, D., in: Microbial Technology (eds.), Peppler, H. J., Perlman, D., vol. 1, p. 241, New York-San Francisco-London, Academic Press 1979
24. Hahn, F. E. (ed.): Antibiotics, Berlin-Heidelberg-New York, Springer 1979
25. Nara, T., in: Ann. Rep. Ferment. Processes (eds.), Perlman, D., Tsao, G. T., vol. 2, p. 223, New York-San Francisco-London, Academic Press 1978
26. Opferkuch, W.: Immun. Infekt. *6*, 133 (1978)
27. Wagner, W. H.: Immun. Infekt. *6*, 180 (1978)
28. Shibata, M., Uyeda, M., in: Ann. Rep. on Ferment. Processes (eds.), Perlman, D., Tsao, G. T., vol. 2, p. 267, New York-San Francisco-London, Academic Press 1978
29. Fuska, J., Proksa, B., in: Adv. Appl. Microbiol. (ed.), Perlman, D., vol. 20, p. 259, New York-San Francisco-London, Academic Press 1976
30. Beale, A. J., Harris, R. J. C., in: Microbiol. Technol., Current State, Future Prospects (eds.), Bull, A. T., Ellwood, D. C., Ratledge, C., p. 151, Cambridge-London-New York, Cambridge University Press 1979
31. Baron, S.: ASM News *45*, 358 (1979)
32. Wang, D. I. C., et al. (eds.): Fermentation and Enzyme Technology, pp. 39, 238, New York-Chichester-Toronto, John Wiley 1979
33. Demain, A. L.: Naturwissenschaften *67*, 582 (1980)
34. Nickol, G. B., in: Microbial Technol. (eds.), Peppler, H. J., Perlman, D., vol. 2, p. 155, New York-San Francisco-London, Academic Press 1979
35. Perlman, D., in: Microbiol Technol. (eds.), Peppler, H. J., Perlman, D., vol. 2, p. 173, New York-San Francisco-London, Academic Press 1979
36. Sebek, O. K., Perlman, D., in: Microbial Technol. (eds.), Peppler, H. J., Perlmann, D., vol. 1, p. 483, New York-San Francisco-London, Academic Prèss 1979
37. Wiseman, A., in: Topics in Enzyme and Fermentation Biotechnology (ed.), Wiseman, A., vol. 1, p. 172, Chichester, Ellis Horwood 1977
38. Rehm, H.-J. (ed.): Industrielle Microbiologie, p. 353, Berlin-Heidelberg-New York, Springer 1980
39. Mosbach, K. (ed.): Methods in Enzymology, vol. 44, New York-San Francisco-London, Academic Press 1976
40. Atkinson, B. (ed.): Biochemical Reactors, pp. 176, 191, London, Pion Ltd. 1974
41. Abbott, B. J., in: Ann. Rep. Ferment. Processes (eds.), Perlman, D., Tsao, G. T., vol. 2, p. 91, New York-San Francisco-London, Academic Press 1978
42. Jack, T. R., Zajic, J. E., in: Adv. Biochem. Eng. *5*, 125 (1977)
43. Bohak, Z., Sharon, N. (eds.): Biotechnological Applications of Proteins and Enzymes, New York-San Francisco-London, Academic Press 1977
44. Chang, T. M. S. (ed.): Biomedical Applications of Immobilized Enzymes and Proteins, New York-London, Plenum Press 1977
45. Lakshminarayanaiah, N. (ed.): Membrane Electrodes, p. 335, New York-San Francisco-London, Academic Press 1976
46. Yamada, K. (ed.): Japan's Most Advanced Industrial Fermentation Technology and Industry, Part I, p. 10, Tokyo, Int. Tech. Inform. Inst. 1977
47. Arima, K.: Develop. Ind. Microbiol. *18*, 79 (1977)
48. Yamada, K.: Biotechnol. Bioeng. *19*, 1563 (1977)
49. Hirose, Y., Okada, H., in: Microbial Technol. (eds.), Peppler, H. J., Perlman, D., vol. 1, p. 211, New York-San Franscisco-London, Academic Press 1979
50. Terui, G., Enatsu, T.: J. Ferment. Technol. *40*, 120 (1974)
51. Fukumura, T.: Proc. Symp. Amino Acid Nucleic Acid. Japan, *5*, 495 (1974)
52. Abbott, B. J.: Adv. Appl. Microbiol. *20*, 203 (1976)
53. Wandrey, C., Flaschel, E.: Adv. Biochem. Eng. *12*, 147 (1979)
54. Lockwood, L. B., in: Microbial Technol. (eds.), Peppler, H. J., Perlman, D., vol. 1, p. 355, New York-San Francisco-London, Academic Press 1979

55. Griffith, W. L., Compere, A. L.: Develop. Ind. Microbiol. *18*, 723 (1977)
56. Ingram, J. M., Blackwood, A. C.: Adv. Appl. Microbiol. *13*, 267 (1970)
57. Suomalainen, H., Letonen, M.: Dechema Monographs, Biotechnology, vol. 82, p. 207, Weinheim-New York, Verlag Chemie 1978
58. Ninet, L., Renaut, J., in: Microbial Technol. (eds.), Peppler, H. J., Perlman, D., Vol. 1, p. 529, New York-San Francisco-London, Academic Press 1979
59. Florent, J., Ninet, L., in: Microbial Technol. (eds.), Peppler, H. J., Perlman, D., vol. 1, p. 497, New York-San Francisco-London, Academic Press 1979
60. Perlman, D., in: Microbial Technol. (eds.), Peppler, H. J., Perlman, D., vol. 1, p. 521, New York-San Francisco-London, Academic Press 1979
61. Martin, C. K. A., Perlman, D.: Biotechnol. Bioeng. *18*, 217 (1976)
62. Beegle, C.: Develop. Ind. Microbiol. *20*, 97 (1978)
63. Ignoffo, C. M.: Develop. Ind. Microbiol. *20*, 105 (1978)
64. Ignoffo, C. M., Anderson, R. F., in: Microbial Technol (eds.), Peppler, H. J., Perlman, D., vol. 1, p. 1, New York-San Francisco-London, Academic Press 1979
65. Nakao, Y., in: Microbial Technol. (eds.), Peppler, H. J., Perlman, D., vol. 1, p. 311, New York-San Francisco-London, Academic Press 1979
66. Kang, K. S., Cottrell, J. W., in: Microbial Technol. (eds.), Peppler, H. J., Perlman, D., vol. 1, p. 417, New York-San Francisco-London, Academic Press 1979
67. Ratledge, C., in: Economic Microbiology (ed.), Rose, A. H., vol. 2, p. 263, New York-San Francisco-London, Academic Press 1978
68. Margaritis, A., et al.: Develop. Ind. Microbiol. *20*, 623 (1979)
69. Sternberg, M., in: Adv. Appl. Microbiol. *20*, 135 (1976)
70. Zajic, J. E., Kosaric, N., Brosseau, J. D.: Adv. Biochem. Eng. *9*, 57 (1978)
71. Perlman, D., Bayan, A. P., Giuffre, N. A.: Adv. Appl. Microbiol. *6*, 27 (1964)
72. Videla, H. A., Arvia, A. J.: Biotechnol. Bioeng. *17*, 1529 (1975)
73. Pace, G. W.: Adv. Biochem. Eng. *15*, 41 (1980)
74. Mitchell, R. (ed.): Water Pollution Microbiology, New York-Chichester-Toronto, John Wiley 1978
75. Ackerman, R. A., Fialkoff, S. L.: Develop. Ind. Microbiol. *19*, 499 (1978)
76. Cork, D. J., Cusanovich, M. A.: Develop. Ind. Microbiol. *20*, 591 (1979)
77. Timmerman, M. W.: Develop. Ind. Microbiol. *20*, 285 (1979)
78. Moo-Young, M., Chahal, D. S., Vlach, D., in: Bioconversion of Cellulosic Substances into Energy, Chemicals, and Microbial Protein (ed.), Ghose, T. K., p. 457, ITT Delhi and SFIT Zürich (1978)
79. Rosenberg, S. L.: Enzyme Microbial Technol. *2*, 185 (1980)
80. Grossbard, E. (ed.): Straw Decay and its Effect on Disposal and Utilization, New York-Chichester-Toronto, John Wiley 1979
81. Gaden, E. L., et al. (eds.): Enzymatic Conversion of Cellulosic Materials, Technology and Application, New York-Chichester-Toronto, John Wiley 1976
82. Wilke, C. R. (ed.): Cellulose as a Chemical and Energy Resource, New York-Chichester-Toronto, John Wiley 1975
83. Kirk, T. K., Higuchi, T., Chang, H. M. (eds.): Lignin Biodegradation, Microbiology, Chemistry and Application, vols. 1 and 2, West Palm Beach, CRC Press 1980
84. Vanek, Z., Hostalek, Z. (eds.): Genetics of Industrial Microorganisms, Amsterdam, Elsevier 1973
85. Erickson, R., Young, F. E.: Develop. Ind. Microbiol. *19*, 245 (1978)
86. Dulbecco, R.: Microbiol. Rev. *43*, 443 (1979)
87. Klingmüller, W. (ed.): Genmanipulation und Gentherapie, Berlin-Heidelberg-New York, Springer 1976
88. Hopwood, D. A.: Develop. Ind. Microbial. *18*, 9 (1977)
89. Elander, R. P., Chang, L. T., Vaughan, R. W., in: Ann. Rep. Ferment. Processes (eds.), Perlman, D., Tsao, G. T., vol. 1, p. 1, New York-San Francisco-London, Academic Press 1977
90. Gresser, I. (ed.): Interferon 1979, vol. 1, New York-San Francisco-London, Academic Press 1979
91. Vilček, J., Gresser, I., Merigan, T. C. (eds.): Regulatory Functions of Interferons, Ann. New York Acad. Sci., vol. 350, New York, The New York Acad. Sci. 1980

92. Lee, S. L., Scott, A. I.: Develop. Ind. Microbiol. *20*, 381 (1979)
93. Reinert, J., Bajay, Y. P. X. (eds.): Plant Cell, Tissue, and Organ Culture, Berlin-Heidelberg-New York, Springer 1977
94. Fiechter, A. (ed.): Adv. Biochem. Eng., vols. 16 and 18, Berlin-Heidelberg-New York, Springer 1980
95. Malmberg, R. L.: Develop. Ind. Microbiol. *19*, 197 (1978)
96. Wilson, G.: Adv. Biochem. Eng. *16*, 1 (1980)
97. Reinhard, E., Altermann, A. W.: Adv. Biochem. Eng. *16*, 49 (1980)
98. Wagner, F., Vogelmann, H., in: Plant Tissue Culture and its Biotechnological Applications. (eds.) W. Barz, E. Reinhard, M. H. Zenk, p. 245, Berlin-Heidelberg-New York, Springer 1977
99. Fiechter, A., in: Handbook of Biotechnology (eds.), Rehm, H., Reed, G., Weinheim-New York, Verlag Chemie (in Press)
100. Spier, R. E.: Adv. Biochem. Eng. *14*, 119 (1980)
101. Smith, R. V., Acosta, D., Rosazza, J. P.: Adv. Biochem. Eng. *5*, 69 (1977)
102. Acton, R. T., Lynn, J. D.: Adv. Biochem. Eng. *7*, 85 (1977)
103. Rothblat, G. H., Cristofalo, V. J. (eds.): Growth, Nutrition, and Metabolism of Cells in Culture, New York-San Francisco-London, Academic Press 1972
104. Andersen, K., Shanmugam, K. T., Valentine, R. C.: Develop. Ind. Microbiol. *19*, 279 (1978)
105. Newton, W., Postgate, J. R., Rodriguez-Barrueco, C. (eds.): Recent Develop. in Nitrogen Fixation. New York-San Francisco-London, Academic Press 1977
106. Subba Rao, N. S. (ed.): Recent Adv. in Biolog. Nitrogen Fixation, London, Edward Arnold 1980
107. Robson, R. L., Postgate, J. R.: Ann. Rev. Microbiol. *34*, 183 (1980)
108. Maier, R. J. et al.: Develop. Ind. Microbiol. *19*, 193 (1978)
109. Hollaender, A. et al. (eds.): Genetic Engineering for Nitrogen Fixation, New York-London, Plenum Press 1977
110. Döbereiner, J. et al. (eds.): Limitations and Potentials for Biological Nitrogen Fixation in the Tropics, New York-London, Plenum Press 1978
111. Torma, A. E.: Adv. Biochem. Eng. *6*, 1 (1977)
112. Schwartz, W. (ed.): Conference Bacterial Leaching 1977 GBF, Braunschweig-Stöckheim, Weinheim-New York, Verlag Chemie 1977
113. Karaivko, G. I. et al. (eds.): The Bacterial Leaching of Metals from Ores, p. 62, England, Technicopy 1977
114. Gale, N. L., Wixson, B. G.: Develop. Ind. Microbiol. *20*, 259 (1979)
115. Olson, B. et al.: Develop. Ind. Microbiol. *20*, 275 (1979)

Chemistry and Spectroscopy of Rare Earths

Christian Klixbüll Jørgensen[1] **and Renata Reisfeld**[2]

1 Département de Chimie minérale, analytique et appliquée, University of Geneva, CH 1211 Geneva 4, Switzerland

2 "Enrique Berman" professor of Solar Energy, Department of Inorganic and Analytical Chemistry, Hebrew University, Jerusalem, Israel

Table of Contents

Christian Klixbüll Jørgensen and Renata Reisfeld

1 Challenge to the Periodic Table

The rare earths are, strictly speaking, the sesquioxides M_2O_3 of seventeen elements M now known to have the atomic numbers $Z = 21$, 39 and all the integers from 57 to 71. Because the metallic elements are readily available since about 1950, many people call the elements themselves "rare earths" and sometimes use the pleonasm "rare-earth oxides". Before the Z values were established by Moseley in 1913, it was uncertain how many elements (now called lanthanides) follow lanthanum, and there was much to be said for the arguments by Crookes [1] that the lanthanides constitute a category intermediate between the distinct elements (defended by Lavoisier) and what we would call isotopes (hardly differing in chemical properties, though having different atomic weights). Interestingly enough, non-radioactive isotopes were found in 1914 as lead 208 in thorium minerals and lead 206 in uranium minerals, to be compared with the average atomic weight 207.09 of the isotope mixture in conventional lead minerals. Crookes [1] also suggested a slow build-up of elements via hydrogen and helium, with higher concentrations of those elements having particular thermodynamic or kinetic stability under dramatic conditions of high temperature, as elaborated today by astrophysicists [2]. Contrary to the large planets (such as Jupiter and Saturn), the small planets (such as the Earth) consist almost exclusively of elements with Z above 7, formed by the collapse of the first-generation stars. Our Sun is, at the earliest, a second-generation star (only half as old as the time 10^{10} years elapsed since the Big Bang) and 0.15 percent of the Earth's crust consist of elements with Z above 30, originally expelled as dust from particularly violent supernova explosions. Of the average crust, 5 ppm (or g/ton) is scandium, 28 ppm yttrium, and 120 ppm lanthanides (to be compared with 16 ppm lead or 12 ppm of thorium). The "rare earths" are not especially rare; if we except promethium (having only radioactive isotopes) the two rarest lanthanides (thulium and lutetium) are more abundant than familiar elements such as mercury, bismuth and silver. However, the great difference is that the lithophilic rare earths are dispersed in silicates and other oxide-type minerals, whereas the chalkophilic elements (including lead) are concentrated in conspicuous sulfide minerals, or even as noble metals.

It is mainly for historical reasons that *scandium* is included in the rare earths, since Nilson found it 1879 in erbium- and ytterbium concentrates. The unit cell parameter a_0 of cubic Sc_2O_3 is only 9.85 Å, to be compared with In_2O_3 10.15, Lu_2O_3 10.391, Er_2O_3 10.547, Tl_2O_3 10.57, Y_2O_3 10.602, Dy_2O_3 10.665 and Gd_2O_3 10.812 Å. Nobody consider indium and thallium (discovered by Crookes 1861) as lanthanides, these two are readily reduced to the metallic state, and show several chalkophilic tendencies. Seen from the point of view of Mendeleev, scandium is considered a triumph for the Periodic Table, since its properties were predicted, together with those of gallium, germanium, rhenium (and in a sense, also hafnium). Yttrium is neither any problem; the specific example of a_0 values given above can be extended to nearly all chemical properties (complex formation constants, solubilities, etc.) being intermediate between holmium and dysprosium ($Z = 67$ and 66). Such regularities created the opinion that ionic size is the only significant parameter in rare-earth chemistry (here, we speak deliberately about the a_0 values known with a far better precision than the derived ionic radii of cations and of the oxide anion). This way of thinking is still quite appropriate for aqueous solutions, whereas non-aqueous solvents and solid-state chemistry can show unexpected deviations. As pointed out by Goldschmidt, the "lanthanide contraction" is predominant in Lu(III) being smaller than Y(III), whereas Zr(IV) and Hf(IV) are extremely similar, and Ta(V) marginally larger (and also less oxidizing) than Nb(V).

It is interesting that pain-staking chemical techniques of *fractional recrystallization* of salts [3] succeeded in isolating all the lanthanides (again except promethium) before 1907. These 15 consecutive elements with closely similar chemistry were considered a superlative transition-group, by extrapolation from the behaviour of iron, cobalt, nickel and copper, which are remarkably similar (especially when they remain bivalent). Mendeleev constructed this, and two other tetrades (ruthenium, rhodium palladium and silver) and (osmium, iridium, platinum and gold) in part because the highest oxidation state defines, in principle, the column number in the Periodic Table, and Cu(II) and Au(III) are frequent. However, during the tremendous pressure for putting $18 = 2 \times 8$ in the short version (space requirements in text-books?) Mendeleev finally cut off the coinage metals (to form a side-group to the alkaline metals) and left three triades. Undoubtedly, somebody has argued that the $(2^4 - 1) = 15$ lanthanides are a hypertriade to be compared with $(2^2 - 1)$.

It is not always recognized that the Periodic Table exists in two closely similar versions [4], the chemical and the spectroscopic version. Obviously, the former has historical priority, but the chemical analogies are not always clear-cut. Thus, the "invalid" similarities between barium and lead, between indium and bismuth, between silver and thallium, and perhaps even between tellurium and osmium (as discussed by Retgers at the end of last Century) or tin and platinum, are more convincing than the legitimate similarities between beryllium and mercury, or between nitrogen and bismuth. Nowadays, many students consider the Periodic Table as a corollary of the *Aufbau principle* [5] of electron configurations. However, there is not always a clear-cut relation [6] between the configuration to which the groundstate of the neutral gaseous atom belongs, and the chemistry of the element considered. Thus, not only beryllium, magnesium, calcium, strontium, barium and radium, but also *helium* (but not the heavier noble gases), zinc, cadmium, *ytterbium* and mercury are "spectroscopic alkaline-earths" (in spite of all chemical evidence to the contrary for the two italicized elements) with closed shells, the last of which is a s orbital. The Aufbau principle was not at all clarified in 1923 (as can be seen from the July 1923 issue of "Naturwissenschaften" commemorating the tenth anniversary of the 1913 model of the hydrogen atom) but Stoner proposed the following year that each nl-shell can accommodate at most $(4l + 2)$ electrons. It is evident from the long Kossel isoelectronic series [7] that a d-shell is filled with ten electrons between $K = 18$ and 28 (K being the number of electrons in the corresponding monatomic entity M^{+z} with the charge $z = (Z - K)$, to be compared with compounds [8] having the oxidation state z), $K = 36$ and 46, or $K = 68$ and 78, and that the 4f shell is filled with 14 electrons from $K = 54$ to 68 (for ionic charges at least +2) and, as we know now, the 5f shell from $K = 86$ to 100. Hund [5] believed that the groundstate of the gaseous lanthanide atoms always belongs to [54] $4f^q5d6s^2$. However, this is only true [9] for La, Ce, Gd and Lu, whereas the eleven other atoms have groundstates belonging to [54] $4f^{q+1}6s^2$. It is discussed below how the conditional oxidation state M[II] can be defined when the number of 4f electrons is q = $(Z - 56)$ whereas it is M[III] when q = $(Z - 57)$.

Connick [10] was the first to emphasize that the lanthanides are not trivalent because of the presence of 4f electrons as such, but that the difficulty of oxidizing M(III) to M(IV) and the difficulty of reducing M(III) to M(II) are related to ionization energies and the numerical values of chemical effects such as hydration energy differences [8,11,12]. A closer analysis [13] shows that the small average radius of the partly filled 4f shell induces a propensity toward a constant oxidation state (though it is much more difficult to predict that it is M(III) in actual compounds) by the huge difference (even in condensed matter) between the ionization energy and the electron affinity of the 4f shell. Between 1949 and 1956, a rather vigorous controversy took place regarding the transthorium elements [4,6,14] between two (equally unreasonable) extreme points of view: that these elements ought to be trivalent because they contain 5f electrons; and that the high oxidation states in the beginning (from thorium to plutonium) shows that 5f electrons are not present, at least not before neptunium. It was shown [14] from absorption spectra of non-metallic compounds that the 5f shell contains $(K - 86)$ electrons, where $K = (Z - z)$ is the difference between the atomic number and the oxidation state. On the other hand, the comparison between the 5f and the 4f groups give a result very similar to 4d vs. 3d, with higher (and more varying) oxidation states [15] in the beginning of the 5f and 4d groups, and it is now known that mendelevium ($Z = 101$) is more readily reduced to Md(II) than thulium, and that it is very difficult to oxidize the $5f^{14}$ nobelium(II). The tendency toward lower oxidation states at the end is also found in the 4d group, where it is more difficult to oxidize silver(I) than copper(I).

Though the difference between the chemical and spectroscopic versions of the Periodic Table is less gruesome than in the case of helium, we have to accept the fact that nearly all neodymium compounds are $4f^3$ and hence Nd(III), though the groundstate [9] of the gaseous atom belongs to [54] $4f^4 6s^2$ whereas it belongs to [86] $5f^3 6d 7s^2$ in the uranium atom, in spite of U(VI) and U(IV) compounds being far more frequent than U(III).

Although the theoretical description of d-group compounds originated in magnetic measurements (in particular, paramagnetism indicating positive values of the spin quantum number S, whereas diamagnetism is diagnostic for S zero) and that Van Vleck proved the presence of $(Z - 57)$ electrons in the 4f shell of M(III) compounds, the magnetic properties of a given M(III) in various compounds are so stereotypic that magnetochemistry has played a minor rôle in the rationalization of chemical bonding in lanthanides. The many interesting details of absorption spectra [8,16] of non-metallic compounds have been far more instructive, and created a strong need for high-purity lanthanides. The streneous separation procedures used earlier were reviewed by Prandtl [17] in 1938. Since about 1942, Spedding and his group in Ames, Iowa developed ion-exchange resin elution techniques for separating kg to ton quantities in high purity, and the prices today for 99.9 (or better) % oxides vary from the order of magnitude of gold (for Tm_2O_3 and Lu_2O_3) to well below the price of silver. After the war, reliable spectrophotometers for the visible and near ultra-violet regions became widely available, and the chemical investigations fluorished in an unprecedented manner.

Since 1960, fifteen Rare Earth Research Conferences have been held, and the proceedings [18–35] of these, and three related conferences, comprise a remarkable variety of physical, chemical and biological studies. In the 8. edition of Gmelin's Handbuch der anorganischen Chemie, a slim volume (Systemnummer 39) published 1938 and treating history of the elements and occurrence in minerals was followed since 1973 by more than twelve volumes treating extensively the physical properties, methods of separation, compounds etc. For our purposes, a chapter "Comparison of Properties of Atoms and Ions Along the Lanthanide Series" in Volume *39B1* may be cited [36]. Four volumes of a treatise "Handbook on the Physics and Chemistry of Rare Earths" edited by Karl A. Gschneidner and LeRoy Eyring have been published in 1978 and 1979. Four times a year, a newsletter "RIC News" appears; it can be obtained (free of charge) from Dr. K. A. Gschneidner, Rare-Earth Information Center, Energy and Mineral Resources Research Institute, Iowa State University, Ames, Iowa 50011, U.S.A.

2 Aqua Ions and Oxides

Corresponding to the pronounced lithophilic character (a "hard" anti-base, i.e. Lewis acid, in Pearson's classification [37–39] having much stronger affinity to "hard" than to "soft" bases) the most frequent neighbour atoms to a given lanthanide are oxygen and fluorine. It is not unexpected that many unidentate ligands, such as chloride, bromide, ammonia and cyanide, are not able to replace the water molecules directly coordinated to the lanthanide in aqueous solution. However, the situation is slightly different in the case of the rather strong Brønsted bases NH_3 and CN^- which, almost instantaneously, establish protonation equilibria with water, and maintain a (small) concentration of OH^- sufficiently high to precipitate the hydroxides, as is well-known from the chemistry of aluminium(III) and iron(III).

Bidentate and multidentate ligands can form strong complexes [40,41] with trivalent rare earths, mainly because of the favourable change of entropy related to the "chelate effect". An extreme case are multidentate "crown ethers". Such neutral molecules have a large affinity [42,43] even to alkaline and alkaline-earth elements normally much more interested in water than other neutral ligands, and complexes have also been studied [44,45] of rare-earth elements. As almost universally true, perchlorate does not normally enter aqua ions, but nitrate (which seems to be bidentate with lanthanides) shows an ambiguous behaviour. The soft anti-base palladium(II) forms perceptible nitrate complexes [46] and the excited state of the uranyl ion [47] forms much stronger nitrate (and bromide) complexes than the groundstate. Quite old observations [49] show nitrate complex formation by moderate change of the absorption bands of coloured lanthanides in 0.5 to 2 molar NO_3^-, whereas chloride [50] has no effect below 5 molar. As long some 10 volume % water (say 5 molar H_2O) is added to non-aqueous solvents, the same situation prevails, whereas almost anhydrous solvents (such as methanol and ethanol) show chloro complexes at much lower chloride concentration. Formation of thiocyanate complexes [51] is also promoted by decreasing water content.

2.1 Coordination Number *N*

Many crystal structures of hydrated lanthanide salts show simultaneous bonding of water molecules and one or more anions to M(III). It is not possible to infer the constituents of the first coordination sphere from the composition of salt hydrates.

Thus, $GdCl_3$, 6 H_2O has two chloride ions and all the six water molecules bound to gadolinium, whereas MCl_2, 6 H_2O (for M = cobalt and nickel) has both the two chlorides and only four water molecules bound to M(II). In 1939, Helmholz [52] established that the ennea-hydrate of neodymium(III) bromate has a crystal structure $[Nd(OH_2)_9] (BrO_3)_3$ with a distinct aqua ion having three water molecules (or strictly speaking oxygen nuclei) in an equilateral triangle in a plane, parallel to the triangular surfaces of a surrounding trigonal prism of six oxygen nuclei. This is a rather rare type of coordination with $N = 9$ (though it was later found in the rhenium(VII) hydride complex ReH_9^{-2}) but reminds one about $N = 12$ in hexagonal close-packed metals, also having a trigonal prism, but the polygon in the middle plane being a regular hexagon. However, the ethylsulfates (which had been important in fractional crystallization, because of reasonably large solubility differences between two consecutive lanthanides were also shown to be $[M(OH_2)_9] (C_2H_5OSO_3)_3$. By the way, the highly soluble bromates form a striking contrast to the almost insoluble, amorphous iodates $M(IO_3)_3$ and $Th(IO_3)_4$ probably being cross-linked by each iodate bound to several M(III) or Th(IV). The trigonal kind of $N = 9$ is also found in all $M(OH)_3$ and in the anhydrous MCl_3 for M = La to Gd, whereas starting with M = Dy, the YCl_3 type has a highly distorted $N = 6$. Anhydrous lanthanum chloride has played a large rôle for spectroscopic studies [53] of (finely resolved) absorption (and fluorescence emission) at low temperature of the various incorporated M(III), including $5f^3$ uranium(III) [54] and the subsequent trans-uranium elements.

In the writers' opinion, it is not perfectly well decided whether the lanthanide aqua ions *in solution* invariantly are $M(OH_2)_9^{+3}$ like in the crystalline bromates and ethylsulfates (which can be prepared of all 15 lanthanides, and of yttrium), or whether Spedding and Habenschuss [55,56] are right in assuming a change-over to a lower N (say, 8) somewhere between samarium and gadolinium. Fidelis and Mioduski [57] have recently produced a fascinating argument for the second alternative. In the beginning of the lanthanides, the large majority of the aqua ions seem indeed to have $N = 9$. Freed [58] found in 1931 that the four characteristic strong absorption bands (due to $4f \rightarrow 5d$ inter-shell transitions) of cerium(III) aqua ions in solution have the same positions (in the ultra-violet) as when substitutionally incorporated in $[La(OH_2)_9] (C_2H_5OSO_3)_3$. There has been some suspicion [50,59] that the first, much weaker, band at 33700 cm^{-1} (297 nm) is due to a minority aqua ion with differing symmetry (and probably lower N) since its intensity increases strikingly with the temperature. On the other hand, Krumholz [60] showed almost perfect agreement between the spectrum of Nd(III) in cold perchloric acid, and of solids containing $Nd(OH_2)_9^{+3}$.

It is beyond discussion that many equilibrium constants of mono-complexes of multidentate ligands (as a function of Z in the lanthanides, yttrium having a pseudo-Z somewhere between 66 and 67) go through some discontinuity before gadolinium [40,61,62]. However, as always, when discussing the change of free energy during the reaction of an aqua ion and a ligand, it is conceivable that the sudden change of properties takes place in the mono-complex, and not in the aqua ion. Thus, the exceedingly sharp (and weak) transition (in the yellow) to $^5D_0$ from the groundstate $^7F_0$ in the aqueous solution of europium(III) ethylenediaminetetra-acetate is represented [63] by two lines (of which the relative intensity changes reversibly as a function of the temperature between 25 and 80 °C). This effect is almost certainly due to two mono-complexes of which the composition differs by one molecule of water (for instance 4 and 3; and the ligand may be further on be quinquedentate or sexidentate).

Though it is undecided whether $N = 9$ is invariant for aqua ions in solution, or whether the lanthanides following samarium tend toward $N = 8$, it is nowadays perfectly clear that there is no strong energetic motivation to adapt a definite N. At this point, the situation is very different from the d-groups. The absorption bands in the visible, rationalized by "ligand field" theory [8,13,64] as transitions from essentially non-bonding d-orbitals to one (quadratic), two (octahedral) or three (tetrahedral) anti-bonding d-like orbitals, show clearly that octahedral $M(OH_2)_6^{+z}$ occur for $3d^1$ Ti(III), $3d^2$ V(III), $3d^3$ V(II) and Cr(III), $3d^5$ Mn(II) and Fe(III) both having groundstates with $S = {}^5/_2$, $3d^6$ Fe(II) with $S = 2$ and Co(III) with $S = 0$, $3d^7$ Co(II), $3d^8$ Ni(II), $4d^6$ Ru(II) and Rh(III) and $5d^6$ Ir(III) (the last three examples diamagnetic, by having the three roughly non-bonding orbitals fully occupied). The diamagnetic $4d^8$ palladium(II) aqua ion [46] seems to be quadratic $Pd(OH_2)_4^{+2}$ though it is not perfectly excluded that it has a fifth water molecule coordinated perpendicular on the Pd(II) O_4 plane, making the complex tetragonal-pyramidic like $Cu(NH_3)_5^{+2}$ and the isoelectronic species TeF_5^-, IF_5 and XeF_5^+. There is a strong correlation between groundstate S values and stereochemistry in d^8 systems; nickel(II) complexes with $S = 1$ are usually octahedral (and much less frequently tetrahedral) whereas the diamagnetic species can be quadratic (e.g. $Ni(CN)_4^{-2}$), rectangular (e.g. with two bidentate sulfur-bound ligands) or tetragonal-pyramidic (such as $Ni(CN)_5^{-3}$). It is necessary to consider each case without prejudices; thus, it seems that the vanadyl ion is $OV(OH_2)_4^{+2}$ without any tautomeric rearrangement to $V(OH)_2^{+2}$ in spite of V(IV) being isoelectronic with Ti(III), and the amphoteric behaviour of the small beryllium(II) corresponds to formation of $Be(OH)_4^{-2}$ at high pH, whereas the large lead(II) produces $Pb(OH)_3^-$, possibly pyramidal like $SnCl_3^-$, $SbCl_3$, $TeCl_3^+$, SeO_3^{-2}, BrO_3^-, IO_3^- and XeO_3. Nuclear magnetic resonance (especially involving ^{17}O) has demonstrated tetrahedral $Be(OH_2)_4^{+2}$ and octahedral $M(OH_2)_6^{+z}$ for M = Mg(II), Al(III), Zn(II) and Ga(III) in solution. Hence, the large majority of aqua ions (outside the rare earths) for which the constitution has been established, are octahedral. Negative evidence for the scandium(III) aqua ion was presented by Geier [65] showing much more rapid exchange of water (in the 10^{-8} to 10^{-7} sec range for all the rare earth aqua ions at 12 °C) in contrast to Al(III) and Ga(III), having interesting corollaries for "ligand field" theory [66]. However, the rapid water exchange of rare-earth aqua ions only suggests that N is above 6 (in spite of the existence of ScF_6^{-3}) but not whether it is 7, 8, 9 or 10.

Since the time of Alfred Werner, there has been an exaggerated expectation of the first-neighbour atoms *always* forming regular polyhedra. There is no doubt that a regular tetrahedron is so favoured for $N = 4$ and a regular octahedron for $N = 6$ that it is the exceptions (e.g. a square or a trigonal prism) that need extended discussion. However, a cube for $N = 8$ is hardly known among monomeric complexes. Werner assumed $Mo(CN)_8^{-4}$ to be cubal, but it is tetragonal anti-prismatic (or alternative, low symmetries) in crystalline salts. Any model of ligands moving on a sphere, having ligand-ligand repulsions monotonically decreasing with increasing distance R at least as strongly as (1/R) predicts that a tetragonal anti-prism is more stable than a cube (though it may not always be the absolutely most stable configuration among all conceivable for $N = 8$). Careful analysis [67,68] shows that the energy differences (as far goes ligand-ligand repulsion) are extremely small, allowing smooth rearrangements (what Al Cotton calls "fluxional" behaviour) and it is expected that the rigidity of the stereochemistry almost vanishes for N above 6. Extended crystals can have cubal $N = 8$ such as fluorite-type CaF_2 and CeO_2 because of stronger Madelung potential. On the other hand, there is only a marginal difference [13] between the Madelung potentials for tetrakaidecahedral (cuboctahedral) $N = 12$ known from cubic close-packed metals (and from the strontium site in the cubic perovskite $SrTiO_3$) and for regular icosahedral $N = 12$ (which is not likely to occur in crystals, because an infinitely extended lattice cannot have five-fold axes). There is no evidence for angular directivity in the chemical bonding of rare earths. If any was observed, it would probably be described in the L.C.A.O. model as due to the empty 5d orbitals (or 4d in the case of yttrium).

Another argument of Alfred Werner is that one can evaluate N for an aqua ion by comparison with other complexes of unidentate ligands (e.g. ammonia). This argument is remarkably successful for the octahedral complexes of central atoms containing a partly filled d-shell, but has to be treated with some circumspection (we know very little about the stereochemistry, and influence of the time-scale, in the case [69] of copper(II) aqua ions in solution, but some crystals contain tetragonally elongated $Cu(OH_2)_6^{+2}$ whereas no crystal structure is known to involve $Cu(NH_3)_6^{+2}$ and the highest complex seems to be $Cu(NH_3)_5^{+2}$) and it is quite uncertain outside the d-groups. Vibrational Raman spectra confirm that $Zn(NH_3)_4^{+2}$ is indeed tetrahedral [70] whereas (at least the major part) of zinc(II) aqua ions are octahedral. Though copper(I), silver(I) and mercury(II) are known [70] to bind two ammonia molecules strongly in aqueous ammonia, this is by no means convincing evidence for the stereochemistry of the aqua ions.

In the case of lanthanides with a partly filled 4f-shell, the narrow absorption bands (having close similarity to atomic spectral lines) have been determined with great precision for aqua ions (in the presence of non-complexing perchlorate or chloride anions) mainly because Beer's law is valid, if the spectrophotometer supplies sufficiently monochromatic radiation. Prandtl and Scheiner [71] made a catalog of narrow bands in 1934, and work by D. C. Stewart and other spectroscopists at the Argonne National Laboratory was reviewed by Carnall [72]. Whereas the intensities of some of the transitions can vary conspicuously in differing complexes, the band positions are remarkably invariant, as first pointed out by Gladstone in 1857. An unexplained trend is that some aqua ions, such as $4f^2$ Pr(III), $4f^5$ Sm(III), $4f^9$ Dy(III), $4f^{12}$ Tm(III) and $4f^{13}$ Yb(III) on the whole tend to give broader bands than the narrow peaks of $4f^3$ Nd(III), $4f^6$ Eu(III), $4f^7$ Gd(III), $4f^{10}$ Ho(III) and $4f^{11}$ Er(III) aqua ions. Figure 1 gives the quantum numbers of J-levels below 42000 cm^{-1} of 11 trivalent lanthanides.

2.2 Short Internuclear Distances in Binary Oxides

The 4f group does not only differ from the d-groups in the sense of freely varying N and lack of evidence for angular directivity of the chemical bonding, but also by a

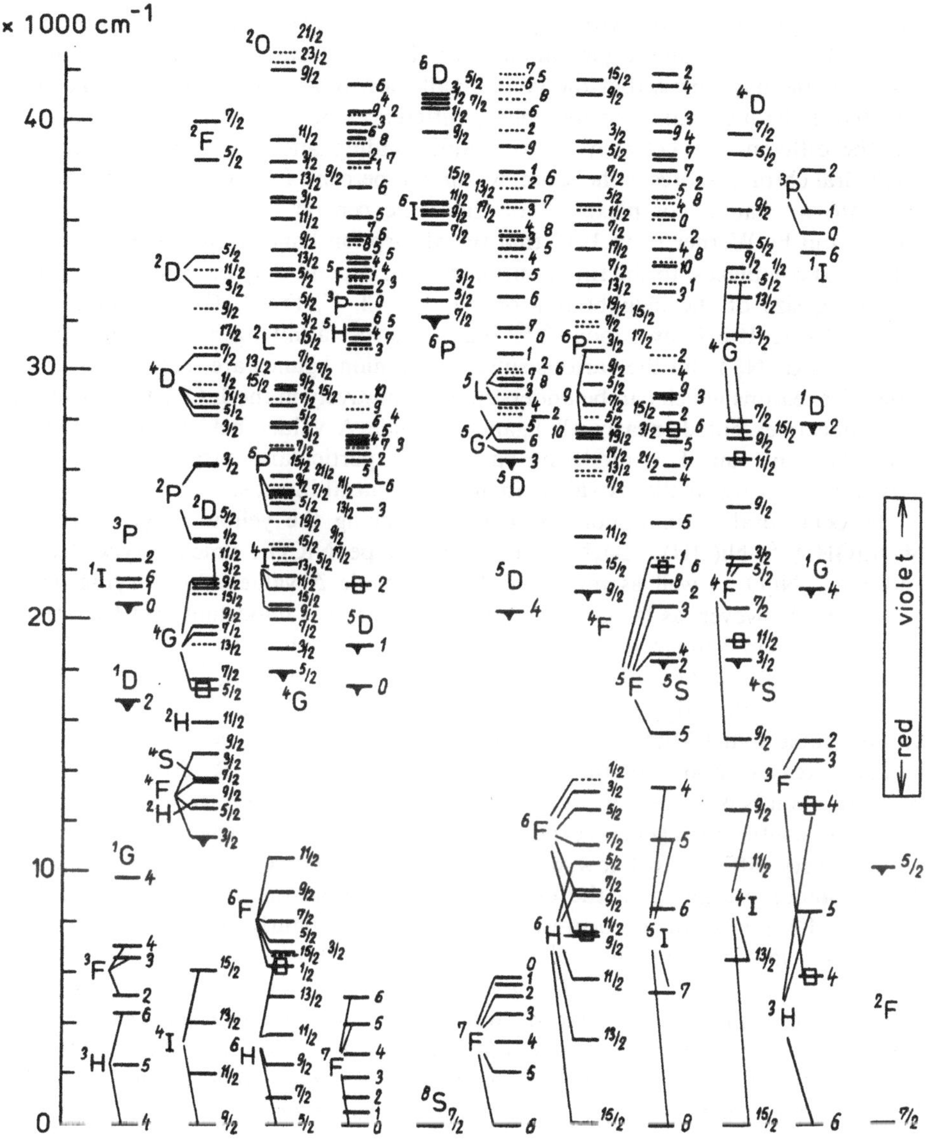

Fig. 1. Energy Levels (J indicated at the right hand) of the trivalent lanthanides (except cerium and promethium) in the unit 1,000 cm^{-1} as a function of the number q of 4f electrons. Excited levels frequently showing luminescence are indicated by a black triangle. The excited levels corresponding to hypersensitive transitions from the groundstate are marked with a square. In cases where the quantum numbers S and L are reasonably well-defined, the Russell-Saunders terms are given at the left. Calculated energy levels [225] are shown as stippled lines. This figure is an extended and modified version of a diagram [53] resuming the work of a generation of rare earth spectroscopists. The empty area (shaped like a wine leaf) in the middle of the figure (q = 5 to 9) corresponds to a large gap between the highest level of a term having the same S_{max} as the groundstate, and the lowest (frequently fluorescent) level of the lowest term with (S_{max} — 1)

much stronger interaction with adjacent oxide anions than with water molecules. One of the most striking arguments against the electrostatic model of the "ligand field", i.e. the hypothesis that the (very small) non-spherical part of the Madelung potential is the mechanism of the energy differences between the five d-orbitals, is that the difference Δ between the two anti-bonding and the 3 other orbitals in octahedral chromophores is the same, within ten percent, for oxo, hydroxo, oxalato, malonato and aqua complexes [73] of the same central atom. It is was already pointed out by Werner [74] in 1912 that the absorption band positions of octahedral cobalt(III) complexes, to the first approximation, depend only on what is essentially decreasing size of the neighbour atom (I, Br, Cl, F, O, N. C) regardless of its electric charge. It is known today [8,13] that some special nitrogen-containing ligands (such as azide N_3^-) and the various sulfur-containing ligands are scattered over this series (for reasons which can be understood qualitatively in the L.C.A.O. model) but if cyanide is excepted, nearly all ligands provide Δ values between 0.75 and 1.3 times the aqua ion. This moderate variation is particularly surprising in view of most anions giving smaller Δ values than most neutral ligands.

The octahedral chromophore $Ni(II)O_6$ producing the yellow to green colours of $Ni(OH_2)_6^{+2}$, $Ni(OH)_2$, $Ba_2Ni(OH)_6$, calcite-type $NiCO_3$, ilmenite-type $NiTiO_3$ and $Mg_{1-x}Ni_xO$ as well as undiluted NiO, provide Δ values also varying within 15 percent. However, as far goes the mixed crystals, it is important that Mg(II) and Ni(II) have very similar ionic radii. Whereas undiluted Cr_2O_3 is green (and has $\Delta = 16600\ cm^{-1}$, to be compared with $17450\ cm^{-1}$ for $Cr(OH_2)_6^{+3}$) many gem-stones are red. such as ruby $Al_{2-x}Cr_xO_3$ (the asymptotic value for small x is $\Delta = 18000\ cm^{-1}$) and the cubic spinel $MgAl_{2-x}Cr_xO_4$. Here, Al(III) is distinctly smaller than Cr(III), and Δ increases because of the smaller internuclear distance R (direct hydrostatic pressure [75] on NaCl-type NiO, where the atoms are on special positions, by R being half the a_0, confirms the general expectation that Δ increases 5% when R decreases 1%) but it is very difficult to know whether the Cr-O distance is fully adapted to the Al-O distance prevailing in the crystal, or whether a compromise is established, where Cr(III) pushes the six nearest oxide neighbours slightly away. A large number of mixed oxides coloured by $3d^3$ Cr(III) or $3d^8$ Ni(II) were studied by Schmitz-DuMont and Reinen [76,77] as far goes the well-resolved reflection spectra, and it is particularly interesting [8] that the ilmenite-type $Cd_{1-x}Ni_xTiO_3$ has an asymptotic value of $\Delta = 6000\ cm^{-1}$ for low x, only 70% of $\Delta = 8500\ cm^{-1}$ for $Ni(OH_2)_6^{+2}$, corresponding to the very long R adapting to the high ionic radius of Cd(II). However, the much smaller $\Delta = 4800\ cm^{-1}$ was reported [78] in the perovskite $BaNi_{0.05}Ca_{0.45}Te_{0.5}O_3$.

The crystallographic studies provide a time-average picture of the electronic density of the average unit cell [13,79]. If twinning and other effects of large-scale disorder are avoided, the parameters of the unit cell are extremely well determined. If atoms on special positions are considered (all the constituents in NaCl, CaF_2, and cubic perovskites) their distances are geometrical fractions of the unit cell parameters. Atoms on general positions have internuclear distances (such as Pt-Cl in cubic K_2PtCl_6) determined from relative intensities in the X-ray diffraction pattern, and tend to have an order of magnitude less precision than the unit cell parameters, especially when the two atoms have a large and a small *Z*. The vibrational motion (persisting to a large extent even at —273 °C) has mean amplitudes typically around 0.1 Å, and it is frequent in modern structures [79] to give "thermal ellipsoids" detecting strong deviations from the vibrational amplitudes, due to local disorder (mobile atoms, differing content of adjacent unit cells, etc.). It is not always easy to have confidence in the distance between a lanthanide and a light atom (such as oxygen) and an interesting case of a purported europium(IV) in zeolite (based on anomalous short R) has been discussed [80]. Sinha [81] has compiled R-values in a large number of rare-earth compound crystalline structures.

Below 1800 °C, the rare earths M_2O_3 are known in three types, hexagonal *A*, monoclinic *B* and cubic *C*. At lower temperatures (say by dehydrating the hydroxides below 800 °C) it is possible to make the *C*-type of even the largest lanthanides, but it is not perfectly established whether this is a question of kinetics of metastable phases, or thermodynamic stability. Strong calcination provides the *A*-type between M = La and Pm. The *B*-type is most readily realized for M = Sm and Eu, but is frequently found for Gd. $N = 7$ for *A*, two triangles perpendicular on the hexagonal axis carrying the seventh oxide. The *B*-type is very complicated with non-equivalent M sites all having $N = 7$. For comparison, it may be mentioned that N is also seven for the monoclinic baddeleyite ZrO_2 and HfO_2 (whereas CeO_2, ThO_2 and UO_2 are $N = 8$ fluorite-type). The chronological prototype of the cubic *C*-type is the mineral bixbyite $Mn_{2-x}Fe_xO_3$ but it has been reported for all the 17 rare earths, In, Tl, Pu, Am, Cm, Bk and Cf (though some are metastable). Being cubic, it can be considered a super-structure of fluorite (but this analogy is far less evident than the mixed oxides discussed below in 2.3). Three-quarters of M are situated on a site with the low local symmetry C_2 and one-quarter on a site having a centre of inversion S_6 (sometimes written C_{3i}) both having $N = 6$. The structure has been much studied [82–84] mainly as function of M, and has recently [85] been refined for Y_2O_3 showing the Y-O distances 2.243, 2.268 and 2.337 Å (each twice) for the C_2 site, and all six 2.288 Å for the S_6 site. If the oxygen nuclei were on fluorite idealized sites, all these distances would be 2.2965 Å. The S_6 site can be considered as a cube having lost two ligating oxygen atoms on a body-diagonal, but the C_2 site is quite far from a cube having lost the oxides situated on one face-diagonal. Anyhow, these two instances of $N = 6$ are far removed from a regular octahedron.

The dispersion of internuclear distances R in rare earths is unusually large. Thus, three kinds of Sm in *B*-Sm_2O_3 have their seven R scattered [81] between 2.25 and 2.76 Å, in spite of the average value 2.46 Å not being significantly different for the three kinds. In this case, the average distances are not much below the M-O distances to the nine water molecules reported for Nd(III) bromate and Pr(III), Y(III) and Er(III) ethylsulfates:

M:	Pr	Nd	Y	Er	
Six R (in Å):	2.47	2.47	2.37	2.37	(1)
Three R:	2.65	2.51	2.55	2.52 .	

X-ray diffraction of a liquid (like electron diffraction of a gas [86]) is able to provide manifolds of internuclear distances, though no explicit information is obtained about the angular distribution. Habenschuss and Spedding [56] studied 3.2 to 3.8 molar MCl_3 solutions, finding $N = 9$ for M = La, Pr and Nd, an integrated signal intensity 8.8 for Sm and 8.3 for Eu, and 8 for Tb to Lu. They report the R values to be (in Å):

La: 2.580	Pr : 2.539	Nd: 2.513	Sm : 2.474	Eu: 2.450
Tb: 2.409	Dy: 2.396	Er : 2.369	Tm: 2.358	Lu: 2.338 (2)

These values are comparable to the six short distances (to the trigonal prism) in Eq. (1) suggesting that the three long distances (to the equatorial triangle) are anomalous. It is noted that they are less different in the bromate [52] than in the ethylsulfates. Because of the large experimental difficulties, X-ray diffraction of aqueous solutions is normally performed on very high concentrations, and preferentially on anions with high *Z*, such as iodide. It is an important question whether the 10 molar chloride anions present render the distances in Eq. (2) less repr sentative for aqua ions in dilute solution. Habenschuss and Spedding [56] find the closest M-Cl distances around 4.9 Å for the aqua ions with $N = 9$ and 4.8 Å for $N = 8$. This is, obviously, an outer-sphere ion-pair (for which the equilibrium formation constant measured by classical physico-chemical techniques is close to 1) and it is well-known [50] that the narrow absorption bands in the visible are not modified by the formation of such ion-pairs. However, if a small concentration of direct M-Cl bonds are present in the strong chloride solutions, the R would be only 0.3 to 0.4 Å longer than the values in Eq. (2) and might be difficult to detect superposed the signal from the eight (or seven, for that matter) water molecules, though chlorine has about twice as many electrons as oxygen.

It is beyond doubt [50,87] that 10 molar hydrochloric acid would form $NdCl(OH_2)_x^{+2}$ and probably some $NdCl_2(OH_2)_x^+$ but the activity coefficient vary quite dramatically [88,89] in this solvent. It seems almost certain today [39] that anhydrous alcoholic solutions [90] of $NdCl_3$ contain species such as $NdCl(CH_3OH)_x^{+2}$. The 3.2 to 3.8 molar solutions of MCl_3 seem to have a less extreme chloride activity than corresponding solutions of H_3O^+ and Cl^-. For instance, NaCl is remarkably soluble in 3M YCl_3 but far less soluble in 9M HCl. However, it is always a problem with strong salt solutions that the water activity (as directly measured from the decreased vapour pressure) may favour species containing less water, if the differences of free energy are small. Thus, 3 to 5 molar sodium perchlorate dehydrate [91] the blue octahedral nickel(II) triethylenetetramine complex Ni(trien) $(OH_2)_2^{+2}$ to yellow Ni(trien)$^{+2}$. By the same token, the high chloride concentration prevailing in the solutions described by Eq. (2) may readily shift the relative concentrations of $M(OH_2)_9^{+3}$ and of $M(OH_2)_8^{+3}$ having comparable free energy.

N is also 9 for $Y(OH)_3$ having [92] six Y-O distances 2.403 and three 2.437 Å, not showing the large difference for the ethylsulfate in Eq. (1), where the six distances actually are given as short as 2.37 Å. The two sets of distances [93] are practically identical, 2.950 and 2.953 Å, in $LaCl_3$, but differ more and more in the isotypic $NdCl_3$, $EuCl_3$ and $GdCl_3$. In the gadolinium chloride, the six distances are 2.822 Å and the three 2.918 Å. For comparison, it may be noted [81] that the four M-O distances in the tetragonal oxychloride MOCl decrease smoothly from 2.39 Å for M = La to 2.25 Å for Ho, about 0.1 Å shorter than the six distances in Eq. (1), whereas the M-Cl distances vary from 3.18 to 3.04 Å, more than 0.2 Å longer than in the anhydrous MCl_3.

The *A*-type La_2O_3 was slightly revised by Müller-Buschbaum and Schnering [94] showing that the hexagonal crystals occur in two forms, one according to a model suggested by Linus Pauling with one "axial" La-O distance 2.48 Å, three (as short as) 2.39 Å and three 2.68 Å; and another (easier to prepare) with some statistical disorder of the oxide positions. This illustrates clearly the unexpected difficulties frequently encountered, when trying to locate as relatively electron-poor atoms as oxygen.

Hence, there seems to be a crystallographic fundament for believing in intrinsically shorter R, and more pronounced covalent bonding in binary oxides than in aqua ions. The observed indication for the latter situation is mainly the nephelauxetic effect (discussed below in Section 5) discovered by Hofmann and Kirmreuther in 1910, and elaborated by Ephraim and collaborators 1926–29, that the absorption bands representing essentially *J*-levels of a partly filled 4f shell [16] indicate energy differences some 3% smaller in the case of Nd(III) and some 1.6% for Er(III) when comparing oxides with aqua ions. This is not a trivial effect of lower electronegativity of the ligands, since anhydrous iodides, bromides and chlorides show a much less pronounced nephelauxetic effect than the oxides. The part of the covalent bonding in rare earths expressed as nephelauxetic shifts is probably a quite minor contribution compared with the covalent bonding due to empty orbitals (such as 5d and 6s in the L.C.A.O. hypothesis) and a risk remains that the tiny effect on the partly filled 4f shell is not proportional to the covalent bonding involving empty orbitals.

2.3 Mixed Oxides Without Discrete Anions

Syncrystallized mixtures of rare earths (as one may prepare by calcining oxalates precipitated from intermediate solutions in the process of fractional recrystallization) generally choose between the *A*, *B* or *C* type according to their average ionic radius. However, Roth and Schneider [82] found that lanthanum mixed with equal molar amounts of the smallest lanthanides formed (non-cubic) perovskites $LaMO_3$ (M = Er, Tm, Yb, Lu) exactly as do the even smaller M = Al, Cr, Fe, Ga, . . . Rittershaus and one of us [95] got interested in this question following a study of reflection spectra by low temperature of various mixed oxides containing rare earths [96]. Rather than heating mixtures of finely divided oxides in very hot furnaces, we preferred to coprecipitate the mixed hydroxides from (rapidly agitated) aqueous solutions of (preferentially) nitrates with ammonia. This does not leave any foreign cations in the final product obtained by heating to 800 or 650 °C (or even less). It turned out that the perovskites $LaErO_3$ and $LaYbO_3$ decompose to other mixed oxides above 900 °C. Besides other well-known types (such as *C* and fluorite), a new type $NdYO_3$ (only characterized by its Debye powder diagram) was consistently obtained in about forty compositions [95] with or without quadrivalent titanium or zirconium.

The oxides containing the same element (cerium, praseodymium or terbium) as M(III) and M(IV) have been investigated thoroughly by LeRoy Eyring [97]. Some of these oxides, such as Pr_7O_{12}, Pr_9O_{16}, Pr_5O_9, $Pr_{11}O_{20}$ and Pr_6O_{11} are relatively strictly stoichiometric, and show definite crystal structures, whereas others are disordered fluorites with wider range of composition. Prandtl believed that praseodymium could become partially quinquevalent in mixed oxides, but Marsh [98] showed that $Y_{1-x}Pr_xO_{1.5+0.5x}$ could not be oxidized beyond Pr(IV), and arguments derived from photo-electron spectra [99] show that Pr(V) is inconceivable in condensed matter. The electron affinity of such a species (being the ionization energy of Pr(IV) corrected for small Franck-Condon effects) would be sufficient to oxidize, not only fluoride, but also argon atoms and caesium(I).

Many minerals are disordered fluorites. Thorianite $Th_{1-x}M_xO_{2-0.5x}$ is able to contain quite high concentrations of trivalent lanthanides, besides Ce(IV), and laboratory samples can be given up to $Th_{0.46}La_{0.54}O_{1.73}$ being a colourless solid lacking, on the average, more than one of the eight oxide anions surrounding a given cation. The type CaF_2 itself can be considered as a super-structure of CsCl systematically lacking half of the cations in a definite pattern. Some minerals may be non-stoichiometric by containing excess anions on the "empty caesium" sites, such as yttrofluorite $Ca_{1-x}Y_xF_{2+x}$ and UO_{2+x} (losing its cubic symmetry by oxidizing to a definite stoichiometry, such as U_4O_9). Other fluorites may show charge compensation with constituents having comparable ionic radii (it is very frequent in rocks to have a given amount of Na(I) + Si(IV) replaced by an equivalent amount of Ca(II) + Al(III) having the same sum of oxidation states) such as trivalent M(III) in $Ca_{1-x}M_xF_{2-x}O_x$ or U(VI) forming a rhombohedral UY_6O_{12} isotypic with Pr_7O_{12} containing (unless the 4f electrons are itinerant in this black material, like the 3d electrons in the cubic spinel magnetite Fe_3O_4) four Pr(III) per three Pr(IV). However, a much more frequent source of non-stoichiometric behaviour in fluorites is *anion deficit*. Besides ThO_2, hafnium(IV) and zirconium(IV) oxides form fluorite-type $Zr_{0.9}Y_{0.1}O_{1.95}$ (the Nernst lamp mass [100]) with highly mobile oxide ions transporting electric current above 600 °C, much like the mobile iodide ions in AgI and $RbAg_4I_5$ at much lower temperatures) and $Zr_{0.9}Mg_{0.1}O_{1.9}$ used for crucibles. Whereas pure ZrO_2 is not a fluorite, definite concentrations of added Y(III) or Mg(II) allow retention of cubic symmetry from 0° to 2000 °C.

When traces of praseodymium or terbium are added to thorium when coprecipitating mixed hydroxides [95], a strong purple colour (due to electron transfer bands) indicates Pr(IV) and Tb(IV) in the calcined oxide, whereas similar traces produce a chamois colour in (otherwise pale yellow) CeO_2 and a strong orange colour in Y_2O_3 and other rare earths (brownish in Nd_2O_3). Though mixed oxides of zirconium with Pr or Tb are greyish, it seems from the a_0 values [95] that the great majority is trivalent.

It is possible to introduce moderate quantities of the small titanium(IV) in rare earths, but larger quantities yield the cubic pyrochlore-type $M_2Ti_2O_7$. *Pyrochlore* (originally a mineral with highly complicated composition) is a super-structure of fluorite, having $N = 8$ for the large M(III), being a cube compressed along one of the body-diagonal (three-fold axis) such as in the carefully investigated [101] $Er_2Ti_2O_7$ where two Er-O = 2.182 Å and the six other distances are 2.471 Å. On the other hand, $N = 6$ for Ti(IV) being approximately a regular octahedron, with all distances 1.955 Å. Like in so many other crystal structures involving rare earths, it is noted how large a spreading M-O distances can show. When preparing mixed oxides [95] containing zirconium, it is quite interesting that $Sm_{0.5}Zr_{0.5}O_{1.75}$ prepared at 800 °C is a disordered fluorite, (like $Dy_{0.5}Zr_{0.5}O_{1.75}$ at all temperatures), but rearranges to a pyrochlore $Sm_2Zr_2O_7$ at 1000 °C. Further heating above 2400 °C disorders this pyrochlore again to a fluorite.

The *perovskites* MZO_3 (and related fluorides like $KNiF_3$) are not all cubic, as reviewed by Khattak and Wang [102] but in the oxides, the sum of the oxidation states of M and Z is 6, such as $RbNbO_3$, $SrTiO_3$ and $LaAlO_3$. In the cubic perovskites, $N = 6$ for Z situated in a regular octahedron, and $N = 12$ for M in a cuboctahedron. The twelve distances can be rather different in low-symmetry perovskites, whereas other, only slightly distorted, such as $LaAlO_3$ show a reversible phase transition at (in the example 435 °C) higher temperature to cubic symmetry. The magnetic coupling between Z containing a partly filled d-shell, such as Cr(III) and Fe(III), can be quite strong [102,103] and the electric conductivity, including phase-transitions between semi-conducting and metallic systems, can be quite unexpected with Z = titanium or cobalt. The photo-electron spectra of $LaCoO_3$ has been studied [104] at various temperatures.

The prototype of the cubic *garnet* is the dark red mineral $Fe_3Al_2Si_3O_{12}$ containing Fe(II) in the idealized formula, but for rare-earth chemistry [102] the typical case is the colourless diamond-like $Y_3Al_5O_{12}$. Three of the five small cations, here Al(III) and in the mineral Si(IV), are situated on tetrahedral sites (called *d*), two octahedral sites (called *a*) whereas the three large cations, e.g. Y(III) are on a *c*-site with $N = 8$ and very low local symmetry, in spite of the cubic symmetry of the whole crystal. Substituted garnets have great importance in magneto-optical devices.

Since the mixed oxides discussed here have the oxygen atoms bridging in an indefinite lattice, one should not call perovskites such as $BaTiO_3$, $LaAlO_3$, $GdFeO_3$ and $DyGaO_3$ titanates, aluminates, ferrites and gallates, because such words insinuate finite anions. The opposite situation occurs in $CaCO_3$ in which both crystalline modifications (calcite having $N = 6$ and aragonite with $N = 9$) contain monomeric carbonate anions. A third, rare modification, hexagonal vaterite, may not exist as pure $CaCO_3$, but it is isotypic with lanthanide borates MBO_3 ($N = 9$). Bevan and Summerville [105] have further reviewed mixed oxides of rare earths, and Felsche [48] the rather complicated silicates.

2.4 Salts of Oxygen-containing Anions

The many mixed oxides not containing definite discrete anions might encourage the opinion that the rare earths tended toward such building of oxide bridges in order to keep N so much higher than the oxidation state. However, this is not always true. For instance, orthophosphate is a quite strong base, pK of $HOPO_3^{-2}$ being 12, and it does not always precipitate straight forward salts. Thus, calcium tends to form apatites $Ca_5(PO_4)_3X$ with $X = OH^-, F^-, \ldots$ Nevertheless, one of the most frequent rare-earth minerals is *monazite* MPO_4 containing mainly the larger M(III) from lanthanum to samarium. This mineral was for many years extracted mainly for its thorium content (for use in candoluminescent Auer mantles to be discussed in Section 7) present due to charge compensation by Th(IV) and SiO_4^{-4}. The almost unique ortho-niobates and -tantalates occur in the isotypic *fergusonite* $M(NbO_4)_{1-x} \cdot (TaO_4)_x$. Under normal conditions, iso-polyniobates are formed even from strongly alkaline solution. The tetragonal type *xenotime* is also MPO_4, but this time with a major concentration of yttrium, and of lanthanides smaller than neodymium. This mineral is isotypic with *zircon* $ZrSiO_4$ and thorite $ThSiO_4$. Whereas the silicate and phosphate monomeric anions are tetrahedral, the cations Y(III), Zr(IV) and Th(IV) have $N = 8$ in the rather low point-group D_{2d}. These phosphates, the corresponding arsenates and orthovanadates have been important for developing the angular overlap model [106-108] relating the tiny energy differences (total spreading below 800 cm^{-1} or 0.1 eV) between the seven 4f orbitals to the anti-bonding influence of the chromophore consisting of the ligating neighbour atoms. As discussed in Section 7, the red cathodoluminescence of $Y_{1-x}Eu_xVO_4$ was for many years of great importance for colour television. The mixed orthovanadates are most readily prepared by pyrolyzing an intimate mixture of the appropriate rare earths with p.a. ammonium metavanadate NH_4VO_3. Though the tendency toward polymerization is not as great as of niobates and tantalates, it is easy to form orange dekavanadates $V_{10}O_{28}^{-6}$ from aqueous solution and brick-red V_2O_5 is not very soluble in water (around pH = 2) in contrast to strongly hygroscopic P_4O_{10}.

The double nitrates $Mg_3M_2(NO_3)_{12}$, 24 H_2O can be recrystallized in strong nitric acid and serve to separate the lighter lanthanides M = La, Pr, Nd and Sm (where Ce has been removed after oxidation to the quadrivalent state). Judd noted [109] that the fine-structure of the absorption band belonging to each *J*-level of M(III) was so peculiar that it looked as if the chromophore was icosahedral with $N = 12$ (which is almost unheard about, outside boron chemistry). The crystal structure [110] of the cubic crystals confirmed entirely Judd's proposal, it is indeed $[Mg(OH_2)_6]_3 \cdot [M(O_2NO)_6]_2$, 6 H_2O with the twelve oxygen atoms provided by the six bidentate nitrate ligands forming approximately a regular icosahedron, circumscribed by a regular octahedron formed by the six nitrogen nuclei. A comparable structure is found in orange $(NH_4)_2Ce(NO_3)_6$ (and the analogous colourless thorium compound) actually containing $M(O_2NO)_6^{-2}$ with M = Ce(IV) or Th(IV). The high $N = 12$ is accompanied by relatively long M-O distances to the bidentate ligands (rationalizing why the nephelauxetic effect is intermediate between the fluorides and the aqua ions). Thus, the average Ce-O distance in the cerium(III) salt [110] is 2.64 Å, almost 0.15 Å longer than extrapolated from the aqua ions in Eq. (1).

Other double nitrates have $N = 10$ such as $M(O_2NO)_5^{-2}$. Since carbonate is isoelectronic with nitrate, it is likely that $M(O_2CO)_5^{-7}$ and $Th(O_2CO)_5^{-6}$ obtained by dissolving the smallest lanthanides M = Er, Tm, Yb and Lu, and thorium(IV), in concentrated carbonate solutions (where the uranyl ion seems to form $UO_2(O_2CO)_3^{-4}$ isoelectronic with $UO_2(O_2NO)_3^-$ known from certain crystal structures) are similar. Sinha [81] mentions several examples of $N = 10$ in mixed complexes of nitrate involving, for instance, bidentate dipyridyl in $Mdip_2(O_2NO)_3$.

Bünzli and collaborators [111] continued previous work [44,45] on crown-ethers and found $N = 11$ for the adduct of $Eu(NO_3)_3$ with a quinquedentate ether, the Eu-O distances distributed between 2.43 and 2.69 Å, the average of the six distances to the bidentate nitrate ligands being 0.13 Å smaller than to the ether oxygens. $N = 11$ occurs also [112] in the lanthanum nitrate hexahydrate $[La(O_2NO)_3(OH_2)_5]H_2O$ and [113] in the anion $Sm(O_2NO)_5(OH_2)^{-2}$. $N = 12$ occurs [45] in the adduct of $Nd(NO_3)_3$ with a sexidentate crown-ether, the six nitrate distances (all close to 2.60 Å) being significantly shorter than to the ether oxygen atoms.

Oxygen atoms of organic anions tend to have shorter distances than do neutral organic molecules, and Sinha [81] has compiled many results, though with the major motivation of establishing the shape of the coordination polyhedra, which are now known to be much less characteristic than in the d-groups. A typical case [114] is the oxalate $Nd_2(C_2O_4)_3$, 10 H_2O where $N = 9$. The six distances to three (1,2-)-bidentate oxalate oxygens between 2.46 and 2.57 Å have an average only 0.01 Å longer than to the three coordinated water molecules. Contrary to monomeric complexes such as $Rh(C_2O_4)_3^{-3}$ where half the carbonyl groups are not coordinated (and show the corresponding vibrational spectrum), each carboxyl group of the oxalate bridges two Nd(III).

2.5 Glasses

Not only oxygen- but also fluoride- and sulfur-containing anions may form vitreous materials incorporating lanthanides. For many years, neodymium (or the technical mixture with praseodymium, still [115] called didymium) has been added to conventional silicate glasses (containing sodium and bivalent ions such as calcium or lead) in order to obtain a mauve plate with extremely sharp absorption bands (which can be used [90] to standardize a spectrophotometer much more accurately than its nominal wave-length, when the positions are changed). Such neodymium glass has a strong absorption band covering the two yellow sodium emission lines, and is used in qualitative analysis to allow the red and violet light emitted by potassium atoms in a gas flame to be detected in the simultaneous presence of sodium, and also with the analogous purpose of protecting glass-workers' eyes against sodium light rendering the incandescent glass less distinct.

Seen from the point of lanthanide spectroscopy, glasses combine some of the characteristics of salts of discrete anions and of mixed oxides with indefinite oxide bridging. The tetrahedral silicate or phosphate constituents of glasses play a rôle similar to zircon or xenotime with $N = 8$ of the larger cation (and it is likely, but difficult to prove, that lanthanides have such an environment in the glass). On the other hand, there is no repeating unit cell in a glass, and there is evidence [116,117] from fluorescence spectra of europium(III) in mixed phosphate-tungstate glasses that the local order may be highly different as a function of the composition.

It is trivially true that any excited state not undergoing radiationless decay by exciting vibrations, or performing photochemical reactions, actually is luminescent. However, it is generally felt thtat luminescence is rather the exception in crystalline or vitreous solids containing lanthanides (especially if the temperature is not very low) and further on, that the most likely candidates [72] are concentrated around the half-filled shell, $4f^6$ Eu(III), $4f^7$ Gd(III) and $4f^8$ Tb(III), and less frequently, $4f^5$ Sm(III) and $4f^9$ Dy(III). This situation is connected [16] with the much larger energy gap in the visible between a given J-level (in the five examples $^5D_0$ at 17250, $^6P_{7/2}$ at 32200, $^5D_4$ at 20500, $^4G_{5/2}$ at 17900 and $^4F_{9/2}$ at 21100 cm^{-1}) and the groundstate, or low-lying J-levels. In the case of gadolinium(III), the emission takes place in the ultra-violet (close to 310 nm). One of us [118-120] analyzed the conditions for radiationless desexcitation, and found it to be essentially determined by the ratio between the energy gap available for the luminescence and the highest normal frequency of vibration of the solid. This ratio indicates roughly the number of phonons needed to degrade the excited state without emission of a photon. The probability of this desexcitation (to be compared with the radiative life-time of the excited state according to the 1917 formula of Einstein) decreases exponentially when the ratio increase, and becomes almost imperceptible [119] above 6 to 8.

It is well-known from fluorescent organic compounds that replacement of the hydrogen atoms by deuterium efficiently counteracts the process of non-radiative desexcitation. The vibrational frequencies are smaller by a factor roughly $\sqrt{2}$, and in particular, the C-H frequencies up to 3500 cm^{-1} are replaced by C-D frequencies up to 2500 cm^{-1}. The same approach to glasses [16,118-120] suggests to replace the silicate and phosphate network-forming anions by other anions with as small a phonon energy as possible. The theory of vibrational frequencies (for approximately harmonic behaviour) opens two alternatives: to increase the atomic masses, and to decrease the force constants. For instance, germanate and tellurite-based glasses combine both alternatives, and the phonon energies are around 900 and 700 cm^{-1}, respectively, to be compared with 1400 cm^{-1} for borate and 1300 cm^{-1} for phosphate glasses. Actually, $4f^{10}$ holmium(III) [121], $4f^{11}$ erbium(III) [122] and $4f^{12}$ thulium(III) [122] emit narrow emission bands (due to transitions from several excited *J*-levels) in such glasses, and their relative intensities (transition probabilities) can be described by the Judd-Ofelt theory (discussed in Section 6). Frequently, luminescense in the near infra-red goes undetected, because it cannot be seen by our eyes. Nevertheless, the transitions from $^4F_{3/2}$ (at 11400 cm^{-1}) in neodymium(III) occur with high yield in many crystalline and vitreous materials, and have acquired a strong technological importance [16] for lasers (to be discussed in Sect. 7). Also, the only excited level of $4f^{13}$ in ytterbium(III) close to 10200 cm^{-1} can be efficiently luminescent, and participate in energy transfer, e.g. from Nd(III) simultaneously present in the glass [123].

Though the luminescence of gadolinium(III) aqua ions in the ultra-violet was discovered by Urbain in 1905, the quenching in aqueous solution is much more extensive than in most glasses, and even the quantum yields of Eu(III) and Tb(III) are quite low. Not unexpectedly, deuterium oxide solutions give higher yields and longer life-times [124-126] and in absorption of Gd(III), coupling with the vibronic spectrum can be observed [126][127].

By comparison with germanate and tellurite glasses, one would not expect glasses formed by mixed fluorides to have particularly low phonon energies. However, it seems that the force constants corresponding to the relatively secondary bonding of the lanthanides to the rest of the glass sometimes can be so small as to be as efficient as germanates. Thus, the luminescence of europium(III) in a barium(II) zirconium(IV) fluoride glass [128] shows the life-times 0.5 m sec for $^5D_2$, 1 m sec for $^5D_1$ and 5 m sec for $^5D_0$. The fluorescence of *three* excited levels is quite unusual; the ratio of the life-times ($^5D_1/^5D_0$) is 0.02 for a tetrafluoroberyllate glass, and as low as 0.0012 for crystalline EuP_5O_{14}. The life-time of $^5D_0$ is 3 m sec for a phosphate glass. Other fluoride-containing glasses have been studied [129-131] and the Judd-Ofelt parameters evaluated. Rutile-type MnF_2 has also very low phonon energies and is excellent for luminescence of incorporated lanthanides [132,133].

Another class of glasses highly effective for lanthanide luminescence contain sulfur replacing the oxygen. The phonon energies are very low, and Ho(III), Er(III) and Tm(III) show high yields of fluorescence of several transitions. The most frequent materials are the mixtures $LaAl_3S_6$ and $LaGa_3S_6$ with aluminium or gallium sulfides [134-136]. The stoichiometric composition of these glasses should not be interpreted as them being compounds in a normal sense of the word. The network-formers are likely to be Al(III) or Ga(III) surrounded by four sulfur atoms. In this connection may be mentioned the colourless ternary sulfide $CdGa_2S_4$ in which small amounts of erbium(III) can be incorporated [137] giving the characteristic $4f^{11}$ absorption spectrum with a pronounced nephelauxetic effect. If the local order is not strongly distorted, it is likely that the rare tetrahedral chromophore $Er(III)S_4$ then occurs.

3 Sulfides and Low-Electronegativity Ligands

The high affinity of lanthanides and yttrium for oxygen-containing ligands, and the experience of aqueous solution chemistry (where the Brønsted base HS^- precipitates the hydroxide) made it unlikely for a long time that sulfides would be readily prepared. However, with the help of the metallic elements and the anhydrous halides now available, it is possible to prepare a large number of binary and ternary sulfides at high temperatures, when avoiding the presence of humidity and oxygen. Flahaut [138,139] has written two magnificient reviews on this subject. Here, we only mention a few details: the cubic type [145] Th_3P_4 with a (low-symmetry) $N = 8$ for Th(IV) and $N = 6$ for P(—III) is found in the metallic Ce_3S_4, in EuM_2S_4 containing [140] europium(II), in the corresponding YbM_2S_4 [M = La to Ho; not Y], in CaM_2S_4 [M = La to Dy], SrM_2S_4 [M = La to Gd], BaM_2S_4 [La, Ce, Pr, Nd] of which reflection spectra [137] show an unusually pronounced nephelauxetic effect, and the sesquisulfides [M = La to Tb] are defect structures $M_{3-0.333}S_4$ which show extended miscibilities with the Th_3P_4 type alkaline-earth ternary sulfides.

The NaCl-type monosulfides MS are physical metals for M = La, Ce, Pr, Nd, Gd, Tb, Dy, Ho, Er, Tm and Lu but semi-conductors (with considerably larger a_0 values) for M = Sm, Eu, and Yb. This choice between two sharply defined alternatives can be rationalized [141,142] by the spin-pairing energy (discussed in Sect. 5). Mixed crystals such as $Nd_xSm_{1-x}S$ or $Th_xSm_{1-x}S$ or $Gd_xSm_{1-x}S$ go metallic at a definite lower limit of x, as has been further investigated [13] by photo-electron spectra.

Binary and ternary selenides and tellurides show essentially the same behaviour [138,139] as the sulfides. The situation is rather special [142] for the TmSe and TmTe containing, on an instantaneous picture [13] 80 or 50 percent, respectively, of the conditional oxidation state $4f^{12}$ Tm[III] and the rest (20 or 50%) $4f^{13}$ Tm[II].

The cathodoluminescent zircon-type $Y_{1-x}Eu_xVO_4$ has to a large extent been replaced by the oxysulfide $Y_{2-x}Eu_xO_2S$. The M_2O_2S (M = La to Lu, including Y) have essentially the *A*-type of La_2O_3. Among $N = 7$ neighbour atoms, three atoms are sulfur, and four oxygen. As a warning about problems with oxygen impurities, what was taken for many years as a modification of Pr_2S_3 is indeed $Pr_{10}OS_{14}$ with a quite definite crystal structure [139].

Hulliger [145,146] reviewed the NaCl-type MP, MAs, MSb and MBi. Though black, it seems that these compounds are low-energy-gap semi-conductors, since their magnetic properties indicate M[III]. The analogous mononitrides MN are almost certainly semi-conductors with about 2 eV energy-gap, but it seems at present impossible to prepare them without impurities making them conducting (like silicon before the war). Seem from the point of view of chemical bonding, the photo-electron spectra [143,147] of mono-antimonides have shown that the ionization energy of the filled Sb5p shell *always* is lower than of the partly filled M4f shell.

We do not have the space here to discuss halides, which have been reviewed by Haschke [148]. It may only be mentioned that though $N = 6$ is not very frequent for the lanthanides and yttrium, it is even much rarer to encounter octahedral coordination. The absorption spectra [149] of MCl_6^{-3} and MBr_6^{-3} show quite unique aspects of high symmetry; the band intensities are unusually low, and many of the electronic transitions are only seen as co-excitations of odd (ungerade) vibrations, lacking the electronic zero-phonon lines (which are the only intense features in other lanthanide spectra). Ryan [150] later succeeded in preparing solutions of MI_6^{-3} in acetonitrile. The mineral elpasolite (K_2NaAlF_6) was found to be isotypic with the berkelium(III) compound $Cs_2NaBkCl_6$. Such hexachloride salts can be made of all the rare earths. The Cartesian axes of this cubic type contain alternatingly NaClMClNaClMCl . . . much like the cubic perovskite $SrTiO_3$ has alternatingly TiOTiO . . ., and each chloride is surrounded by an octahedron consisting of one M and one Na on one axis, and four Cs (having $N = 12$) in a plane, which does not need exactly to contain the chlorine nucleus. The absorption spectra of such elpasolites have been extensively studied [151–154] like a phosphonium salt [155] of $NdCl_6^{-3}$. It remains an open question whether the chlorine nuclei tend to leave the Cartesian axes on an instantaneous picture. $N = 6$ is certainly quite low for the first lanthanides.

For reasons quite similar to sulfur-containing ligands (where the absorption spectra can be studied in 1,2-dichloroethane solution [156] of dithiocarbamates such as $M(S_2CNR_2)_3$ and $M(S_2CNR_2)_4^-$ and where Pinkerton [157] showed trigonal prismatic $N = 6$ in the dithiophosphinates $M(S_2PR_2)_3)$ it was believed for many years that ligands would not bind exclusively to lanthanides with nitrogen atoms. A quite outstanding case is $N = 3$ in the hexamethyldisilylamide $M(N(Si(CH_3)_3)_2)_3$ which are known for the various rare earths [158] like for Cr(III) and for Fe(III) constituting an exceptional case of low N. According to the compilation of Sinha [81] the M-N distance is short (as expected for such a low N), 2.259 Å for Eu, 2.049 Å for Sc and 1.917 Å for Fe. Interestingly enough [158] the Eu(III) compound is orange, and Yb(III) yellow, the same colours as the dithiocarbamates [156] caused by *electron transfer bands*, where the excited state contains one 4f electron more, transferred from the reducing ligands. Again, $EuBr_6^{-3}$ is orange (like anhydrous $EuBr_3$) and $YbBr_6^{-3}$ is yellow [149] whereas EuI_6^{-3} is dark green and YbI_6^{-3} deep pruple, having stronger reducing ligands [159].

The Brønsted basicity of heterocyclic nitrogen-containing ligands (such as pyridine) is much weaker than of aliphatic amines or ammonia. Combined with the chelate effect of bidentate dipyridyl or phenanthroline or tridentate terpyridyl is it feasible to prepare complexes of the latter heterocyclic ligands (normally containing coordinated anions, and sometimes even water molecules) from moderately polar solvents, such as ethanol. Such complexes are reviewed by Sinha [81] who did much of the early work. Under strictly anhydrous circumstances (and in less polar solvents) is it possible to prepare complexes of multidentate aliphatic amines such as ethylenediamine (the tetrakis-complex has $N = 8$) and the tridentate diethylenetriamine, as reviewed by Forsberg [160] and by Thompson [41]. In the review [161] (with 406 references) on (mainly solid) complexes of rare-earth elements with neutral ligands bound by oxygen atoms, there are quite a few cases of simultaneous binding of nitrogen and oxygen atoms. This is a more frequent situation in anions, such as biological α-amino-acids. However, again because of the chelate effect, multidentate synthetic amino-acids developed for complexometric titrations and for masking of undesirable cations, [162] are more studied.

An early case is ethylenediaminetetra-acetate (called EDTA by most authors, but here $enta^{-4}$ avoiding element-like capitals). Schwarzenbach and his collaborators [163] showed that this potentially sexidentate ligand (two tertiary amine-N and four carboxylate groups normally being unidentate because of steric conditions, and not bidentate, like most carbonate and a few acetate complexes) sometimes have a free-dangling carboxylate and provide $N = 5$. Thus, cobalt(III) forms $Co(enta)^-$, $Co(enta)\,OH_2^-$, $Co(enta\ H)\,OH_2$ and $Co(enta)\,OH^{-2}$ where, dependent on pH, the non-ligating carboxylate group may protonate, and the aqua ligand deprotonate to hydroxide. Chromium(III) is only known [163] in the three latter situations. It seems also verified [164] that the stable $Ni(enta)\ OH_2^{-2}$ contains one molecule of ligated water, which can be exchanged, such as $Ni(enta)NH_3^{-2}$ and the cyano complex $Ni(enta)\,CN^{-3}$. A complex rearranging in a few minutes (also the initial form of Cr(III) has a differing absorption spectrum) may involve a quadridentate ligand (free to protonate on two carboxylates) and be $Ni(enta)\,(OH_2)_2^{-2}$. Not all 3d-group complexes are octahedral; both $Mn(enta)\,(OH_2)^{-2}$ and $Fe(enta)\,(OH_2)^-$ have $N = 7$, but with unusually long Mn-N and Fe-N distances. The lanthanides show a similar behaviour; crystalline $[La(enta\ H)\,(OH_2)_4]$, $3\ H_2O$ shows $N = 10$ and a salt of $La(enta)\,(OH_2)_3^-$ $N = 9$.

Under these circumstances the potentially octadentate ligand diethylenetriamine penta-acetate with three nitrogen atoms and five carboxylate groups (called DTPA by most people, but here [165] $denpa^{-5}$) should bind rare earths better than $enta^{-4}$ whereas there should not be a great difference in the 3d-group. This is indeed [166,167] observed from the equilibrium formation constants. The next homolog in this series (corresponding [165] to a well-known set of organic "Christmas tree garland" amines) is the potentially decadentate triethylenetetraminehexa-acetate (called TTHA by Martell [168] but here $trienha^{-6}$ for reasons [165] as above) with four nitrogen atoms attached in a linear butane chain and six carboxylate groups. Indeed, four non-ligated carboxylates are available for protonation [168] in $Ni(trienha\ H_4)$. Interestingly enough, the formation constants [169] for M(III) complexes of $trienha^{-6}$ decrease from La(III) to Ho(III), and then further to Ga(III), whereas Y(III) has the unexpected place [167] close to Pm(III), as far goes $denpa^{-5}$ complex formation constants.

Thorium(IV) [170] has $N = 10$ in $Th(trienha)^{-2}$. The same is true [171] for the pentakis (tropolonate) anion of Th(IV) and U(IV) whereas the neutral tetrakis-(tropolonate) molecules have $N = 8$. The question of N for thorium(IV) [172] and for uranium(IV) [173] has been discussed on basis of formation of mixed complexes of multidentate ligands. It is clear from our arguments related to rare-earth aqua ions that one cannot establish a definite N for the Th(IV) aqua ion this way (though 10 may not be unreasonable). With pK = 4, it is the weakest Brønsted acid among the M(IV) aqua ions (the pK values for M = Zr, Ce and Hf are undoubtedly negative) and already the uranium(IV) aqua ion has pK = 1 (which may surprise, in view of the moderate decrease of ionic radius). Spectroscopic arguments have been given [174] that U(IV) aqua ions have $N = 9$ like Nd(III). It is well-known that many anion complexes, such as the oxalate $U(C_2O_4)_4^{-4}$ have $N = 8$. Though it is easier [175] to make UCl_6^{-2} (and even UBr_6^{-2} and UI_6^{-2}) than the corresponding

MX_6^{-3}, the former species is not detectable in aqueous hydrochloric acid, and the Pearson-hard character of U(IV) can be seen [176] from spectra of $UCl_nBr_{6-n}^{-2}$ in nitromethane, where the equilibrium constants for introducing bromide ligands are quite low.

Another neutral, nitrogen-containing ligand is acetonitrile CH_3CN. At one side, it seems little apt to coordinate rare earths, since MX_6^{-3} can be studied [149,150] in mixtures of acetonitrile and succinonitrile, but on the other side. Bünzli, Yersin and Mabillard [177] have shown from well-resolved fluorescence (and infra-red) spectra that europium(III) perchlorate in acetonitrile forms several species, among which the predominant are $Eu(NCCH_3)_9^{+3}$ and $Eu(OClO_3)(NCCH_3)_8^{+2}$ (though we write chemical formulae on one line, we try to convey as much stereochemical information as feasible by not separating adjacent atoms, when it can be avoided). The same ambiguity occurs with nickel(II) complexes, where octahedral $Ni(NCCH_3)_6^{+2}$ is known, but tetrahedral $NiCl_4^{-2}$ (which is very difficult to maintain in other solvents) can be obtained with a small excess of chloride. One may conclude that chloride is *much* better bound than in water, for reasons that may be connected [39] with the lower local dielectric constant. Acetonitrile shows many other differences from water, for instance, copper(I) and silver(I) dissolve as $M(NCCH_3)_4^+$.

4 Organolanthanide Chemistry

Whereas binding of rare-earth central atoms with nitrogen- or sulfur-ligating atoms have attracted attention only since about 1960, the situation is even more extreme with regard to ligation by carbon atoms. It is worth remembering that alkaline and alkaline-earth metals also form organometallic compounds, though with exception of the small atoms lithium and beryllium, they are not usually typical organic molecules (like organometallic compounds formed by silicon, germanium, tin and lead) but have distinct carbanion characteristics (after all, CH_3^- is isoelectronic [8] with NH_3) and, frequently, unusual coordination numbers N. They have to be made (like isolating the pyrophoric zinc(II) alkyls; or keeping magnesium(II) Grignard reagents in ether solution) under scrupulous exclusion of humidity or solvents with mobile protons capable of protonating carbanions.

Though $N = 3$ is only known [158] for the $M(N(Si(CH_3)_3)_2)_3$ and the vapours of M(III) halides, the organolanthanide compounds are now known to vary from $N = 4$ to $N = 16$. This is not really as surprising, when comparing with organoberyllium compounds. Though Be(II) sticks to $N = 4$ much more firmly than carbon(IV), many dimeric beryllium alkyls $RBeR_2BeR$ have $N = 3$ of Be, and two bridging alkylgroups have carbon with $N = 5$. The monomeric *t*-butyl derivative $(H_3C)_3CBeC(CH_3)_3$ is normal, as far goes carbon, but has linear $N = 2$ of beryllium, like mercury would have. Frequently, carbon has N above 4. The molecule $Li_4(CH_3)_4$ shows $N = 6$, each carbon being bound to three hydrogen atoms and to three lithium situated on a face of a regular tetrahedron Li_4. The amber-yellow (non-metallic) fluorite-type Be_2C has allright $N = 4$ for Be(II) but cubal $N = 8$ for C(—IV).

$N = 4$ occurs [178] in the "classical" lutetium(III) complex of four 1-deprotonated 2,6-dimethylbenzene carbanions with average Lu-C distance 2.45 Å (actually 0.11 Å longer than Eq. (2) gives for Lu(III) aqua ions). There is much evidence for steric hindrance in this complex, but at the same time, it is likely that the two methyl substituents on the phenyl group protects against rapid rearrangements. However, it is clear that this species is far more reactive than the similar $B(C_6H_5)_4^-$. The Yb(III) analog has been prepared, but not with larger lanthanides.

In 1954, Wilkinson and Birmingham [179] prepared cyclopentadienides $M(C_5H_5)_3$ of most rare-earth elements. However, as reviewed by Baker, Halstead and Raymond [180] and by Thompson [41] several of their crystal structures do not correspond to molecules with 15 roughly equal M-C distances, but rather to an irregular packing of planar, pentagonal ions $C_5H_5^-$ frequently bridging two M(III). This situation is very different from the "sandwich" cyclopentadienides of the d-groups [181] such as ferrocene $Fe(C_5H_5)_2$ and the isoelectronic cobalticenium cation $Co(C_5H_5)_2^+$ both having strictly $N = 10$. Adducts with one molecule of cyclohexyl*iso*nitrile such as $(H_5C_5)_3PrCNC_6H_{11}$ are monomeric (with an average 2.77 Å of fifteen M-C(ring) distances, and the sixteenth M-C distance 2.68 Å) and the analogous ytterbium(III) complex, as well as $Yb(C_5H_5)_3$ alone, are green [182] due to an electron transfer band from the reducing ligands in the red (at 3000 cm^{-1} lower wave-number than of YbI_6^{-3}). This coordination with $N = 16$ is well-known [180] from several uranium(IV) complexes $XU(C_5H_5)_3$ (among which the chloride was prepared by Reynolds and Wilkinson in 1956). Ytterbium(III) bis(cyclopentadienide) monochloride [180] is a chloride-bridged dimer $(H_5C_5)_2YbCl_2Yb(C_5H_5)_2$ with apparently less reducing ligands [8], the solid being orange.

Uranium(IV) and thorium(IV) form true "sandwich" cyclooctatetraenide complexes "uranocene" $U(C_8H_8)_2$ and $Th(C_8H_8)_2$ with the average [180] of the sixteen U-C distances 2.647 Å and Th-C distances 2.701 Å. Cerium(III) forms a far more reactive anion $Ce(C_8H_8)_2^-$ with the quite high average 2.74 Å of the sixteen Ce-C distances.

There exist some aryl and acetylene complexes of rare earths, for which there is not much structural evidence available [41]. The two mixed complexes [183] $C_6H_5Yb \cdot (C_5H_5)_2$ and $C_6H_5C{\equiv}CYb(C_5H_5)_2$ are both orange like $CH_3Yb(C_5H_5)_2$ whereas $CH_3Er(C_5H_5)_2$ has the usual pink erbium(III) colour. It is important to note that $Tb(C_5H_5)_3$ is colourless; there are hence no inverted electron transfer to low-lying empty M.O., nor $4f{\rightarrow}5d$ transitions, before in the ultra-violet.

The nephelauxetic effect [137,184] is slightly more pronounced in $Nd(C_5H_5)_3$ and $Er(C_5H_5)_3$ than in the corresponding sesquioxides, but there is no indication of an essentially different kind of covalent bonding in organolanthanide compounds. The "ligand field" treatment of 4f and 5f group "sandwich" compounds [185] does not involve parameters highly different from other M(III) and M(IV) complexes. Green [186] has reviewed the photo-electron spectra of gaseous (vapourized) d- and f-group organometallic compounds. Whereas $5f^2{\rightarrow}5f^1$ ionization can be observed in $U(C_8H_8)_2$ at 6.20 eV and in $U(C_5H_5)_4$ at 6.34 eV, problems of relative signal intensities seem to prevent 4f signals to be detected in the gaseous organolanthanides studied as far. Fortunately, the X-ray induced photo-electron spectra of solid lanthanide compounds [99,143,147,187,188] have been far more informative about the 4f ionization energies.

5 Electron Configurations, Spin-pairing Energy and Nephelauxetic Effect

The atomic spectra of lanthanides are far richer in lines than even the arc spectrum of iron atoms. The identification of the individual *J*-levels, and their classification in electron configurations, has been a very difficult work [9]. The situation is slightly less complicated in gaseous M^{+2} (called M III without parenthesis by atomic spectroscopists, because it is "the third spectrum of the elements"; we reserve Roman numerals *with* parentheses for oxidation states [8] in condensed matter) because the lowest *J*-levels belong to the configuration [54] $4f^q$ with $q = (Z - 56)$ (we except Gd^{+2} with [54] $4f^7 5d$) followed by the two next configurations *[54]* $4f^{q-1}5d$ and *[54]* $4f^{q-1}6s$. A few cases, such as samarium, europium and thulium (having only one excited level of *[54]* $4f^{13}$ at 8774 cm^{-1}) are extensively studied, both as gaseous M^{+2} and as M(II) in solid samples.

The fascinating fact is that the narrow absorption and fluorescence bands of M(III) in solution and in vitreous and crystalline solids have been considered as a kind of atomic spectra for at least 50 years, in spite of the first gaseous M^{+3} to be characterized as [54] 4f was Ce^{+3} (by R. J. Lang in 1936) and twelve of the thirteen expected *J*-levels of *[54]*$4f^2$ in Pr^{+3} by Jack Sugar [189] in 1965. More recently, Kaufman and Sugar [190] extended the analysis of Yb^{+3} and found also the only excited level of *[54]*$4f^{13}$ at 10214 cm^{-1} (above the groundstate) coinciding with the band position of Yb(III) found by Gobrecht [191]. The ten intervening Nd(III) to Tm(III) are well understood today [16] but the evidence [9] available for the ten corresponding M^{+3} is more than fragmentary. It may be mentioned that serious doubts have been expressed [9] about the *[54]* $4f^{11}$ levels of Er^{+3} previously cited [8]. The evidence [9] for M^{+2} can be of some help for the chemical spectroscopists; e.g. 22 *J*-levels have been identified both in *[54]*$4f^3$ of Pr^{+2} and in Nd(III), and it turns out [8] that the *J*-energy differences in Pr^{+2} are very precisely 0.81 times those of the isoelectronic Nd(III).

Actually, the theory for (*S*,*L*)-term and *J*-level distribution in a configuration containing one partly filled shell (which originated with Slater 1929 and was elaborated in the treatise [192] by Condon and Shortley, and further stream-lined by Racah [193]), works much better for $4f^q$ configurations of M(III) than e.g. for monatomic entities containing from six to eight electrons [8,192] where the lowest configuration is $1s^2 2s^2 2p^q$. The straight forward description of S.C.S. (Slater-Condon-Shortley) or Racah parameters as integration of interelectronic repulsion over definite (say Hartree-Fock) radial functions can be critized [194,195] but it remains true that these parameters remain approximately inversely proportional to the average radius of the partly filled shell, but decreased by a dielectric screening effect [196,197] corresponding (to the first approximation) to the multiplication of the integrals of interelectronic repulsion by the factor $(z + 2)/(z + 3)$ where z is the ionic charge. It turns out [8,13,196] that the distance between the baricenter of all states having a definite value of $S = (S_0 - 1)$ (obtained by weighting the term energies by the factor $(2S + 1)(2L + 1)$ indicating the number of mutually orthogonal wave-functions) and of those having $S = S_0$ can be written $2DS$ (for all q values) where the *spin-pairing energy parameter D* is a definite linear combination of S.C.S. (or Racah) parameters of interelectronic repulsion. Relative to the baricenter of all the states of the configuration f^q, the baricenter of all states having a given value of S is

$$\left[\frac{3q(14-q)}{52} - S(S+1)\right] D \tag{3}$$

providing a quantitative expression for one of Hund's rules [5] that when a configuration containing one partly filled shell is able to exhibit differing S values (i.e. q between 2 and $4l$), the groundstate has the highest S possible. In the cases q = 2, 3, 4, 5 and 9 to 12, more than one L value is compatible with the maximum value of S, and the groundstate is more stabilized than given by Eq. (3). A further (small) correction can be introduced for spin-orbit coupling. The S baricenters of Eq. (3) move on a straight line as function of Z, expressing the smoothly increasing (negative) one-electron energy of the f electrons. If the parameter $(E - A)$ describing the latter effect [39] is 3200 cm^{-1}, $D = 6500\ cm^{-1}$, the Racah parameter $E^3 = 500\ cm^{-1}$, and accepting the empirical values of the Landé parameter of spin-orbit coupling, the ionization energy $(q \rightarrow q-1)$ is (in the unit 1000 cm^{-1}) relative to an arbitrary zero-point, for each q value given:

1: 1.3	2: 12.5	3: 20.9	4: 21.6	5: 22.2	6: 30.2	7: 40.3	
8: 0.8	9: 11.9	10: 19.8	11: 20.0	12: 20.0	13: 27.2	14: 37.2	(4)

It is noted that the ionization energy in Eq. (4) is almost the same for (q + 7) as for q (this is the numerical expression for the "effect of half-filled shells") and further on, that two plateaux occur, giving nearly the same ionization energy for q = 3, 4, 5 and 10, 11, 12. This is a general feature of the "refined spin-pairing energy theory" although e.g. the difference between q = 7 and 14 is determined by a delicate balance [198] between D and $(E - A)$. Thus, the ionization energies [199] of gaseous M^{+2} and M^{+3} vary as a function of q (or Z, keeping z constant) qualitatively agreeing with Eq. (4) though the agreement would be better with slightly different D and $(E - A)$.

For the chemist, it is more interesting that the standard oxidation potentials E^0 of M(II) to M(III) aqua ions can be rationalized [8,15,200,201] by Eq. (4). Though one would not really have expected so, the $4f^q \rightarrow 4f^{q-1}5d$ inter-shell transitions of M(II) [202] and, at much higher wave-numbers, of M(III) [203] in CaF_2. Corresponding transitions [9,147] occur in gaseous M^{-2} and M^{+3} at some 10000 cm^{-1} higher energy than for the M(II) and M(III). These examples all refer to the removal of an electron from the 4f shell. Actually, Eq. (4) was originally derived [156] with the purpose of explaining *electron transfer bands* [159] generally moving toward higher wave-numbers along the series

$$4f^6\ \text{Eu(III)} < 4f^{13}\ \text{Yb(III)} < 4f^5\ \text{Sm(III)} < 4f^{12}\ \text{Tm(III)} < \ldots \tag{5}$$

corresponding to Eq. (4) having the opposite sign, going from (q — 1) to q. The original measurements [156] providing Eq. (5) were done on almost anhydrous MBr_3 in anhydrous ethanol, where it is likely that the major species in solution is $MBr(HOC_2H_5)_x^{+2}$. Later [149,150] definite octahedral species MX_6^{-3} (X = Cl, Br and I) were measured in acetonitrile. Eq. (5) also applies to complexes of oxygen-containing anions in aqueous solution [204] (the europium(III) aqua ion has its first electron transfer band [205] at 53200 cm^{-1} because water is so weakly reducing) and to solid mixed oxides [206–208].

The S.C.S. parameters of interelectronic repulsion entering the "ligand field" expressions for the energy levels [8,13,209] of d-group complexes are multiplied by nephelauxetic ratios β between 0.94 (for manganese(II)fluoride) to well below 0.4 (e.g. cobalt(III) and rhodium(III) complexes of sulfur-containing ligands, but also NiF_6^{-2}). The *nephelauxetic effect* (this Greek word meaning "cloud-expanding" was proposed [210] by the late Professor Kaj Barr) is the most pronounced for higher oxidation states and for ligating atoms of lower electronegativity. This is what would expect covalent bonding does to the d-like orbitals, though a major part [8] of the nephelauxetic decrease of S.C.S. parameters seems to be due to an expansion of the radial function of the partly filled shell (corresponding to a weaker central field than in gaseous M^{+z}) and not only to delocalization of the anti-bonding d-like orbitals. The situation is far from the purely covalent bonding between two identical atoms, since β in the L.C.A.O. model is proportional to the square of the electronic density (and hence, to the fourth power of the wave-function amplitude) and would be about 0.25 in such a case.

The nephelauxetic shift of *J*-levels of rare-earth compounds is known since 1910, and it is particular striking in some solids [50,211]. However, a closer analysis is needed [96,149] because the shift (usually a few hundred cm^{-1}) has the same order of magnitude as the "ligand field" effects separating the $(2J + 1)$ states of a given *J*-level to a certain extent [212,213]. In a few fortunate cases, all of these states have been safely identified in a large number of *J*-levels, and the shifts of the *J*-baricenters than give the same result as the more primitive technique [95,96,149] of comparing the intensity baricenters with the aqua ion (preferentially measured in the ethylsulfate at low temperature). One has to evaluate β relative to the aqua ion for all the Nd(III) to Tm(III); only gaseous Pr^{+3} is known [189] for direct comparison with Pr(III).

In the infra-red, most of *J*-level energy difference is due to spin-orbit coupling [191]. Though the Landé parameter should be decreased half as much (proportional to the squared amplitude) as the parameters of interelectronic repulsion in a pure L.C.A.O. description, much evidence [214] is available that the Landé parameter varies less than half of (1 — β) from one compound to another. The difference (22007 — 6415 = 15592 cm^{-1}) between the two *J*-levels $^3P_1$ and $^3F_3$ of Pr^{+3} does not depend on the Landé parameter ζ_{4f} and is decreased [215] to 14954 cm^{-1} (β = 0.959 for $Pr_xLa_{1-x}F_3$) whereas β = 0.958 for $[Pr(OH_2)_9](C_2H_5OSO_3)_3$, 0.945 for $Pr_xLa_{1-x}Cl_3$, 0.942 for $Pr_xLa_{1-x}Br_3$ and close to 0.92 for $BaPr_2S_4$. The distance is 11759 cm^{-1} in the isoelectronic Ce^{+2} or 0.754 times the value of Pr^{+3} to be compared with the ratio 0.81 between Pr^{+2} and Nd(III) mentioned above. It is seen that the spreading between β = 0.96 for PrF_3 and 0.92 for $BaPr_2S_4$ (as a function of the ligating atoms) is far smaller than between β = 0.94 for rutile-type MnF_2 and perovskite $KMnF_3$ down to 0.82 for the green NaCl-type MnS and 0.80 for $Zn_{1-x}Mn_xS$ with $N = 4$.

We already mentioned that the large nephelauxetic effects in binary (and most of the mixed) rare earths do not occur in the d-group. For instance, β = 0.93 for $Mn(OH_2)_6^{+2}$ may be compared with 0.89 for NaCl-type MnO, or β = 0.90 for $Ni(OH_2)_6^{+2}$ with [8] the β values

$Mg_{1-x}Ni_xO$: 0.83 NiO: 0.77 $KNiF_3$: 0.92
$Mg_{1-x}Ni_xTiO_3$: 0.81 $Cd_{1-x}Ni_xTiO_3$: 0.80 $NiTiO_3$: 0.79 (6)

though it must be admitted that the sub-shell energy difference varies much less than $(1 - \beta)$ in these nickel(II) compounds. However, certain mixed oxides [95,96] of erbium(III) have more moderate nephelauxetic effects, such as pyrochlore $Er_2Ti_2O_7$ having [101] two unusually short Er-O distances 2.182 Å but the other six rather long, 2.471 Å. Their average 2.40 Å is significantly larger than the average 2.284 Å of the distances [85] in Y_2O_3 (which can be safely extrapolated to be 0.01 Å shorter in Er_2O_3). We have to admit that aqua ions are also complexes (and there is not such as a thing as a "free ion" in rare earth compounds) and their shortest distance, cf. Eqs. (1) and (2), in Er(III) is 2.37 Å, marginally shorter than in the pyrochlore structure.

For the chemist, the most important conclusion of the nephelauxetic effect in lanthanides is the moderate extent, all known β values being above 0.90. It is likely that all erbium(III) compunds have β between 0.985 and 0.945 (though the J-levels of Er^{+3} are not [9] known). Fortunately, the three contributions to the nephelauxetic effect all pull in the same direction, so the observed β values constitute higher limits to each of the three mechanisms. The delocalization in L.C.A.O. pictures would be as high as $(1 - \beta)/2$, i.e. 2 percent in PrF_3, but all available evidence (from nuclear hyperfine structure of paramagnetic resonance, due to nuclei of ligating atoms, etc.) indicates less delocalization in the 4f (though not in the 5f) group. Hence, it is likely that a modified central field of M(III) gives the main contribution to $(1 - \beta)$, and it was noted above that this quantity would be as large as 0.25, if the central field was changed to that of M^{+2}. However, a third contribution is the chemical enhancement of the Watson effect [196,197] due to a kind of dielectric screening.

There is much evidence available that the Racah parameter $E^1 = (8D/9)$ of Eq. (3) is less influenced by the nephelauxetic effect than the parameter E^3 determining the distances between terms having the highest S possible (like the groundstate) but differing L. This can be clearly seen [8,36,215] on the position of $^1I_6$ of $4f^2$ systems (mainly determined by E^1) relative to $^3P_2$ (situated $42E^3 + 3.5\zeta_{4f}$ above the groundstate $^3H_4$). The energy difference between $^1I_6$ and $^3P_1$ is 205 cm^{-1} in Pr^{+3}, 253 cm^{-1} in $Pr(OH_2)_9^{+3}$ (where its presence [216] was first suspected), 340 cm^{-1} in the chloride and 380 cm^{-1} in the bromide, showing that there is not a common nephelauxetic ratio β multiplying all the J-level energy differences. However, one should not forget the conspicuous Watson and Trees effects [195-197] in the gaseous ions, and the distance between $^1I_6$ and $^3P_1$ has increased to 897 cm^{-1} in Ce^{+2} showing that the weakly covalent effects on Pr(III) can be interpolated between Pr^{+3} and the isoelectronic Ce^{+2}.

As reviewed by Fidelis and Mioduski [57] the formation constant K of a complex with a given ligand (or the distribution coefficient for extraction in another solvent, or an ion-exchange resin) shows a ratio (in the case of two consecutive lanthanides) which provides perceptible variations (from a constant) not only at the half-filled shell (q = 7) Gd(III) but also at the plateaux q = 3 and 4, as well as 10 and 11. These quarter-shell effects can be rationalized [217,218] by the refined spin-pairing energy theory. If D of Eq. (3) is decreased 1% (65 cm^{-1}) by the nephelauxetic effect in a relatively more covalent Gd(III) complex (relative to the aqua ion), the groundstate is 840 cm^{-1} less stabilized (relatively to the baricenter of the $4f^7$ levels) than the Gd(III) aqua ion. Further on, this differentiation is only 600 cm^{-1} in the Eu(III) and Tb(III) complexes. The free energy 240 cm^{-1} available, suffices to decrease K_{Gd}/K_{Eu} by a factor 3.1 at 25 °C ($\Delta \log K = 0.5$) and to increase K_{Tb}/K_{Gd} by the same factor. The plateaux of the quarter-shell effect [57] can be explained if E^3 changes more than E^1, as discussed above, providing non-linear variations, like in Eq. (4).

The M-X distances [57] or the derived ionic radii of M(III) also follow the same pattern, the particularly stabilized Gd(III) being *larger* than interpolated from the neighbour M(III) on both sides. This fact cannot be explained by differential changes in free energy, but might be connected with the tendency pointed out by Katriel and Pauncz [194] that partly filled shells with large spin-pairing stabilizations achieve (via the virial theorem) smaller average radii. This would have the consequence that the filled *5p* shell (essentially situated outside the 4f shell) would expand, a change in the same direction as produced by the lack of one proton on the gadolinium nucleus, providing a less attractive central field for the *5p* electrons.

6 Inductive Quantum Chemistry

It is not always realized that the spectra of monatomic entities were classified by the Hund vector-coupling scheme [5] before they were rationalized by quantum mechanics. Quite generally, the first 20 to 400 energy levels of a monatomic entity have *J*-levels (even or odd parities) fitting them to the electron configurations obtained [8] by modifying the (*nl*)-value of zero (in the case of a partly filled shell), one, two or (in rarer cases) three electrons of the configuration to which the groundstate belongs. This statement is, by no means, equivalent to an anti-symmetrized Slater determinant [192] being a very good approximation to the wave-function of the groundstate. Correlation effects are quite extensive [196]: the correlation energy (defined as the difference between the Hartree-Fock energy and the experimental groundstate corrected for relativistic effects) is higher than the first ionization energy of the sodium atom, and of all neutral atoms with Z above 11; and whereas the squared amplitude of the optimized configuration ($1s^2$) is 0.99 in the groundstate of the helium atom, it is close to 0.8 in the neon atom, and may readily be lower than 0.5 in many atoms with Z above 20. This does not modify the taxocological utility [7,8] of electron configurations.

The development of "ligand field" theory [13] between 1951 and 1963 (when the angular overlap model [106] was established) shows many analogies to the understanding of atomic spectra between 1913 and 1929. At this point, the energy levels of a partly filled 4f shell are so close to spherical symmetry that the tiny differences between the seven one-electron energies of the 4f shell have complicated relations with the energy differences (of the same order of magnitude) between the sub-levels formed by the $(2J + 1)$ states of a given *J*-level. Both the model [72,212,213] of the (very small) non-spherical part of the Madelung potential, and the angular overlap model [13,106,108] have the point in common that the various *J*-levels are assumed to be described by the same set of 4f one-electron energy differences in a given chromophore, but it has turned out that one needs very artificial modifications of the parameters of the Madelung potential, whereas the angular overlap model mostly is successful in transferring the parameters of anti-bonding effects from one chromophore [16] to another having a different symmetry or coordination number N.

Systems containing two or more nuclei have a problem absent from monatomic entities, the fact that two nuclei have one distance R, and that N (three or more) nuclei have $(3N - 6)$ mutually independent distances in the Born-Oppenheimer approximation. Consequently, the energy of a given electronic state corresponds to a potential hypersurface in a $(3N - 5)$ dimensional space. Quantum chemistry (allowing the internuclear distances to vary) is a very difficult problem for more than two or three nuclei. Text-books frequently speak about the "reaction coordinate" corresponding to a geodesic curve on the $(3N - 5)$ dimensional hypersurface, much like the easiest path of hiking in a mountainous landscape. Spectroscopic transitions (such as optical excitations or photo-electron ionization processes) are "vertical" in the sense of keeping the distribution of R values of the original groundstate. It is easy to see that quantum chemistry is much more suitable for "vertical" situations, avoiding the $(3N - 5)$ dimensional nightmares, but at the same time relinquishing the hope of describing chemical reactions. It is perfectly evident from this review that (following G. N. Lewis from 1916) many chemists entertain the illusion that $2N$ electrons can be allotted to N "bonds". Besides trivial cases [219] such as H_2 this is nearly always debatable. In no simple sense are there 18 electrons involved in $M(OH_2)_9^{+3}$ nor 16 in fluorite-type CeO_2, and there are not enough electrons available for such a description of fluorite-type Be_2C and neither enough (outside strongly bound inner shells) in the NaCl-type oxides MgO, MnO, NiO, CdO and EuO which cannot all be fully electrovalent. No atomic spectroscopist would ever have looked for the mechanism keeping two electrons together in a "bond". They are only co-existing by the combined action of the kinetic-energy operator in quantum mechanics [13,219] and of Pauli's exclusion principle for fermions. Outside carbon chemistry, we are rather reluctant to speak about "multiplicity" of bonds. This concept has rarely distinct experimental consequences, with the exception of oxygen atoms (almost lacking proton affinity) in CO and in the vanadyl and uranyl ions, which may appropriately be said to contain triple bonds like N_2.

The rare earths have many unexpected properties. One of the surprises were the high S values, up to $S = {}^7/_2$ for Eu(II), Gd(III) and Tb(IV) corresponding to Hund's rules [5] for monatomic entities, and like $S = {}^5/_2$ for *most* Mn(II) and Fe(III) compounds. *Photo-electron spectra* of solids [66,143,147,187,188,220,221] have allowed the ionization of the groundstate of $4f^q$ to the various J-levels of $4f^{q-1}$ to be observed. The intensities of such signals are proportional [147,188,222] to the square of the coefficient of the ionized J-level in the original groundstate. The first surprise was that in M_2O_3 and MF_3 the ionization energy of M4f is comparable to *X2p* for M = Sm, Eu, Tm and Yb, and distinctly higher for M = Gd (inspite of the half-filled shell), even higher than for M = Lu. In all MSb, the ionization energy of the filled shell Sb5p is lower than of M4f. These classes of compounds, as well as the metallic elements [143,188] (excepting europium and ytterbium exemplifying M[II]) all follow Eq. (4) as a function of q. Already Klemm prepared ytterbium alloys having magnetic moments (and molar volumes) indicating Yb[III], and Wertheim [223] reported a well-resolved $4f^{13} \rightarrow 4f^{12}$ photo-electron spectrum of $YbNi_5$. The situation of higher ionization energy of the $4f$ shell (including weakly anti-bonding orbitals) than of the loosest bound orbitals of the ligands had already been observed in a few iron(III)-

complexes (containing anti-bonding 3d-like orbitals) and is colloquially named [66] the "third revolution in ligand field theory" and has been further analyzed [147, 224].

The second surprise derived from photo-electron spectra of lanthanide compounds was the evaluation of the nephelauxetic ratio β for the *J*-levels of $4f^{q-1}$ obtained by ionization. Though gaseous M^{+3} and M^{+4} are not known, a reliable estimate of the M^{+4} *J*-level positions can be obtained by multiplying the energy differences [225] of the isoelectronic M(III) of the preceding element by 1.20. It turns out [147, 188] that β = 0.92 for *TbSb* and for metallic terbium, and the marginally higher β = 0.91 for TmSb compared to 0.89 for metallic thulium. It is even more unexpected [143] that TmTe containing comparable amounts of Tm[III] and Tm[II] on an instaneous picture show β = 0.91 for the $4f^{11}$ ionized states of the former case, but β = 0.99 for the $4f^{12}$ obtained by ionizing Tm[II]. Said in other words, covalent bonding, invasion of the 4f shell by conduction electrons of metals, or other conceivable processes, do not modify the 4f radial function very much, and especially, the β values allow at most 5% delocalization in L.C.A.O. models.

This question in related to the satellite signals observed in many photo-electron spectra [187, 220, 226] because a given ionization process can bifurcate in various final states, providing photo-electrons with differing kinetic energy. Magnetic properties of metallic elements [227] allow to determine not only the *J*-groundstate of M[III] but also to evaluate "ligand field" effects separating the $(2J + 1)$ states to roughly half the extent of aqua ions (once more showing that the "ligand field" has very little to do with Madelung potentials). The choice between M[II] and M[III] has been related to the ease of reduction in Eq. (5) and the tendency towards M[III] with $q = (Z - 57)$ becomes more pronounced along the series

$$\textit{gaseous } M^0 < MI_2 < MTe < MSe < MS < \textit{solid } M \tag{7}$$

where the gaseous atoms [9] has only four M[III], viz. La, Ce, Gd and Lu, whereas MS have 8 additional M[III], viz. Pr, Nd, Pm, Tb, Dy, Ho, Er and Tm, and metallic M finally a thirteenth M[III] in samarium. It is noted [39, 201] that Eq. (7) is ordered as a function of decreasing M-M distances for enhanced tendency toward M[III]. Contrary to a previous mistake [39] gaseous M^{+2} would be earlier than M^0 in Eq. (7) by having only three M[III], this time without cerium.

Another aspect of inductive quantum chemistry applied to lanthanides, are the relative intensities of internal transitions between *J*-levels of the partly filled 4f shell. Here, the situation is entirely different from isolated monatomic entities. With exception of magnetic dipolar transitions, atomic spectral lines occur only between *J*-levels of opposite parity (i.e. necessarily belonging to two configurations differing in an odd number of electrons with odd *l*) and *J* changes at most by one unit. The oscillator strength *P* (called *f* by many authors) depends in a dramatic way on the detailed radial functions [192, 228] and is, among other factors, proportional to the square of the electric dipole moment of the product of the odd and the even wave-function. In this sense, the transitions between different *J*-levels of *[54] $4f^q$* are strictly forbidden as electric dipolar transitions. Actually, *P* of typical narrow lanthanide absorption bands are between 10^{-7} and 10^{-5} whereas strong spectral lines of monatomic entities have *P* between 0.1 and 1.5.

In 1945, Broer, Gorter and Hoogschagen [229] pointed out that admixtures of configurations having opposite parity of $4f^q$ might occur, due to odd normal modes of vibration, or due to hemihedric deviations of the "ligand field" from having a centre of inversion. The absolute selection rule for these two intensity-giving mechanisms is that J can at most change by 6 units. Considering the "monatomic" configurations $4f^{q-1}5d$ and $4f^{q-1}5g$ at high energy, Judd [230] and Ofelt [231] proposed independently a theory with only three material parameters (Ω_t with t = 2, 4 and 6) to be multiplied by definite matrix elements U_t (frequently called $|U^{(t)}|^2$ by a more precise, but elaborate notation) which can be calculated [225] between each pair of J-levels (and remain invariant, if the J-levels remain at the same distance from all the other J-levels.). Then P is proportional to

$$\frac{h\nu}{(2J_0 + 1)} [\Omega_2 U_2 + \Omega_4 U_4 + \Omega_6 U_6] \tag{8}$$

where $h\nu$ is the transition energy, and J_0 the J-value of the groundstate. Judd and one of us [232] pointed out that Nd(III) shows one, and Ho(III) and Er(III) two *hypersensitive pseudoquadrupolar transitions* with P values strongly increasing with conjugated organic ligands, with bidentate nitrate and carbonate, and, as it turned out later, in gaseous [233] $NdBr_3$ and NdI_3, in the volatile [234] adduct $CsNdI_4$ and the mixture [235] of erbium(III) and aluminium(III) chloride vapour, perhaps containing $Er(Cl_2AlCl_2)_3$. These transitions are called "pseudoquadrupolar" because they follow the selection rules for electric quadrupolar transitions (but the observed P are many thousand times the calculated values) among which the strongest in Russell-Saunders coupling has (J_0, S_0, L_0) of the groundstate going to ($J_0 - 2$, S_0, $L_0 - 2$). Whereas Ω_2 of Eq. (8) is small (and not much larger than the experimental uncertainty) for aqua ions and fluorides, it is very large for the ligands providing strong hypersensitive pseudoquadrupolar transitions, and their selection rule is actually a large value of U_2. On the other hand, Ω_4 and Ω_6 have comparable size for most materials. Judd [236] has recently given thorough arguments for a previous opinion [232] that the physical mechanism behind the large Ω_2 is the break-down of describing the compound (in the classical theory of light propagation) as a homogeneous dielectric.

The Judd-Ofelt theory of absorption band intensities has been reviewed several times [16, 237, 238]. It can also be adapted to transition probabilities of fluorescence [121, 122, 128] using definite U_t matrix elements, but also same Ω_t as for absorption of the same material. By the same token as the positions of the J-levels are essentially determined [215] by three parameters of interelectronic repulsion (the S.C.S. integrals F^2, F^4 and F^6, or Racah's linear combinations E^1, E^2 and E^3) and the Landé parameter ζ_{4f} of spin-orbit coupling, it is surprising that both absorption and emission probabilities are given (with a precision typically around 20% excepting rare deviations such as the unexpected high intensity of $^3H_4 \rightarrow {}^3P_2$ absorption in Pr(III) compounds) by only three Ω_t. It is conceivable that such regularities are connected with the concept of *chemical polarizability* previously discussed [69] in this series.

7 Lasers, Cathodoluminescent Television and Candoluminescence

There has been many, relatively minor, technological applications of rare earths for the last century. The extraction of thorium from monazite sand (to provide candoluminescent Auer mantles $Th_{0.99}Ce_{0.01}O_2$ for gaslight) left huge quantities of the lighter lanthanides from lanthanum to samarium. It is easy to separate cerium(IV) from such mixtures, and this element was used a lot in chemical analysis, and CeO_2 as optical polishing powder. Great quantities of the remaining elements were reduced to an alloy (with or without iron added) "Mischmetall" used for flints in cigarette-lighters, and today in large quantities in steel-making. Permanent magnets [18–35] are made of alloys such as $SmCo_5$.

However, it is fair to say that the most specific uses of the rare earths are related to spectroscopic properties and light emission. The coherent, monochromatic light from *lasers* [16] was predicted by Einstein in 1917 from thermodynamic arguments based on Planck's explanation of the standard continuous spectrum emitted at a definite temperature by an opaque object ("black body"), accepting the idea of photons, but not involving any hypothesis about the electronic structure. It has turned out that many kinds of materials in different physical states may serve as lasers [16, 239]. Among the more obvious are electric discharges in a dilute gas of monatomic entities, where metastable states E_2 are frequent. The first solid-state laser was the ruby $Al_{2-x}Cr_xO_3$ though it is a *three-level* laser. The more favourable case is the *four-level* laser having the energy levels (perhaps with other intervening levels) $E_0 < E_1 < E_2 < E_3$. A strong absorption from the groundstate E_0 to E_3 (or energy transfer [16, 120, 240] from other, strongly coloured species; or photochemical reactions) pump the high-lying E_3 which decays rapidly (and normally non-radiatively) to E_2. The monochromatic light emission takes place from E_2 to E_1. It is necessary that a concentration of E_1 does not build up (especially in continuous rather than in pulsed modes of operation) and that E_1 decays rapidly (radiatively or not) to the groundstate E_0. In a three-level laser, E_1 and E_0 coincide. It is a primordial condition for laser action that *population inversion* occurs, that each (of the, say, $2J + 1$) state of E_2 has a higher concentration (per unit volume) than E_1 (which is not feasible by thermal equilibrium at any temperature). In a three-level laser, a corollary of this imperative requirement is that at (the very) most 49% of the system remain in the groundstate. In the four-level laser, this requirement is much easier to satisfy, because the ratio between the Boltzmann population of E_1 and E_0 is $\exp(-(E_1 - E_0)/kT)$. Hence, *if* E_1 and E_0 are in thermal equilibium, $e^{-10} = 4.54 \cdot 10^{-5}$ is this ratio in the case $(E_1 - E_2) = 10kT$.

Actually, there is a technically important solid-state laser, where such situation occurs at room temperature (where kT is 200 cm^{-1}). The fifth J-level $^4F_{3/2}$ of neodymium(III) at 11400 cm^{-1} has a favourable branching ratio for luminescence (in the near infra-red at 9400 cm^{-1}) terminating at the first excited J-level $^4I_{11/2}$ 2000 cm^{-1} above the groundstate $^4I_{9/2}$. Crystalline lasers [16, 239] such as the garnet $Y_{3-x}Nd_xAl_5O_{12}$, the perovskite $Y_{1-x}NdAlO_3$ or the pentaphosphate $La_{1-x}Nd_xP_5O_{14}$ (where even the undiluted NdP_5O_{14} works) are rather expensive to shape, when retaining high optical quality, and a variety of glass lasers have been proposed. The SHIVA laser system in Livermore, California (in activity since January 1978) consists of ten Nd(III) silicate glass lasers (each weighing several tons) delivering 10^{-9} sec pulses of effect 10^{13} W (10 terawatt) each, for the purpose of inducing thermonuclear fusion in imploding pellets containing a mixture of deuterium and tritium.

Another approach to four-level lasers is to consider *antiferromagnetic coupling* between two lanthanides (with positive S) bridged by an oxygen (or sulfur) atom at short distances. In a system simplified [241,242] to GdOYb, the monomeric groundstates $^8S_{7/2}$ and $^2F_{7/2}$ are coupled (according [13] to Hund vector-coupling) to $(2J_{Gd} + 1) \times (2J_{Yb} + 1) = 64$ states having $S = 3$ and 4. The first excited level $^6P_{7/2}$ of Gd(III) in the ultra-violet couples the same way with the Yb(III) ground level, producing 64 states, but this time having $S = 2$ and 3, to a good approximation. Let these states close to 32000 cm^{-1} be E_2 of the four-level laser. The inversion of population is readily obtained, if E_1 cannot be formed by absorption from the groundstate (as is true for the noble-gas monohalide gas lasers, where the excited E_2 corresponds to a potential minimum, but the luminescence to E_1 produces a dissociative curve for the diatomic molecule). Such states E_1 can be reached around 10500 cm^{-1} (after fluorescence in the blue-green) constructed from $^8S_{7/2}$ of Gd(III) and $^2F_{5/2}$ of Yb(III) providing $8 \cdot 6 = 48$ states. Actually, green fluorescence has already been observed [243] in $Yb_{2-x}Gd_xO_3$ and in [244] $Yb_{1-x}Gd_xPO_4$ (but this, of course, is not a sufficient condition for laser action). By the way, absorption bands in the green [245] have been observed in pure Yb_2O_3 due to simultaneous excitation of two adjacent Yb(III) by the same photon. Their oscillator strength is about 10^{-9}, thousands of times weaker than the transitions (at half the wave-number) in one Yb(III), and correspond to the 36 states belonging to two simultaneously excited $^2F_{5/2}$ systems.

When evaluating new laser materials, it is helpful to know the Judd-Ofelt parameters from Eq. (8) and in particular be able to apply them to luminescent transitions from E_2 to E_1 as done for many glasses [121,122,128,134–136,246] avoiding much more extensive work on explicit laser properties. Another aspect is predicting the multi-phonon desexcitation [16,118–120] being the major competitive process decreasing (sometimes dramatically) the quantum yield of luminescence. *Energy transfer* is also of great importance for lasers, the constructive side being the pumping of E_3 of a four-level laser (maintaining a quasi-stationary concentration of the light-emitting state E_2) and the less desirable side of cross-relaxation [16,239] between two Nd(III), or between two different lanthanides, being the more likely at higher concentrations. The antiferromagnetic coupling between Gd(III) and Yb(III) mentioned above as a technique of obtaining population inversion, may also be involved in the easy energy transfer [123] from Nd(III) to Yb(III) in germanate and tellurite glasses. The various theoretical treatments of energy transfer have been reviewed [16,120,240].

The conditions for laser materials are, to a great extent, also valid for good glasses for fluorescent concentrators for collecting *solar energy* to feed photovoltaic cells. A major economic problem [247,248] is the huge investment in the silicon used, and it would help a lot, if the amount of silicon needed can be decreased by an order of magnitude. Goetzberger and Greubel [249] analyzed the action of trapping isotropic fluorescent radiation (to an extent of 70 to 80%) by total reflection inside a plate with large surface area, and having silicon around the rim. Such a concentrator has been further discussed [250,251] and was realized [252] as a uranyl glass plate (having the electron transfer band starting at 20500 cm^{-1}). Though the light with lower wave-number going through the plate can be used for the heating of water etc., it is an amelioration [253] to come closer to the energy gap of the silicon cell (9000 cm^{-1}) and allow energy transfer from the excited uranyl ion to Nd(III) or Ho(III). It would be particularly attractive to use near infra-red luminescence of Yb(III), as recently studied [123] as an acceptor for energy transfer in low-phonon glasses. Batchelder, Zewail and Cole [254] described planar fluorescent concentrators of plastic containing fluorescent organic dye-stuffs, and Goetzberger and Wittwer [255] also elaborated this concept. We write a review [256] in a volume of "Structure and Bonding" devoted to "Solar Energy Materials".

When Crookes liberated electrons with high kinetic energy in the vacuum of his tube, he detected *cathodoluminescence* of the wall exposed to the impact of the electrons (which was later shown by Röntgen to emit X-rays). When placing various mixed oxides on the inside of the wall, the cathodoluminescence increased strongly, and with a spectroscope, Crookes [1, 3] detected narrow emission bands due to traces of lanthanides. Working intensively (and with great imagination) on the enigmatic problem of rare earths, he used the presence of definite bands as an argument for new elements in mixtures obtained by fractional crystallization (like he used the spectral line of thallium in the green to discover the element, and like Bunsen and Kirchhoff discovered rubidium and caesium by their spectral lines). The weak point is that though traces of some lanthanides can be detected, there is no proportionality between their concentration and the cathodoluminescent intensity, and problems of quenching at higher concentrations, and energy transfer to other species, are dramatic.

The greatest impact of rare earths on daily life (at the moment) is due to Urbain [257] reporting in 1909 the narrow red emission bands of cathodoluminescent mixed oxides containing small amounts of europium(III) (this element was discovered by Demarçay in 1901). As we mentioned above, the red colour in television is due to the cathodoluminescence of $Eu_xY_{1-x}VO_4$ or $Eu_xY_{2-x}O_2S$ (showing a more saturated red colour because of enhancement of the hypersensitive pseudoquadrupolar transition from $^5D_0$ to $^7F_2$ and because of a slightly more pronounced nephelauxetic effect). The red lines had previously been ascribed by Crookes [3] to some of his supposed elements (such as victorium and incognitum). Urbain [258] has reviewed the complicated questions around the recognition of elemental character of several of the rare earths. It may also be interesting to compare [259] with the circumstances surrounding the discovery by Marignac of ytterbium in 1878 and gadolinium in 1880. Blasse [260] has reviewed the general subject of cathodoluminescence and (photo-)-luminescence in solids (called "phosphors") containing rare earths. Whereas usually only one excited *J*-level (in a few cases two) is sufficiently long-lived to provide detectable luminescence in the latter case, cathodoluminescence frequently shows a much richer spectrum of emission bands at unexpected high wave-numbers (which is a technological disadvantage in the case of Eu(III) then emitting a pinkish light, due to admixture of green and blue light) corresponding to high-lying (and, in part, unidentified) excited levels. There is, obviously, much more work to be done on cathodoluminescence, which is mainly due to secondary electrons with moderate kinetic energy ejected in the solid by the passage of the exceedingly rapid, primary electrons. Hence, there is a certain relation between cathodoluminescent behaviour and electroluminescence, the light emission by solids, when (say 220 V alternating) electric current is passed.

The absorption lines of dilute gases (including flames and stellar atmospheres) containing neutral atoms were shown by Bunsen and Kirchhoff to coincide with a part (of the perhaps much more numerous) emission lines of the same atom. Since Gladstone had noted the similarity in 1857 between the narrow absorption bands of rare earths and the Fraunhofer lines of the solar spectrum, Bunsen and Bahr showed before 1864 that incandescent rare earths may emit narrow bands roughly at the positions of the absorption bands, and in particular that pink Er_2O_3 is brilliant emerald-green, when heated in a flame. At the end of last Century, the *candoluminescence* got great technological importance for gas-light, and Auer von Welsbach (who worked [3] most of his life with rare earths, and in 1885 separated didymium in Pr and Nd) invented the mantle, where a textile is impregnated with a concentrated nitrate solution of one or more metallic elements, and pyrolyzed in a flame to an "inorganic textile" of microcrystalline oxide, which should be as weakly conducting of heat as possible, in order to keep the temperature and the yield of visible light as high as feasible. Auer mantles of pure ThO_2 emit a rather weak, pink light, whereas admixtures of cerium in small quantities provide a brilliant greenish-white light (early optimized at $Th_{0.99}Ce_{0.01}O_2$) but in higher concentrations a dull, yellow light. The mechanism is related to the theorem of Kirchhoff; very little infra-red radiation is emitted, where the oxide is transparent, and the emitivity approaches that of the standard opaque object in the visible [16]. The strong absorption in the visible (not perceptible at room temperature) seems to be due to an electron transfer band of cerium(IV) which has shifted (or broadened), or perhaps to an electron transfer band due to transfer of a 4f electron from Ce(III) to Ce(IV), as known from dark blue CeO_{2-x}. There were very few studies of narrow-band candoluminescence of trivalent lanthanides in the meantime. It is possible [261,262] to make Auer mantles of most of the mixed oxides previously obtained by calcining co-precipitated hydroxides [95] and whereas (presumably quadrivalent) praseodymium and terbium produce broad continuous spectra (and orange light), several emission bands are seen in the orange light emitted by Nd(III), mauve Ho(III), brilliant green Er(III) and deep purple Tm(III) (corresponding to the colour complementary to that of the sesquioxides). There has been some discussion [263,264] as to whether the candoluminescence of Auer mantles involves a component of chemoluminescence from reactive species of the flame recombining on the oxide surface, but it seems now established [265] that it is nearly exclusively Kirchhoffian, in the sense of emitting only in the absorption bands (some new bands of Nd(III) are due to thermal population of $^4I_{11/2}$). By far the strongest bands are those corresponding to hypersensitive pseudo-quadrupolar transitions.

8 References

1. Crookes, W.: Die Genesis der Elemente (Zweite deutsche Auflage von W. Preyer). Braunschweig: Friedrich Vieweg und Sohn 1895
2. Trimble, V.: Rev. Mod. Phys. *47*, 877 (1975)
3. Weeks, M. E.: The Discovery of the Elements (7. ed). Easton, Penn.: J. Chem. Educ. Publ. 1968
4. Jørgensen, C. K.: J. chim. physique *76*, 630 (1980)
5. Hund, F.: Linienspektren und Periodisches System der Elemente. Berlin: Julius Springer 1927

6. Jørgensen, C. K.: Angew. Chem. *85*, 1; Int. Ed. *12*, 12 (1973)
7. Jørgensen, C. K.: Adv. Quantum Chem. *11*, 51 (1978)
8. Jørgensen, C. K.: Oxidation Numbers and Oxidation States. Springer: Berlin, Heidelberg, New York 1969
9. Martin, W. C., Zalubas, R., Hagan, L.: Atomic Energy Levels, the Rare-earth Elements, NSRDS-NBS *60*. Washington, D. C.: National Bureau of Standards 1978.
10. Connick, R. E.: J. Chem. Soc. (Suppl.) 235 (1949)
11. Jørgensen, C. K.: Chimia *23*, 292 (1969)
12. Jørgensen, C. K.: Comments on Inorganic Chemistry (Gütlich, P., Sutin, N., eds.) *1*, 123 (1981)
13. Jørgensen, C. K.: Modern Aspects of Ligand Field Theory. Amsterdam: North-Holland 1971
14. Jørgensen, C. K.: Mol. Phys. *2*, 96 (1958)
15. Jørgensen, C. K.: Chem. Phys. Letters *2*, 549 (1968)
16. Reisfeld, R., Jørgensen, C. K.: Lasers and Excited States of Rare Earths. Springer: Berlin, Heidelberg, New York 1977
17. Prandtl, W.: Z. anorg. Chem. *238*, 321 (1938)
18. 1. R.E.R.C. (Lake Arrowhead, California, October 1960) Rare Earth Research. Kleber, E. V. (ed.). New York: Macmillan 1960
19. 2. R.E.R.C. (Glenwood Springs, Colorado, September 1961) Rare Earth Research. Nachman, J. F., Lundin, C. E. (eds.). New York: Gordon and Breach 1961
20. 3. R. E. R. C. (Clearwater, Florida, April 1963) Rare Earth Research *2*. Vorres, K. S. (ed.). New York: Gordon and Breach 1963
21. 4. R.E.R.C. (Phoenix, Arizona, April 1964) Rare Earth Research *3*. Eyring, L. (ed.). New York: Gordon and Breach 1964
22. 5. R.E.R.C. (Ames, August 1965). Six volumes: *1* (spectra); *2* (solid state); *3* (chemistry); *4* (solid state); *5* (metallurgy) and *6* (solid state). AD-627221 to AD-627226 (also CONF-650804). Springfield, Virginia: National Technical Information Service 1965
23. Symposium in New York, September 1966. Advances in Chemistry Series *71*: Lanthanide and Actinide Chemistry. Fields, P. R., Moeller, T. (eds.). Washington, D. C.: American Chemical Society 1966
24. 6. R.E.R.C. (Gatlinburg, Tennesee, May 1967). CONF-670501. Springfield, Virginia. National Technical Information Service 1967
25. 7. R.E.R.C. (Coronado, California, October 1968). CONF-681020 (two volumes). Springfield, Virginia: National Technical Information Service 1968
26. Les Eléments des Terres Rares (Paris and Grenoble, May 1969). Colloques Internationaux C.N.R.S. No. 180 (two volumes). Paris: CNRS, 15 Quai Anatole France 1969
27. 8. R.E.R.C. (Reno, Nevada, April 1970). Lindstrom, R. (ed.). Reno: Metallurgy Research Center, U.S. Bureau of Mines 1970
28. 9. R.E.R.C. (Blacksburg, Virginia, October 1971). Taylor, L. (ed.). Blacksburg: Department of Chemistry, Virginia Polytechnic Institute and State University 1971
29. NATO Advanced Study Institute on Analysis and Applications of Rare Earth Materials (Kjeller, Norway, August 1972). Oslo: Universitetsforlaget, Blindern 1972
30. 10. R.E.R.C. (Carefree, Arizona, May 1973). CONF-730402 (in two)
31. 11. R.E.R.C. (Traverse City, Michigan, October 1974). Eick, H. A. (ed.). East Lansing. Michigan: Department of Chemistry, Michigan State University, 1974
32. La Spectroscopie des éléments de transition et des éléments lourds dans les solides (Lyon, June 1976). Gaume, F. (ed.). Colloques Internationaux C.N.R.S. No. 255 Paris: CNRS, 15 Quai Anatole France 1977
33. 12. R.E.R.C. (Vail, Colorado, July 1976). Lundin, C. E. (ed.). Denver: University of Denver, 1976
34. 13. R.E.R.C. (Oglebay Park, West Virginia, October 1977). The Rare Earths in Modern Science an Technology. McCarthy, G. J., Rhyne, J. J. (eds.). New York: Plenum 1978
35. 14. R.E.R.C. (Fargo, North Dakota, June 1979). The Rare Earths in Modern Science and Technology *2*. McCarthy, G. J., Rhyne, J. J., Silber, H. B. (eds.). New York: Plenum 1980
36. Jørgensen, C. K.: Gmelin Handbuch der anorganischen Chemie: Seltenerdelemente, *B1*, 17 (Springer-Verlag: 1976)
37. Pearson, R. G.: J. Am. Chem. Soc. *85*, 3533 (1963)

38. Pearson, R. G.: J. Chem. Educ. *45*, 581 and 643 (1968)
39. Jørgensen, C. K.: Handbook on the Physics and Chemistry of Rare Earths (eds. K. A. Gschneidner and L. Eyring) *3*, 111, Amsterdam: North-Holland 1979
40. Moeller, T., Martin, D. F., Thompson, L. C., Ferrus, R., Feistel, G. R., Randall, W. J.: Chem. Rev. *65*, 1 (1965)
41. Thompson, L. C.: Handbook on the Physics and Chemistry of Rare Earths (eds. K. A. Gschneidner and L. Eyring) *3*, 209, Amsterdam: North-Holland 1979
42. Lehn, J. M.: Structure and Bonding *16*, 1 (1973)
43. Truter, M. R.: Structure and Bonding *16*, 71 (1973)
44. Bünzli, J. C. G., Wessner, D.: Helv. Chim. Acta *64*, 581 (1981)
45. Bünzli, J. C. G., Klein, B., Wessner, D.: Inorg. Chim. Acta *44*, L147 (1980)
46. Jørgensen, C. K., Parthasarathy, V.: Acta Chem. Scand. *A32*, 957 (1978)
47. Marcantonatos, M. D., Deschaux, M., Celardin, F.: Chem. Phys. Letters *69*, 144 (1980)
48. Felsche, J.: Structure and Bonding *13*, 99 (1973)
49. Quill, L. L., Selwood, P. W., Hopkins, B. S.: J. Am. Chem. Soc. *50*, 2929 (1928)
50. Jørgensen, C. K.: Mat. fys. Medd. Danske Vid. Selskab *30*, no. 22 (1956)
51. Barnes, J. C., Day, P.: J. Chem. Soc. 3886 (1964)
52. Helmholz, L.: J. Am. Chem. Soc. *61*, 1544 (1939)
53. Dieke, G. H.: Spectra and Energy Levels of Rare Earth Ions in Crystals. New York: Interscience 1968
54. Gruber, J. B., Morrey, J. R., Carter, D. G.: J. Chem. Phys. *71*, 3982 (1979)
55. Spedding, F. H., Rard, J. A., Habenschuss, A.: J. Phys. Chem. *81*, 1069 (1977)
56. Habenschuss, A., Spedding, F. H.: J. Chem. Phys. *70*, 2797 and 3758 (1979) and *73*, 442 (1980)
57. Fidelis, I. K., Mioduski, T. J.: Structure and Bonding *47* (1981)
58. Freed, S.: Phys. Rev. *38*, 2122 (1931)
59. Heidt, L. J., Berestecki, J.: J. Am. Chem. Soc. *77*, 2049 (1955)
60. Krumholz, P.: Spectrochim. Acta *10*, 274 (1958)
61. Geier, H., Karlen, U.: Helv. Chim. Acta *54*, 135 (1971)
62. Powell, J. E.: Handbook of the Physics and Chemistry of Rare Earths (eds. K. A. Gschneidner and L. Eyring) *3*, 81, Amsterdam: North-Holland 1979
63. Geier, G., Jørgensen, C. K.: Chem. Phys. Letters *9*, 263 (1971)
64. Jørgensen, C. K.: Adv. Chem. Phys. *5*, 33 (1963)
65. Geier, G.: Ber. Bunsenges. *69*, 617 (1965)
66. Jørgensen, C. K.: Chimia *27*, 203 (1973) and *28*, 6 (1974)
67. Muetterties, E. L., Wright, C. M.: Quart. Rev. *21*, 109 (1967)
68. Hoffmann, R., Beier, B. F., Muetterties, E. L., Rossi, A. R.: Inorg. Chem. *16*, 511 (1977)
69. Jørgensen, C. K.: Topics Current Chem. *56*, 1 (1975)
70. Bjerrum, J.: Chem. Rev. *46*, 381 (1950)
71. Prandtl, W., Scheiner, K.: Z. anorg. Chem. *220*, 107 (1934)
72. Carnall, W. T.: Handbook on the Physics and Chemistry of Rare Earths (eds. K. A. Gschneidner and L. Eyring) *3*, 171, Amsterdam: North-Holland 1979
73. Hartmann, H., Schläfer, H. L.: Angew. Chem. *66*, 768 (1954)
74. Werner, A.: Annal. Chem. *386*, 31 (1912)
75. Smith, D. W.: J. Chem. Phys. *50*, 2784 (1969)
76. Schmitz-Du Mont, O., Reinen, D.: Z. Elektrochem. *63*, 978 (1959)
77. Reinen, D.: Structure and Bonding *6*, 30 (1969)
78. Reinen, D.: Theor. Chim. Acta *5*, 312 (1966)
79. Truter, M. R.: Annual Reports *76C* (for 1979) 161. London: Royal Society of Chemistry 1980
80. Jørgensen, C. K.: J. Am. Chem. Soc. *100*, 5968 (1978)
81. Sinha, S. P.: Structure and Bonding *25*, 69 (1976)
82. Roth, R. S., Schneider, J.: J. Res. Nat. Bur. Stand. *A64*, 309 and 317 (1960)
83. Templeton, D. H., Dauben, C. H.: J. Am. Chem. Soc. *76*, 5237 (1974)
84. Staritzky, E.: Analyt. Chem. *28*, 2023 (1956)
85. Faucher, M., Pannetier, J.: Acta Cryst. *B36*, 3209 (1980)
86. Hargittai, I.: Topics Current Chem. *96*, 43 (1981)

87. Malkova, T. V., Shutova, G. A., Yatsimirskii, K. B.: Russ. J. Inorg. Chem. *9*, 993 (1964)
88. Bjerrum, J., Halonin, A. S., Skibsted, L. H.: Acta Chem. Scand. *A29*, 326 (1975)
89. Skibsted, L. H., Bjerrum, J.: Acta Chem. Scand. *A32*, 429 (1978)
90. Bjerrum, J., Jørgensen, C. K.: Acta Chem. Scand. *7*, 951 (1953)
91. Jørgensen, C. K.: Acta Chem. Scand. *11*, 399 (1957)
92. Christensen, A. N., Hazell, R. G., Nilsson, A.: Acta Chem. Scand. *21*, 481 (1967)
93. Morosin, B.: J. Chem. Phys. *49*, 3007 (1968)
94. Müller-Buschbaum, H., Schnering, H. G. von: Z. anorg. Chem. *340*, 232 (1965)
95. Jørgensen, C. K., Rittershaus, E.: Mat. fys. Medd. Danske Vid. Selskab *35*, no. 15 (1967)
96. Jørgensen, C. K., Pappalardo, R., Rittershaus, E.: Z. Naturforsch. *A19*, 424 (1964) and *A20*, 54 (1965)
97. Eyring, L.: Handbook on the Physics and Chemistry of Rare Earths (eds. K. A. Gschneidner and L. Eyring) *3*, 337, Amsterdam: North-Holland 1979
98. Marsh, J. K.: J. Chem. Soc. 15 (1946)
99. Jørgensen, C. K.: Structure and Bonding *13*, 199 (1973)
100. Möbius, H. H.: Z. Chem. *2*, 100 (1962) and *4*, 81 (1964)
101. Knop, O., Brisse, F., Castelliz, L., Sutarno: Canad. J. Chem. *43*, 2812 (1965)
102. Khattak, C. P., Wang, F. F. Y.: Handbook on the Physics and Chemistry of Rare Earths (eds. K. A. Gschneidner and L. Eyring) *3*, 525, Amsterdam: North-Holland 1979
103. Landolt-Börnstein Tables: II, *9*: Magnetic Properties I. Berlin: Springer 1962
104. Orchard, A. F., Thornton, G.: J. Electron Spectr. *22*, 271 (1981)
105. Bevan, D. J. M., Sommerville, E.: Handbook on the Physics and Chemistry of Rare Earths (eds. K. A. Gschneidner and L. Eyring) *3*, 401, Amsterdam: North-Holland 1979
106. Jørgensen, C. K., Pappalardo, R., Schmidtke, H. H.: J. Chem. Phys. *39*, 1422 (1963)
107. Kuse, D., Jørgensen, C. K.: Chem. Phys. Letters *1*, 314 (1967)
108. Linares, C., Louat, A., Blanchard, M.: Structure and Bonding *33*, 179 (1977)
109. Judd, B. R.: Proc. Roy. Soc. (London) *A241*, 122 (1957)
110. Zalkin, A., Forrester, J. D., Templeton, D. H.: J. Chem. Phys. *39*, 2881 (1963)
111. Bünzli, J. C. G., Klein, B., Chapuis, G., Schenk, K. J.: Inorg. Chem., submitted
112. Eriksson, B., Larsson, L. O., Niinistö, L., Valkonen, J.: Inorg. Chem. *19*, 1207 (1980)
113. Burns, J. H.: Inorg. Chem. *18*, 3044 (1979)
114. Hansson, E.: Acta Chem. Scand. *24*, 2969 (1970) and *27*, 2852 (1973)
115. Weyl, W. A.: Coloured Glasses. London: Dawson's of Pall Mall 1959
116. Reisfeld, R., Mack, H., Eisenberg, A., Eckstein, Y.: J. Electrochem. Soc. *122*, 273 (1975)
117. Mack, H., Boulon, G., Reisfeld, R.: Proceedings of the International Conference on Luminescence, Berlin, July 1981 (in print at North-Holland, Amsterdam)
118. Reisfeld, R.: Structure and Bonding *13*, 53 (1973)
119. Reisfeld, R.: Structure and Bonding *22*, 123 (1975)
120. Reisfeld, R.: Structure and Bonding *30*, 65 (1976)
121. Reisfeld, R., Hormadaly, J.: H. Chem. Phys. *64*, 3207 (1976)
122. Reisfeld, R., Eckstein, Y.: J. Chem. Phys. *63*, 4001 (1975)
123. Reisfeld, R., Kalisky, Y.: Chem. Phys. Letters *80*, 178 (1981)
124. Kropp, J. L., Windsor, M. W.: J. Chem. Phys. *45*, 761 (1966)
125. Haas, Y., Stein, G.: J. Phys. Chem. *75*, 3668 and 3677 (1971)
126. Haas, Y., Stein, G.: Chem. Phys. Letters *11*, 143 (1971) and *15*, 12 (1972)
127. Stavola, M., Friedman, J. M., Stepnoski, R. A., Sceats, M. G.: Chem. Phys. Letters *80*, 192 (1981)
128. Blanzat, B., Boehm, L., Jørgensen, C. K., Reisfeld, R., Spector, N.: J. Solid State Chem. *32*, 185 (1980)
129. Lucas, J., Poulain, M.: ref. 34, p. 259 (1978)
130. Miranday, J. P., Jacoboni, C., De Pape, R.: Rev. chim. min. (Paris) *16*, 277 (1979)
131. Reisfeld, R., Katz, G., Spector, N., Jørgensen, C. K., Jacoboni, C., DePape, R.: J. Solid State Chem., in press
132. Hegarty, J., Imbusch, G. F.: ref. 32, p. 199 (1977)
133. DiBartolo, B. (ed.): Luminescence of Inorganic Solids. New York: Plenum Press 1978
134. Reisfeld, R., Bornstein, A.: Chem. Phys. Letters *47*, 194 (1977)
135. Reisfeld, R., Bornstein, A., Bodenheimer, J., Flahaut, J.: J. Luminescence *18*, 253 (1979)

136. Reisfeld, R.: under preparation
137. Jørgensen, C. K., Pappalardo, R., Flahaut, J.: J. chim. physique *62*, 444 (1965)
138. Flahaut, J.: Les éléments des terres rares. Paris: Masson 1969
139. Flahaut, J.: Handbook on the Physics and Chemistry of Rare Earths (eds. K. A. Gschneidner and L. Eyring) *4*, 1, Amsterdam: North-Holland 1979
140. Hulliger, F., Vogt, O.: Phys. Letters *17*, 238 (1965) and *21*, 138 (1966)
141. Jørgensen, C. K.: Mol. Phys. *7*, 417 (1964)
142. Hulliger, F.: Helv. Phys. Acta *41*, 945 (1968)
143. Campagna, M., Wertheim, G. K., Bucher, E.: Structure and Bonding *30*, 99 (1976)
144. Sovers, O. J., Yoshioka, T.: J. Chem. Phys. *51*, 5330 (1969)
145. Hulliger, F.: Structure and Bonding *4*, 83 (1968)
146. Hulliger, F.: Handbook on the Physics and Chemistry of Rare Earths (eds. K. A. Gschneidner and L. Eyring) *4*, 153, Amsterdam: North-Holland 1979
147. Jørgensen, C. K.: Structure and Bonding *22*, 49 (1975)
148. Haschke, J. M.: Handbook on the Physics and Chemistry of Rare Earths (eds. K. A. Gschneidner and L. Eyring) *4*, 89, Amsterdam: North-Holland 1979
149. Ryan, J. L., Jørgensen, C. K.: J. Phys. Chem. *70*, 2845 (1966)
150. Ryan, J. L.: Inorg. Chem. *8*, 2053 (1969)
151. Schwartz, R. W.: Mol. Phys. *30*, 81 (1975)
152. Serra, O. A., Thompson, L. C.: Inorg. Chem. *15*, 504 (1976)
153. Morrison, C. A., Leavitt, R. P., Wortman, D. E.: J. Chem. Phys. *73*, 2580 (1980)
154. Urland, W.: Chem. Phys. Letters, in press
155. Gruber, J. B., Menzel, E. R., Ryan, J. L.: J. Chem. Phys. *51*, 3816 (1969)
156. Jørgensen, C. K.: Mol. Phys. *5*, 271 (1962)
157. Pinkerton, A. A., Meseri, Y., Rieder, C.: J.C.S. Dalton 85 (1978)
158. Bradley, D. C., Ghotra, J. S., Hart, F. A.: J.C.S. Dalton 1021 (1973)
159. Jørgensen, C. K.: Progress Inorg. Chem. *12*, 101 (1970)
160. Forsberg, J. H.: Coord. Chem. Rev. *10*, 195 (1973)
161. Koppikar, D. K., Sivapullaiah, P. V., Ramakrishnan, L., Soundararajan, S.: Structure and Bonding *34*, 135 (1978)
162. Martell, A. E., Calvin, M.: Chemistry of the Metal Chelate Compounds. New York: Prentice-Hall 1952
163. Schwarzenbach, G.: Helv. Chim. Acta *32*, 839 (1949)
164. Jørgensen, C. K.: Acta Chem. Scand. *9*, 1362 (1955) and *10*, 887 (1956)
165. Jørgensen, C. K.: Inorganic Complexes. London: Academic Press 1963
166. Anderegg, G., Nägeli, P., Müller, F., Schwarzenbach, G.: Helv. Chim. Acta *42*, 827 (1959)
167. Moeller, T., Thompson, L. C.: J. Inorg. Nucl. Chem. *24*, 499 (1962)
168. Bohigian, T. A., Martell, A. E.: J. Am. Chem. Soc. *89*, 832 (1967)
169. Yingst, A., Martell, A. E.: J. Am. Chem. Soc. *91*, 6927 (1969)
170. Bohigian, T. A., Martell, A. E.: Inorg. Chem. *4*, 1264 (1965)
171. Muetterties, E. L.: J. Am. Chem. Soc. *88*, 305 (1966)
172. Bogucki, R. F., Martell, A. E.: J. Am. Chem. Soc. *90*, 6022 (1968)
173. Carey, C. H., Martell, A. E.: J. Am. Chem. Soc. *89*, 2859 (1967)
174. Folcher, G., Lambard, J., Kiener, C., Rigny, P.: J. chim. physique *75*, 875 (1978)
175. Ryan, J. L., Jørgensen, C. K.: Mol. Phys. *7*, 17 (1963)
176. Jørgensen, C. K.: Acta Chem. Scand. *17*, 251 (1963)
177. Bünzli, J. C. G., Yersin, J. R., Mabillard, C.: Inorg. Chem., submitted
178. Cotton, S. A., Hart, F. A., Hursthouse, M. B., Welch, A. J.: J.C.S. Chem. Comm. 1225 (1972)
179. Wilkinson, G., Birmingham, J. M.: J. Am. Chem. Soc. *76*, 6210 (1954)
180. Baker, E. C., Halstead, G. W., Raymond, K. N.: Structure and Bonding *25*, 23 (1976)
181. Warren, K. D.: Structure and Bonding *27*, 45 (1976)
182. Pappalardo, R., Jørgensen, C. K.: J. Chem. Phys. *46*, 632 (1967)
183. Ely, N. M., Tsutsui, M.: Inorg. Chem. *14*, 2680 (1975)
184. Nugent, L. J., Laubereau, P. G., Werner, G. K., Vander Sluis, K. L.: J. Organometallic Chem. *27*, 365 (1971)
185. Warren, K. D.: Structure and Bonding *33*, 97 (1977)

186. Green, J. C.: Structure and Bonding *43*, 37 (1981)
187. Jørgensen, C. K., Berthou, H.: Mat. fys. Medd. Danske Vid. Selskab *38*, no. 15 (1972)
188. Cox, P. A., Baer, Y., Jørgensen, C. K.: Chem. Phys. Letters *22*, 433 (1973)
189. Sugar, J.: J. Opt. Soc. Am. *55*, 1058 (1965)
190. Kaufman, V., Sugar, J.: J. Opt. Soc. Am. *66*, 439 (1976)
191. Gobrecht, H.: Ann. Physik *28*, 673 (1937); *31*, 181, 600 and 755 (1938)
192. Condon, E. U., Shortley, G. H.: Theory of Atomic Spectra (2. ed.). London: Cambridge University Press 1953
193. Racah, G.: Phys. Rev. *76*, 1352 (1949)
194. Katriel, J., Pauncz, R.: Adv. Quantum Chem. *10*, 143 (1977)
195. Jørgensen, C. K.: Israel J. Chem. *19*, 174 (1980)
196. Jørgensen, C. K.: Orbitals in Atoms and Molecules. London: Academic Press 1962
197. Jørgensen, C. K.: Solid State Phys. *13*, 375 (1962)
198. Jørgensen, C. K.: ref. 35, p. 425 (1980)
199. Sugar, J., Reader, J.: J. Chem. Phys. *59*, 2083 (1973)
200. Nugent, L. J., Baybarz, R. D., Burnett, J. L., Ryan J. L.: J. Phys. Chem. *77*, 1528 (1973)
201. Johnson, D. A.: Adv. Inorg. Radiochem. *20*, 1 (1977)
202. McClure, D. S., Kiss, Z.: J. Chem. Phys. *39*, 3251 (1963)
203. Loh, E.: Phys. Rev. *147*, 332 (1966)
204. Barnes, J. C.: J. Chem. Soc. 3880 (1964)
205. Jørgensen, C. K., Brinen, J. S.: Mol. Phys. *6*, 629 (1963)
206. Blasse, G., Bril, A.: J. Chem. Phys. *45*, 2350 and 3327 (1966)
207. Barnes, J. C., Pincott, H.: J. Chem. Soc. (A) 842 (1966)
208. Blasse, G.: Structure and Bonding *26*, 43 (1976)
209. Jørgensen, C. K.: Progress Inorg. Chem. *4*, 73 (1962)
210. Schäffer, C. E., Jørgensen, C. K.: J. Inorg. Nucl. Chem. *8*, 143 (1958)
211. Boulanger, F.: Ann. Chim. (Paris) *7*, 732 (1952)
212. Hellwege, K. H.: Ann. Physik *4*, 95, 127, 136, 143, 150 and 357 (1948)
213. Prather, J. L.: Atomic Energy Levels in Crystals, Nat. Bur. Stand Monograph no. 19. Washington, D. C.: National Bureau of Standards 1961
214. McLaughlin, R. D., Conway, J. G.: J. Chem. Phys. *38*, 1037 (1963)
215. Jørgensen, C. K.: Int. Rev. Phys. Chem. (eds.: Buckingham, D. A., Thomas, J. M., Trush, B. A.) *1*, 225 (1981)
216. Jørgensen, C. K.: Acta Chem. Scand. *9*, 540 (1955)
217. Jørgensen, C. K.: J. Inorg. Nucl. Chem. *32*, 3127 (1970)
218. Nugent, L. J.: J. Inorg. Nucl. Chem. *32*, 3485 (1970)
219. Ruedenberg, K.: Rev. Mod. Phys. *34*, 326 (1962)
220. Jørgensen, C. K.: Structure and Bonding *24*, 1 (1975) and *30*, 141 (1976)
221. Jørgensen, C. K.: Fresenius Z. Analyt. Chem. *288*, 161 (1977)
222. Cox, P. A.: Structure and Bonding *24*, 59 (1975)
223. Wertheim, G. K.: Chem. Phys. Letters *72*, 518 (1980)
224. Ferreira, R.: Structure and Bonding *31*, 1 (1976)
225. Carnall, W. T., Fields, P. R., Rajnak, K.: J. Chem. Phys. *49*, 4412, 4424, 4443, 4447 and 4450 (1968)
226. Wendin, G.: Structure and Bonding *45*, 1 (1981)
227. Wallace, W. E., Sankar, S. G., Rao, V. U. S.: Structure and Bonding *33*, 1 (1977)
228. Sobelman, I. I.: Atomic Spectra and Radiative Transitions. Springer: Berlin, Heidelberg, New York 1979
229. Broer, L. J. F., Gorter, C. J., Hoogschagen, J.: Physica *11*, 231 (1945)
230. Judd, B. R.: Phys. Rev. *127*, 750 (1962)
231. Ofelt, G. S.: J. Chem. Phys. *37*, 511 (1962)
232. Jørgensen, C. K., Judd, B. R.: Mol. Phys. *8*, 281 (1964)
233. Gruen, D. M., DeKock, C. W.: J. Chem. Phys. *45*, 455 (1966)
234. Liu, C. S., Zollweg, R. J.: J. Chem. Phys. *60*, 2384 (1974)
235. Papatheodorou, G. N., Berg, R. W.: Chem. Phys. Letters *75*, 483 (1980)
236. Judd, B. R.: J. Chem. Phys. *70*, 4830 (1979)
237. Peacock, R. D.: Structure and Bonding *22*, 83 (1975)

238. Henrie, D. E., Fellows, R. L., Choppin, G. R.: Coord, Chem. Rev. *18*, 199 (1976)
239. Weber, M. J.: Handbook on the Physics and Chemistry of Rare Earths (eds. K. A. Gschneidner and L. Eyring) *4*, 275, Amsterdam: North-Holland 1979
240. Powell, R. C., Blasse, G.: Structure and Bonding *42*, 43 (1980)
241. Jørgensen, C. K.: ref. 35, p. 607
242. Jørgensen, C. K.: Proceedings of International Conference on Lasers (New Orleans, December 1980) published by the Society for Optical and Quantum Electronics, McLean, Virginia
243. Feofilov, P. P., Trofimov, A. K.: Optika i Spektroskopia *27*, 538 (1969)
244. Nakazawa, E.: J. Luminescence *12*, 675 (1976)
245. Schugar, H. J., Solomon, E. I., Cleveland, W. L., Goodman, L.: J. Am. Chem. Soc. *97*, 6442 (1975)
246. Reisfeld, R.: Proceedings of International Conference on Lasers (New Orleans, December 1980) published by the Society for Optical and Quantum Electronics, McLean, Virginia
247. Kelly, H.: Science *199*, 634 (1978)
248. Reisfeld, R.: Naturwissenschaften *66*, 1 (1979)
249. Goetzberger, A., Greubel, W.: Appl. Phys. *14*, 123 (1977)
250. Reisfeld, R., Greenberg, E., Kisilev, A., Kalisky, Y.: "Collection and Conversion of Solar Energy for Photoconverting Systems" at Third International Conference on Photochemical Conversion and Storage of Solar Energy, Colorado, August 1980 (ed. J. S. Connolly)
251. Jørgensen, C. K.: ibid., "Quantum Harvesting and Energy Transfer"
252. Reisfeld, R., Neuman, S.: Nature *274*, 144 (1977)
253. Reisfeld, R., Kalisky, Y.: Nature *283*, 281 (1980)
254. Batchelder, J. S., Zewail., A. H., Cole, T.: Appl. Optics *18*, 3090 (1979); and "Role of Electro-Optics in Photovoltaic Energy Conversion", Society of Photo-Optical Instrumentation Engineers *248*, 105 (1980)
255. Goetzberger, A., Wittwer, V.: Advances in Solid State Physics, pp. 427–451. Wiesbaden: Vieweg 1979
256. Reisfeld, R., Jørgensen, C. K.: Structure and Bonding *49* (1981)
257. Urbain, G.: Ann. chim. phys. (Paris) *18*, 222 and 289 (1909)
258. Urbain, G.: Chem. Rev. *1*, 143 (1925)
259. Jørgensen, C. K.: Chimia *32*, 89 (1978) and *34*, 381 (1980)
260. Blasse, G.: Handbook on the Physics and Chemistry of Rare Earths (eds. K. A. Gschneidner and L. Eyring) *4*, 237, Amsterdam: North-Holland 1979
261. Jørgensen, C. K.: Chem. Phys. Letters *34*, 14 (1975)
262. Jørgensen, C. K.: Structure and Bonding *25*, 1 (1976)
263. Ivey, H. F.: J. Luminescence *8*, 271 (1974)
264. Corredor, A., Tsong, I. S. T., White, W. B.: ref. 34, p. 573
265. Jørgensen, C. K., Bill, H., Reisfeld, R.: Proceedings of the International Conference on Luminescence, Berlin, July 1981 (in print at North-Holland, Amsterdam)

Systematization and Structures of the Boron Hydrides

Lawrence Barton

Department of Chemistry, University of Missouri-St. Louis, St. Louis, Missouri 63121, U.S.A.

Table of Contents

1 Introduction

The chemistry of the boron hydrides and their derivatives is indeed complex. Boron is known to form compounds in which its coordination number ranges from 2 to 8, cages having 4 to 20 vertices and linear and ring systems of varying complexity. It has been pointed out many times that the capacity of boron to catenate and form self-bonded complex molecular networks is a property exhibited by no other element except carbon. The challenges presented to chemists by the boranes and their derivatives are wide in scope, and have had a major effect on chemistry in general. Some of these aspects will be mentioned in the following, others will be left for the reader to find in the various footnotes and references. Let it be suffice to say that the general area of research on boron-hydrogen compounds has been recognized by the award of the Nobel Prize twice in the last five years: To William N. Lipscomb[1] for his contributions to the understanding of the bonding and structure of the boranes; and to Herbert C. Brown[2] who first developed reagents which allowed more facile preparation of the boranes and later exploited the remarkable applications of boranes in organic synthesis.

It was an interesting coincidence that the first of these awards was made in 1976, 100 years after the birth of Alfred Stock. As Lipscomb[1] has said "Alfred Stock established boron chemistry". Indeed, Stock,[3,4,5] whose initial interests in boron hydride chemistry were stimulated by the efforts of William Ramsey to make diborane(6), gave up his research in the area for a while. Stock as a very young chemist had been greatly impressed by an extraordinary experimental lecture by Ramsey and when Ramsey later relayed to him that the area of boron hydrides had been "thoroughly investigated in his laboratories" Stock abandoned the line of research. Fortunately for all chemists and chemistry, Stock returned to the study of the boranes when he had his own laboratory in 1912. Sidgewick[6] wrote in his classic text in 1950: "All statements about the hydrides of boron earlier than 1912, when Stock began to work upon them, are untrue". Stock[6] developed techniques for the handling of the poisonous and unpredictable explosive air and water sensitive gases, B_2H_6, B_4H_{10}, B_5H_9, B_5H_{11} and B_6H_{10}. These techniques and the equipment used to study them form the basis of the high vacuum techniques used today.[7]

The dilemma of the structure of diborane(6) puzzled chemists for more than a quarter of a century after its composition and molecular formula had been established and this stimulated debate which had important consequences for chemistry in general. Some of these consequences have only been recognized quite recently by many chemists. The electron deficiency, which resulted in the inability of recently by many chemists. The electron deficiency, which resulted the inability of chemists to account for all the bonds in the ethane-like structure which was postulated from early electron diffraction data[8], led to suggestions of resonance forms involving one-electron bonds, hyperconjugated and no-bond structures and these were generally found to be unsatifactory[9].

Indeed the correct structure, shown in Fig. 1, was predicted by Dilthey[10] in 1921. It seems that this structure was not considered seriously until the early 1940's when the infrared spectral studies on diborane(6) by Bell and Longuet-Higgins[11] and also by Price[12] indicated that the bridge structure was probably correct. This

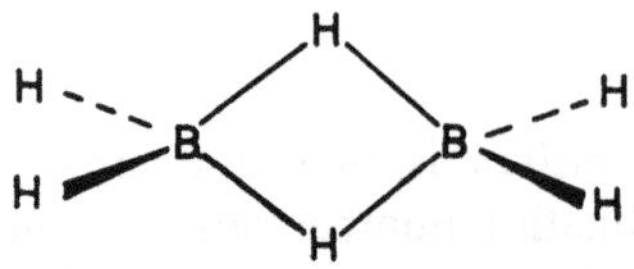

Fig. 1. The Structure of Diborane(6)

was confirmed in 1951 by the electron diffraction studies of Hedberg and Schomaker.[13] This work was soon followed by the pioneering low temperature X-ray diffraction work of Lipscomb which established the structures of essentially all the simple boron hydrides and some of the more complex ones.[14]

The questions concerning the bonding in boranes posed by what were perceived to be their unusual structures are still being discussed today and the resulting research has had major consequences. In 1970, Linus Pauling[15] stated that "the formulas of the boranes do not conform in any simple way to chemical valence theory".

This survey presents a historical review of the classification of boranes; a description of the modern approach follows, and then a survey of the structures of the known boranes completes the presentation.

2 Historical Review of the Systematics and Classification of Boron Hydrides

Stock[16] originally attempted to classify and name the boron hydrides using arguments based on analogies with the hydrocarbons. Although this was not entirely successful owing to the unavailability of structural data, some aspects of his classification have persisted through all the subsequent ones. Stock suggested that there were two series of compounds represented by the empirical formulas B_nH_{n+4} and B_nH_{n+6}.

Members of the first of these series, those represented by the general formula B_nH_{n+4}, were referred to as the "stable" boranes and included B_2H_6, B_5H_9, B_6H_{10} and $B_{10}H_{14}$. The second group, with the general formula B_nH_{n+6}, were called "unstable" boranes and included B_4H_{10}, B_5H_{11} and B_6H_{12}. These definitions clearly have meaning. The boranes B_5H_9 and $B_{10}H_{14}$ are thermally quite stable and can be handled at temperatures above 100 °C without rapid decomposition whereas B_4H_{10} and B_5H_{11} decompose, apparently spontaneously, at room temperature.

The ideas of Stock were extended by Parry and Edwards in 1959.[17] The work was based on a knowledge of most of the structures of the then-known species and of the structures of reaction products and intermediates. These workers noted that the two series, described by Stock, were related by the BH_3 moiety, i.e., the series B_nH_{n+4} is obtained by removing a BH_3 group from B_nH_{n+6}. Further, a new series B_nH_{n+2} is obtained by removal of a second BH_3 group, and in addition another series B_nH_{n+8} should be available if BH_3 is added to B_nH_{n+6}. Although these two new series were unknown at the time, the predictions were correct in that such series are now well established in terms of their electronic structure. In fact, before the end of 1959, species which are electronically placed in these two new series were known. These included the polyhedral borane $[B_{10}H_{10}]^{2-}$ [18] and the bis-ligand adduct $B_5H_9[N(CH_3)_3]_2$ [19].

Another significant aspect of the approach of Parry and Edwards was based on the types of reactions most easily participated in by the various classes of boranes. Species which contain a BH_2 unit (which included B_2H_6 and the series B_nH_{n+6}) all reacted with nucleophiles to generate either molecular or ionic fragments. These two reaction modes were described as symmetrical and non-symmetrical cleavage respectively. The two modes of cleavage are illustrated in Figure 2 for B_4H_{10}.

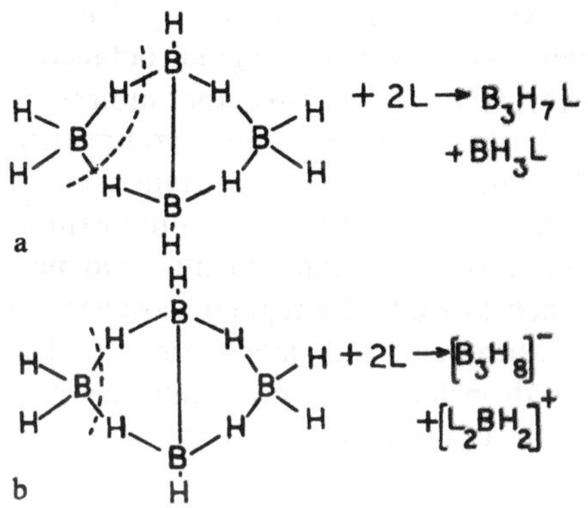

Fig. 2. Scheme Showing the Two Modes of Cleavage of B_4H_{10} by Bases; **a** symmetrical cleavage; **b** non-symmetrical cleavage)

The terms symmetrical and non-symmetrical are based on cleavage of diborane(6) to form two BH_3 groups and to form $[BH_2]^+$ and $[BH_4]^-$ moieties, respectively. These two alternative modes of cleavage are available to all boranes containing BH_2 groups which include B_2H_6, B_4H_{10}, B_5H_{11}, B_6H_{12} and B_9H_{15}. Although there are reports of non-symmetric cleavage of boranes not containing BH_2 groups, for example B_5H_9, these examples are rare and the products have very limited stability.[20]

Another feature of this classification is the Brønsted acidity of the bridge hydrogens in members of the B_nH_{n+4} class. At the time Parry and Edwards raised these issues, only $B_{10}H_{14}$ had been unambiguously shown to demonstrate this property[21] although there were some indications of this from the cleavage of B_4H_{10} with NH_3.[22] It is now recognized that this property is a general one of both the "stable" and "unstable" classes of boranes.

Another type of reaction used in the classification by Edwards and Parry was the evolution of molecular hydrogen by members of the B_nH_{n+6} series to yield the corresponding member of the B_nH_{n+4} series. The former species are clearly less stable than the latter. For example, B_5H_{11} forms B_5H_9 as one of its major decomposition products; however, this property has not been used with success to distinguish the two classes of boranes.

In 1966 Ditter, Spielman and Williams,[23] in a report in which they described the syntheses of B_9H_{15} and B_8H_{12}, extended the proposal of Parry and Edwards that the structural building block for boranes are BH_3 groups. Thus borane frameworks, represented by $(BH)_n$, were described to require either two or three BH_3 units. Species containing two BH_3 groups conform to Stock's B_nH_{n+4} series and those containing three BH_3 groups conform to the B_nH_{n+6} series. These workers denoted all boranes by one general expression $(BH)_n(BH_3)_x$ where x = 2 for "stable" boranes and x = 3 for "unstable" boranes. Thus it was possible to systematize all existing boranes according to the formula $(BH)_n(BH_3)_x$ and to assume that species stable enough to be isolated conformed to values of x = 2 and x = 3. Species not known which were represented by the general formula were assumed to exist as very unstable or transient species.

Other important developments in borane chemistry in the 1960's are pertinent to this discussion. Previously, Longuet-Higgins and Roberts[24,25] studied theoretically the bonding and stability of two solid state boron structures. It had been recognized since the X-ray structural work of Allard[26] and also Pauling[27] in the 1930's that several hexaborides of general formula MB_6 existed as regular octahedra of boron atoms. Longuet-Higgins and Roberts[24] showed that the molecular orbital bonding description of the $[B_6]^{2-}$ ion indicated a "closed shell" arrangement of high stability. They predicted that the bivalent metal borides of this type, should be insulators and when the metal ion has a higher charge the crystal should exhibit metallic conductivity. These predictions were in accord with experimental evidence. These workers used the same theoretical approach to study the regular icosahedron of boron atoms; the dominant structural feature of solid-state boron[28]. The results of the study[25] indicated that the icosahedron has thirteen bonding orbitals available for holding the polyhedron together in addition to the twelve outward-pointing equivalent orbitals on the separate boron atoms. Thus, it was concluded that if a borane of formula $B_{12}H_{12}$ were to be prepared, it would be stable only as the dianion $[B_{12}H_{12}]^{2-}$. This prediction was verified in 1960 when Hawthorne and Pitochelli[29] prepared stable salts of the $[B_{12}H_{12}]^{2-}$ ion, one year after they had prepared salts of the anion $[B_{10}H_{10}]^{2-}$[18]. Both anions were highly stable and existed as regular polyhedra, the latter having the structure of a bicapped Archimedian antiprism[30] and $[B_{12}H_{12}]^{2-}$ the regular icosahedron[31]. Soon Muetterties and his coworkers prepared salts of the analogous anions $[B_{11}H_{11}]^{2-}$[32], $[B_9H_9^{\bullet}]^{2-}$, $[B_8H_8]^{2-}$, $[B_7H_7]^{2-}$[33] and $[B_6H_6]^{2-}$[34] and the field of polyhedral borane chemistry was born. The relationship between these species and the neutral boron hydrides was not immediately recognized; however, this was recognized in the 1970's and is described later in this article.

The discovery of the carboranes, in the early 1960's, also provided important indications of a wider extent of systematization of the boranes than had previously been envisioned. The first ones to be discovered were the species $C_2B_3H_5$, $C_2B_4H_6$ and $C_2B_5H_7$[35] whose structures were deduced to be those of the regular polyhedra containing 5, 6 and 7 vertices, namely the trigonal bipyramid, the octahedron and the pentagonal bipyramid, respectively. These discoveries were followed by that of the species $C_2B_{10}H_{12}$[36], having a regular icosahedral structure[37]. The species $C_2B_4H_6$ and $C_2B_{10}H_{12}$ are clearly related to the polyhedral borane anions $[B_6H_6]^{2-}$ and $[B_{12}H_{12}]^{2-}$ in that they are isoelectronic and isostructural. If the C atoms in the carboranes are replaced by B^-, one has the identical pair of species. The existence of the polyhedral anions and the *closo*-carboranes[40a] is pertinent in that this class of species, represented as B_nH_{n+2}, was predicted by Parry and Edwards in 1959.

The preparation of $C_2B_4H_8$[38] and the identification of its structure as that of a pentagonal pyramid[39] as indicated in Fig. 3, represented the first member of the *nido*[40b)] class of carboranes. The term *nido* (derived from the Greek for nest), was originally applied to all open-carboranes; however, as we shall see below, this was modified later.

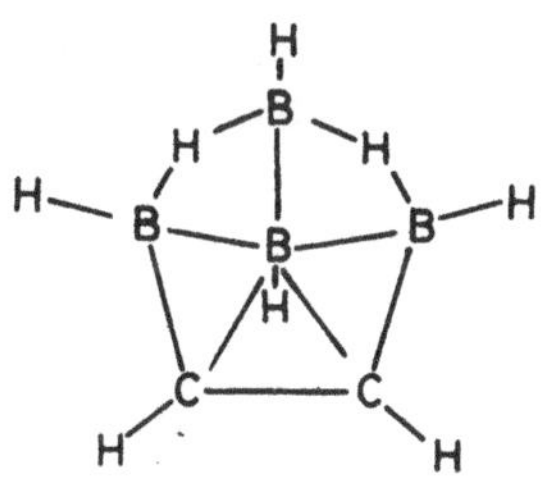

Fig. 3. The Structure of $C_2B_4H_8$

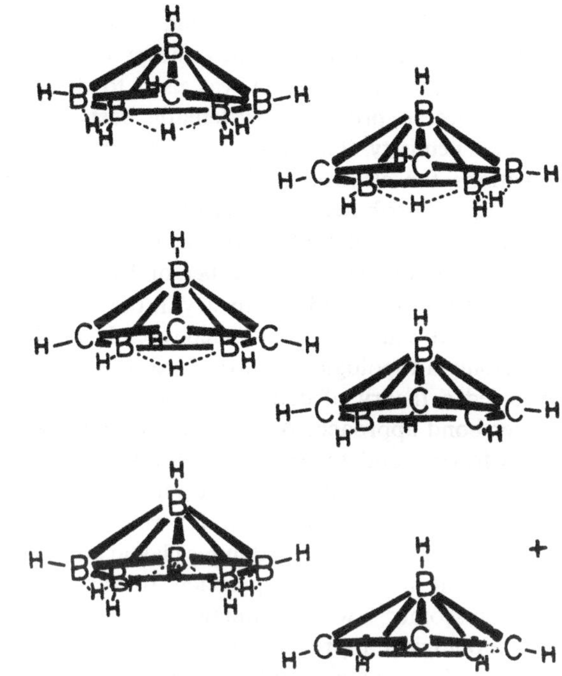

Fig. 4. The Series of *nido*-Boranes Isoelectronic with B_6H_{10}

The Structure of $C_2B_4H_8$ is very similar to that of its isoelectronic analogue B_6H_{10} and recently the whole series of species B_6H_{10}, CB_5H_9, $C_2B_4H_8$, $C_3B_3H_7$, $C_4B_2H_6$, and $[C_5BH_6]^+$ was completed[41)] (see Fig. 4). All of these species belong to Stock's stable class of boranes, B_nH_{n+4}, i.e., $C_mB_nH_{n+4}$. The discovery of a new type of "*nido*"-borane, $C_2B_7H_{13}$, in 1966 by Tebbe, Garrett and Hawthorne,[42)] represented the isolation of the first member of the series $C_mB_nH_{n+6}$, analogous to Stock's unstable series of boranes, B_nH_{n+6}. The structure of this species is an open polyhedron, analogous to the boranes containing BH_2 groups[43)]. Thus the carboranes were shown to the represented in three series, analogous to the three series of boranes.

The contributions of theory to the understanding of the structures of the boron hydrides cannot be over-emphasized. Once the bridge structure was accepted for diborane(6), Longuet-Higgins[44)] described the interaction of the bridge hydrogen 1s orbitals with two appropriate boron orbitals to yield three molecular orbitals including a bonding orbital which he named a "three-center bond"[45)]. As Lipscomb and his coworkers made available the structures of the simple boranes B_4H_{10},

B_5H_9 and B_5H_{11}, Longuet-Higgins[45] applied the three-center bond theory to these systems. The essence of the approach was to utilize all of the orbitals in these electron-deficient systems to generate an electronic structure involving delocalization of the electrons in the boron framework of the borane.

In 1954 Eberhardt, Crawford and Lipscomb[46] published a landmark paper which applied the three-center bond concept to all the known boron hydrides. For the higher boranes the concept was used to describe three-center boron-boron bonds and thence to provide a systematic basis for structural and bonding aspects of borane chemistry. The existence of unknown boranes was speculated by the use of equations for electron and orbital balance[47,48]. These equations, the so-called styx rules, distributed the electrons among the possible types of orbitals available, viz. s = number of bridge hydrogens t = number of B—B—B bonds, y = number of B—B bonds and x = number of BH_2 groups. This paper has had a profound effect on boron hydride chemistry in the ensuing 25 years.

Lipscomb's ideas were extended when he published a paper with R. E. Dickerson[49] describing a topological approach for the generation of structures of boranes which involved a theory of connectivities of various bonding patterns within the three-center bond approach. A geometrical theory was developed to avoid overcrowding of hydrogens and to preserve appropriate bond angles about boron atoms in the icosahedral fragments[50]. Indeed, for several years it was believed that all open-structured boranes (except B_5H_9) were fragments of a regular icosahedron. Lipscomb's work is documented in his classic book and is not detailed here[14].

A recent topological approach to three-center valence structures by Epstein and Lipscomb[51] was formulated to exclude open three-center B—B—B bonds. The topological approach was extended to describe the course of ionic substitution reactions by Epstein[52] and more recently by Rudolph and Thomson[53]. Throughout the 1970's advances in the understanding of the electronic structures of boranes continued to be made by Lipscomb[54] and his coworkers. These include predictions concerning the existence and structures of borane polyhedra with more than 12 vertices[55] and further refinement of the bonding descriptions of the boranes and carboranes, especially those for which it is impossible to write single simple valence structures[56-60].

3 The Current Approach to the Classification of Boranes

The decade of the 1970's saw the beginings of a unified approach to the structure and bonding in boranes and carboranes which was soon applied to other cluster systems including metalloboranes, polynuclear metal carbonyls, organometallic compounds, main-group element clusters and even hydrocarbons. What are colloquially referred to as "Wade's Rule", after K. Wade[61] who first applied these ideas to clusters in general, were developed independently by Williams[62], Wade, and Rudolph[63]. Several reviews and articles[64-71] on the subject have appeared wherein the overall scheme is applied to boranes, carboranes, metalloboranes and organometallic species; however, the current treatment will emphasize boranes and will only describe derivatives when appropriate parent borane examples are not available.

Boranes are recognized as clusters of boron atoms which represent triangular-faced polyhedra. It was Williams[62] in 1970 who pointed out in a classic paper that the structures of all boranes are derived from those of the closed polyhedral borane anions, or carboranes, with vertices ranging from 5 to 12. These regular polyhedra are shown in Fig. 5 and represent the structures of the *closo*-boranes (or carboranes) from which all other borane structures are derived.

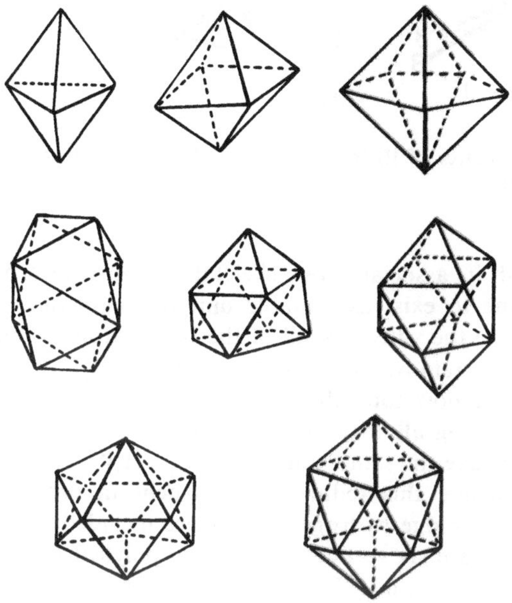

Fig. 5. The Regular Polyhedra Containing 5 to 12 Vertices

The *closo*-boranes all have the general formula $[B_nH_n]^{2-}$ where n represents the number of vertices in the regular polyhedron[72]. Borane species are well characterized for values of n = 6 to 12. The carborane $C_2B_3H_5$, isoelectronic with the unknown species $[B_5H_5]^{2-}$, completes the series 5 to 12. For the *closo*-boranes and *closo*-carboranes containing n vertices in the cage, there are n + 1 pairs of electrons bonding the cage atoms together. Thus each B—H unit supplies a pair of electrons for cage bonding in addition to a pair of electrons in the *exo*-polyhedral B—H sigma bond. The electrons possessed by the cage in addition to the two electrons per B—H bond are referred to as *skeletal* electrons. This general rule which will be described later applies to all boranes, i.e. all electrons except two per boron atom, are included in the skeletal electron count. Thus a species of general formula $[B_nH_n]^{2-}$ contains n + 1 skeletal electron pairs. Longuet-Higgins in 1954 predicted that the B_6 octahedron should contain 7 bonding molecular orbitals[24] and that the $B_{12}H_{12}$ unit contains 13 icosahedral bonding orbitals[25] and thus should exist as the dianion and Lipscomb predicted the analogous result for $[B_{10}H_{10}]^{2-}$ [73] and also predicted the existence of $[B_5H_5]^{2-}$ [74].

The polyhedra which represent the *closo*-boranes also serve as the basis for the structures of the *nido*-, *arachno*- and *hypho*-boranes. The term *nido*, as mentioned before, is derived from the Greek work for nest, and refers to the B_nH_{n+4} class of boranes, all of which have nest-like structures obtained by removing the highest coordination vertex from a *closo*-borane. Examples of this are shown in Fig. 6 for B_6H_{10} and $[B_7H_7]^{2-}$.

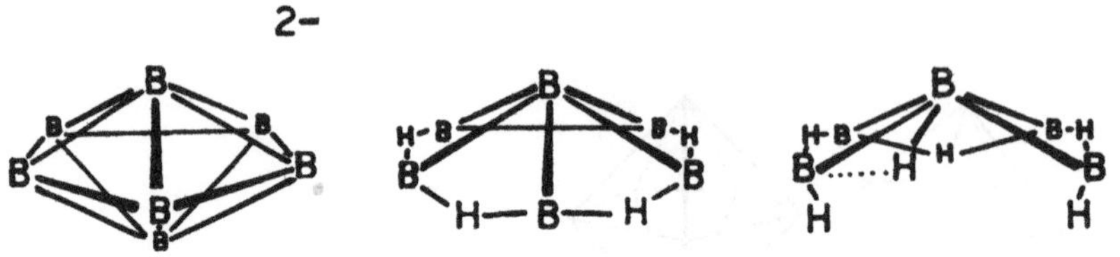

Fig. 6. The Relationship Between the Structures of $[B_7H_7]^{2-}$, B_6H_{10} and B_5H_{11} (each B atom has a terminal H atom which is not shown)

On the removal of a BH unit (with its associated two skeletal electrons) in order for the more open structured borane to exist as a stable unit two electrons (as two H atoms) are supplied to complete the valence shell of the boron atoms. Thus the stable neutral borane, formed by removal of a BH unit from $[B_7H_7]^{2-}$ is B_6H_{10}, after formal protonation. If we now total the skeletal electrons, using the principle described previously of including all electrons except one *exo*-polyhedral pair per boron atom, we find that we have the same number as in $[B_7H_7]^{2-}$, i.e., 16. This clearly follows however we include this redundant statement in order to emphasize that electrons derived from bridge hydrogen atoms are included in the skeletal electron count. In B_6H_{10} there are six vertices in the open polyhedron and eight pairs of skeletal bonding electrons, i.e., n + 2 skeletal bonding pairs where n is the number of vertices in the cluster. Thus we generalize that *nido*-boranes, whose formula is represented by B_nH_{n+4}, contain n + 2 skeletal electron pairs.

Removal of two :BH units from $[B_7H_7]^{2-}$, i.e., removal of a :BH unit from B_6H_{10}, accompanied by the formal addition of two H$^\cdot$ atoms and protonation yields the species B_5H_{11} which is a member of the *arachno*-boranes. These structures are more open and flattened than the *nido*-boranes and typically contain BH_2 groups (The term *arachno* is derived from the Greek work for cobweb[75]). The *arachno*-boranes represent Stock's "unstable" boranes with the general formula B_nH_{n+6}. B_5H_{11} contains 26 valence electrons, and if we subtract 10 electrons for five *exo*-polyhedral :B—H units, we are left with 16 electrons for skeletal bonding. It is important to note that this is the same number of skeletal electron pairs as are possessed by the related species B_6H_{10} and $[B_7H_7]^{2-}$. In the case of the *arachno*-species we have included in the skeletal electron count, the electrons associated with the boron framework, electrons in B—H—B bridging molecular orbitals and also one electron for every "extra" hydrogen appearing in a BH_2 group. These are the electrons associated with the *endo*-hydrogens and this point is discussed below. The structural relationship between $[B_7H_7]^{2-}$, B_6H_{10} and B_5H_{11} is illustrated in Fig. 6.

A fourth class of boranes, the *hypho*-boranes, relates to the above species in that the polyhedron, composed of n vertices, contains n + 4 skeletal electron pairs

(The word *hypho* is derived from the Greek for net and is meant to imply an almost flat species[53)]). Removal of :BH from B_5H_{11} and the addition of two H atoms yields B_4H_{12}. This species is unknown although several Lewis base di-adducts of B_4H_8 have been characterized by NMR spectroscopy[76)]. These species are iso-electronic with B_4H_{12} although the static structures may not necessarily be identical. The B_5 [19,77)] and B_6 [78)] *hypho*-boranes have been structurally characterized. For example, the species $B_6H_{10}\ 2\ P(CH_3)_3$, a *hypho*-borane containing n + 4 skeletal electron pairs and derived from the tricapped trigonal prism structure of $[B_9H_9]^{2-}$ [79)] by the removal of three vertices, has a more flattish and open structure than its *nido*- and *arachno*-counterparts. B_6H_{10} contains 16 skeletal electrons and the two Lewis base molecules provide two electrons each giving a total of 20 electrons, i.e., n + 4 pairs of electrons available for skeletal bonding. The structure of $B_6H_{10}\ 2\ P(CH_3)_3$, is shown in Fig. 7.

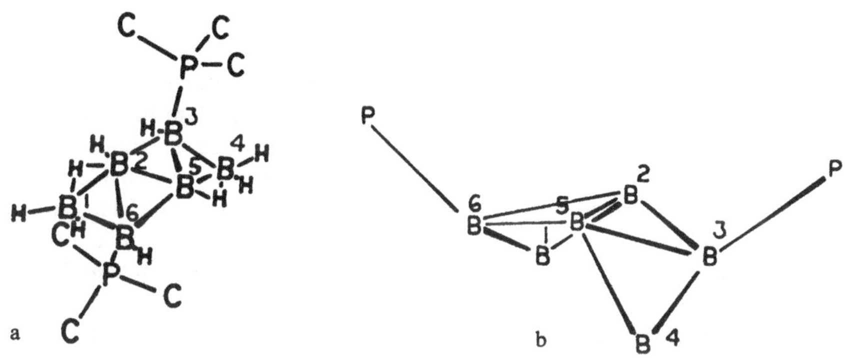

Fig. 7a. Structure of $B_6H_{10}[P(CH_3)_3]_2$ (methyl H's omitted); **b** Conformation of the B_6 Framework in $B_6H_{10}[P(CH_3)_3]_2$

Table 1. Examples of *closo*-, *nido*-, *arachno*-, and *hypho*-Boranes

No. of Skeletal Bonding Pairs	Basic Polyhedron	*closo*-Borane	*nido*-Borane	*arachno*-Borane	*hypho*-Borane
6	trigonal bipyramid	$C_2B_3H_5$	B_4H_8	$[B_3H_8]^-$	—
7	octahedron	$[B_6H_6]^{2-}$	B_5H_9	B_4H_{10}	$B_3H_5L_3$
8	pentagonal bipyramid	$[B_7H_7]^{2-}$	B_6H_{10}	B_5H_{11}	$B_4H_8L_2$
9	dodecahedron	$[B_8H_8]^{2-}$	$(CO)_4FeB_7H_{11}$	B_6H_{12}	$B_5H_9L_2$ $[B_5H_{12}]^-$
10	tricapped trigonal prism	$[B_9H_9]^{2-}$	B_8H_{12}	$[B_7H_{12}]^-$	$B_6H_{10}L_2$
11	bicapped Archimedian antiprism	$[B_{10}H_{10}]^{2-}$	$[B_9H_{12}]^-$	B_8H_{14}	—
12	octadecahedron	$[B_{11}H_{11}]^{2-}$	$B_{10}H_{14}$	n-B_9H_{15} i-B_9H_{15}	B_8H_{16}
13	icosahedron	$[B_{12}H_{12}]^{2-}$	$[B_{11}H_{13}]^{2-}$	$[B_{10}H_{14}]^{2-}$	—

Table 1 shows examples of *closo-*, *nido-*, *arachno-* and *hypho-*boranes containing 6 through 13 skeletal bonding electron pairs. Where neutral borane example are not available, ionic species or heteroboranes are cited.

The Position of Hydrogens and Skeletal Electrons Counting

In the assignment of the number of skeletal electrons we include those involved solely in boron-boron bonding, the bridge hydrogens and those associated with one of the B—H bonds in a BH_2 group (*endo-*hydrogens). The bridge hydrogens and the *endo-*hydrogens are regarded as participating in holding together the boron skeleton whereas the *exo-*hydrogens are not. Thus the former are included in the skeletal-electron count and the latter are not included. These counting rules were developed independently by Wade[61)] and by Rudolph and Pretzer[63)], but it was Shore[78b)] who cited an earlier statement by Lipscomb[80)] to rationalize these rules; Lipscomb stated that "in each of the boron hydrides, based on a single polyhedral fragment, there are two surfaces the surface of the terminal (*exo*) H atoms and a smaller nearly spherical surface containing the B atoms, H bridges, and the extra H atoms (*endo*) of BH_2 groups", and Shore went on to point out that the skeletal electrons are defined as those associated with this inner sphere. Clearly, the electrons bonding the boron atoms to the *exo-*hydrogens, which in a *closo-*borane point outwards from the center of the polyhedron, are not involved in the bonding of the boron skeleton. In *nido-*boranes, the bridging hydrogens are considered to hold together the open face of the polyhedron and the *endo-*hydrogens of BH_2 groups in *arachno-*boranes are situated over incomplete edges or faces. Thus, in counting up the skeletal electrons, all electrons are included except the two in *exo* B—H sigma bonds.

It is easy to see why bridging hydrogens are included in the skeletal electron count. The B—H—B moiety may be regarded as a protonated $[B—B]^-$ unit and indeed it is often convenient to regard the four borane classes formally as $[B_nH_n]^{2-}$, $[B_nH_n]^{4-}$, $[B_nH_n]^{6-}$ and $[B_nH_n]^{8-}$. Protonation of the latter three species yields the easily recognisable *closo-*, *nido-*, *arachno-* and *hypho-*boranes. As mentioned previously, the *endo-*hydrogens of BH_2 groups lie on the surface of an inner sphere which also contains the bridging hydrogens. In some species the bridge and *endo-*hydrogens are almost indistinguishable. For example the bridge and *endo-*hydrogens are exchange-averaged on the NMR time scale in $[B_9H_{14}]^-$ and appear as a single broad resonance in the 220 MHz ^1H spectrum[81)]. *Endo-*hydrogens do, in fact, form pseudo bridges in some cases, e.g., H^2 in B_5H_{11} is situated between the atoms B^2 and B^5 and forms an asymmetric bridge[115)] (see Fig. 17). In $[B_9H_{14}]^-$ the *endo-*hydrogens point towards an extension of the open cavity of the polyhedron whereas the *exo-*hydrogens protrude away from the cavity.

The theoretical basis for the electron counting rules has been discussed elsewhere[63,65,68)]. Wade has formulated a general expression for the determination of the number of electrons available for skeletal bonding for skeletal atoms. Boron has three orbitals available for cluster bonding if it uses its remaining valence shell atomic orbital to accommodate a bonding pair of electrons or a lone pair. Thus B—H supplies two electrons for cluster bonding or $v + x - 2$ where v = the number of electrons in the valence shell of B (i.e., 3), x = the number of electrons supplied by the ligand (in this case H, so $x = 1$). For a BH_2 group, the number of skeletal bonding electrons is 3 + 2 — 2, i.e., 3 and for a BH group to which is coordinated a Lewis base, for example $P(CH_3)_3$ in $B_6H_{10}[P(CH_3)_3]_2$, the number of electrons available for skeletal bonding is 4 since two additional electrons are provided by the ligand.

Recent treatments have described the skeletal bonding[64,68,71] contributions for ligated transition metal moieties in metalloboranes and in transition-metal clusters. Wade indicated the important similarity between boranes and transition-metal clusters, in particular the species $[B_6H_6]^{2-}$ and several octahedral transition-metal clusters; and Mingo[82,83], on the basis of detailed molecular orbital calculations for $[Co_6(CO)_{14}]^{4-}$, has confirmed the validity of these comparisons.

Wade[64d], Corbett[84,85] and others[66,86] have pointed out the relationship between the boranes and other main-group clusters and have demonstrated that the electron-counting schemes and structural predictions are identical. Recent work by Wade and Housecraft[64e] has applied these rules to cyclic and polycyclic hydrocarbons with remarkable success.

Finally, several authors including Lipscomb himself, have pointed out the relationship of the "styx" rules (see above) to Wade's rules[55,87,88]. The sum of the digits in the topological styx formula for a borane structure gives the number of skeletal electron pairs in the molecule since both approaches "factor out" the bonds to exopolyhedral substituents. Thus, in the general formula B_pH_{p+q} used in the styx rules, q values of 2, 4, 6 and 8 are identified with *closo-*, *nido-*, *arachno-* and *hypho-*borane structures, respectively.

4 Structures of the Boranes

This section presents the structure of the boron hybrides and is arranged in accordance with the relationship defined by Wade's Rules and expressed by Williams and Rudolph. Thus for the boranes containing six or more pairs of skeletal bonding electrons, the relationship between the structures of the *closo-*, *nido-*, *arachno-* and *hypho-*species is described. In cases where the parent borane does not exist, examples from heteroboranes with the correctly predicted structure based on Williams' coordination number pattern recognition theory (CNPR) of borane structures will be described [70]. Treated separately will be mono- and diborane species and also species with more than 12 boron atoms. Although there have been several reviews on the structures of the boranes in recent years none have used the current approach[89].

4.1 Borane(3), BH_3

The existence of BH_3 has been established by many groups[90]; however, structural information is only available from theoretical calculations[91] and from a low-temperature matrix isolation study of pyrolysis products of BH_3CO[92]. The molecule is assumed to possess ideal D_3h symmetry with a bond distance of 1.32 Å[91].

It is appropriate to compare the structure proposed for BH_3 with that of the derivative BH_3CO as determined by microwave spectroscopy. The structural parameters in BH_3CO, in which the disposition of the three hydrogens and the CO moieties around the boron atom is roughly tetrahedral, are $R_{B-C} = 1.53$ Å, $R_{B-H} = 1.22$ Å, < HBC = 104° and < HBH = 114° [93].

4.2 Diboranes

4.2.1 Diborane(4), B_2H_4

Only the 1,2-diadducts of diborane(4) are known. A representative structure is that of $B_2H_4(CO)_2$ which has a 1,2-disubstituted ethane-like structure with a center of inversion[94]. Bond distances are: B—B, 1.78 Å; B—C, 1.52 Å; C—O, 1.125 Å, B—H, 1.14 Å; B—H_2, 1.11 Å; and bond angles are B—B—C, 102.3°; B—C—O, 177.5°; H_1—B—C, 108°; H_2—B—C, 100°; B—B—H_1, 115°; B—B—H_2, 117°; B_1—B—H_2, 112°.

4.2.2 Diborane(6), B_2H_6

Although diborane(6) may formally be described as belonging to Stock's classification of stable boranes and thus as a *nido*-species, since it is not structurally derived from a closed polyhedral species, it is usually considered in a class of its own and not part of the general scheme described above in Section 3. The structure of diborane(6) has been determined by-gas-phase electron diffraction[13,95], high-resolution infrared and Raman spectrecopy[96], and by low temperature X-ray diffraction[97]. The structure is shown in Fig. 1 and bond distances are given in Table 2.

Table 2. Structural Parameters for B_2H_6

Bond/Angle	B—H^t	B—H_μ	B—B	H^tBH^t	$H_\mu BH_\mu$
electron diffraction[95]	1.19 Å	1.34 Å	1.77 Å	120°	97°
X-ray diffraction[97]	1.08 Å	1.24 Å	1.78 Å	124°	90°
IR and Raman spectra[96]	1.19 Å	1.32 Å	1.77 Å	121°	96°

The shorter B—H bond distances for the X-ray diffraction data are ascribed to thermal vibration and rigid rotation of the molecule in the crystal[98].

4.3 Borane Clusters Containing 6 to 13 Skeletal-Bonding Electron Pairs (N)

4.3.1 The N = 6 Class; $C_2B_3H_5$, $[B_4H_7]^-$ and $[B_3H_8]^-$

The species $[B_5H_5]^{2-}$ with D_3h symmetry and nine equal B—B distances was predicted to be stable by Lipscomb et al.; however, thus far the species has eluded preparation[74]. Of the dicarbapentaborane analogues, three isomers are possible but the only unsubstituted isomer known is 1,5-$C_2B_3H_5$. The structure, which has been determined by electron diffraction[99], is shown in Fig. 8 and structural parameters are shown inset.

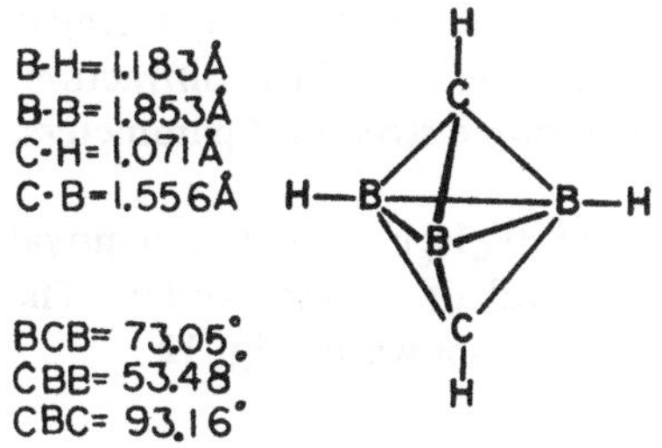

Fig. 8. Structure and Structural Data for 1,5-$C_2B_3H_5$

Although substituted derivatives of the 1,2- isomer are also known, detailed structural data are not available; however, spectral data suggest the trigonal bipyramidal closed polyhedral structure[100]. Theoretical calculations indicate that the 1,5-isomer is 61 kcal/mol lower in energy than the 1,2-isomer[101].

The *nido*-species whose structure is derived by the removal of an axial vertex from $[B_5H_5]^{2-}$ is B_4H_8. This species has not been isolated although it is suggested as an unstable intermediate in several process[102]; however, its Bronsted conjugate base, $[B_4H_7]^-$, is known. Kodama et al.[20,103] have prepared solutions of $[B_4H_7]^-$ and predict the structure shown in Fig. 9 based upon the ^{11}B NMR spectrum. This structure, (i.e., a trigonal pyramid) is clearly one of two possible ones for a *nido*-species with four vertices.

The only *arachno*-species known for a borane where N = 6 are $[B_3H_8]^-$ and B_3H_7L. The structure of $[B_3H_8]^-$ is that predicted for the *arachno*-species derived by removing the two axial vertices from $[B_5H_5]^{2-}$ and is shown in Fig. 10.

The structure, which was determined by Peters and Nordman[104], has $B—B_{base}$ = 1.80 Å, $B—B_{sides}$ = 1.77 Å, $B—H_t$ = 1.06 to 1.20 Å, B_p —H = 1.5 Å, B_{base}—H = 1.2 Å and is consistent with interpretations of the ^1H NMR spectrum of the ion[105].

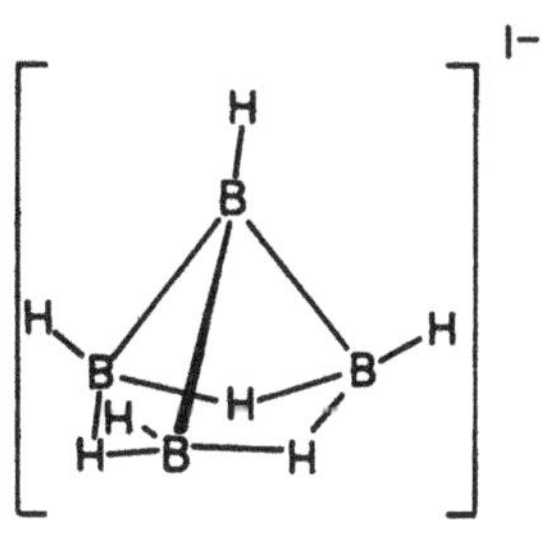

Fig. 9. Proposed Structure of the $[B_4H_7]^-$ Ion

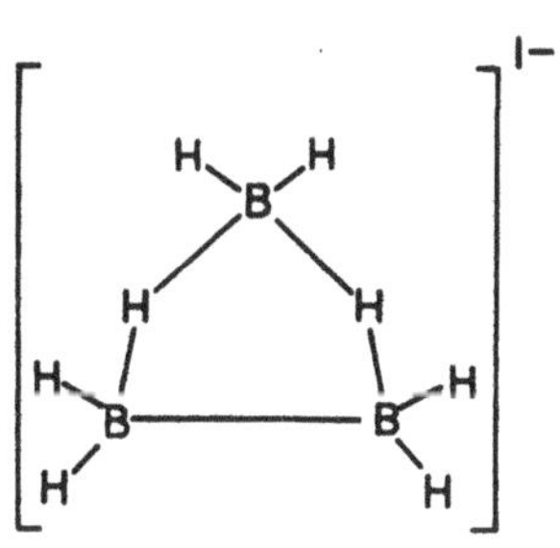

Fig. 10. Structure of the $[B_3H_8]^-$ Ion

4.3.2 The N = 7 Class; $[B_6H_6]^{2-}$, B_5H_9, B_4H_{10} and $B_3H_5L_3$

The hexahydrohexaborate(2—) ion has the expected structure of a regular octahedron. The species has a B—B distance of 1.69 Å and a B—H distance of 1.11 Å determined by X-ray diffraction[106].

The structure of *nido*-pentaborane(9) is obtained by the removal of a single vertex from $[B_6H_6]^{2-}$ and is shown in Fig. 11. Since electron- and X-ray diffraction, and microwave spectroscopy have all been used to determine structural parameters, the data are all summarized in Table 3.

The structure of tetraborane(10) is obtained from $[B_6H_6]^{2-}$ by the removal of two adjacent vertices and from B_5H_9 by the removal of a basal vertex. The resulting species has the expected *arachno*-structure and is shown in Fig. 12.

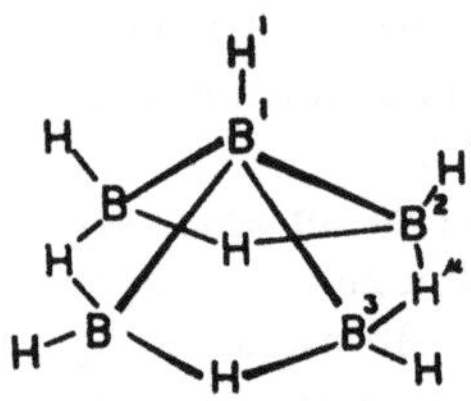

Fig. 11. Structure of B_5H_9

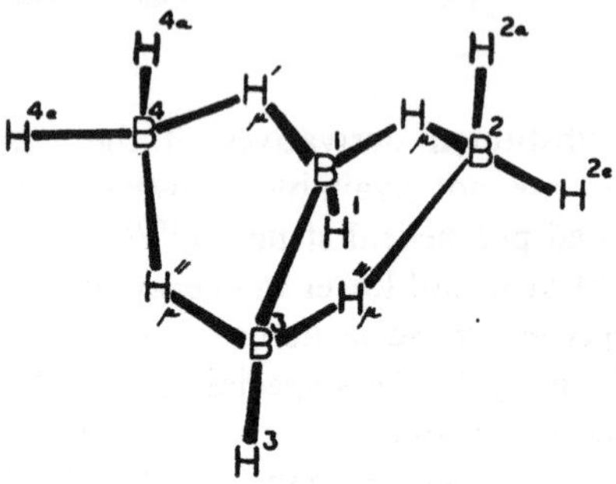

Fig. 12. Structure of B_4H_{10}

Table 3. Structural Parameters for B_5H_9

Bond Distances	Microwave	Electron Diffraction	X-Ray Diffraction
B^1—B^2	1.69	1.70	1.65
B^2—B^3	1.80	1.80	1.75
B^1—H^1	1.22[a]	1.23[a]	1.14
B^2—H^2	1.22[a]	1.23[a]	1.07
B^2—H_μ	1.35	1.36	1.27
Bond Angles			
B^1—B^2—H^2	136°	120 ± 2	130 ± 2°
B^2—H^1—B^3	—	—	64°
B^2—H_μ—B^3	—	—	87°
B^3—B^2—H_μ	—	—	46.4 ± 1°
B^2—B^1—H^1	—	—	131°
B^1—B^2—B^3	—	—	58°
B^3—B^2—H^2	—	—	107°
dihydral angle			
$B^1B^2B^3$—$B^2B^3H_\mu$	196 ± 2°	187 ± 10°	190 ± 5°

[a] The B^1—H^1 and B^2—H^2 distances were assumed to be equal in the two studies, respectively.

The molecule is a butterfly structure, H^{2e} and H^{4e} are the equatorial hydrogens of the BH_2 groups and H^{2a} and H^{4a} are the axial hydrogens. A more up-to-date terminology would refer to H^{2e} and H^{4e} as *exo*-hydrogens and H^{2a} and H^{4a} as *endo*-hydrogens Structural parameters are given in Table 4.

Table 4. Structural Parameter for B_4H_{10} [110)]

Bond Distances (Å)					
B^1—B^2	1,85	B^1—H^1	1.15	B^1—H_μ	1.13
B^2—B^3	1.84	B^3—H^3	1.05	B^1—H_μ^1	1.17
B^3—B^4	1.84	B^2—H^{2a}	1.14	B^3—H_μ^{111}	1.19
B^4—B^1	1.85	B^4—H^{4a}	1.03	B^3—H_μ^{11}	1.14
average	1.84			average	1.16
B^1—B^3	1.71	B^2—H^{2e}	1.09	B^2—H_μ	1.43
B^2—B^4	2.80	B^4—H^{4e}	1.12	B^4—H_μ^1	1.30
		average	1 10	B^2—H_μ^{111}	1.41
				B^4—H_μ^{11}	1.34
				average	1.37

Bond Angles (°)			
H^1—B^1—B^3	118	H^{2e}—B^2 to $B^1B^2B^3$	112
B^1—B^2—B^3	56.6	H^{2a}—B^2 to $B^1B^2B^3$	122
B^1—B^3—B^2	61.7	$B^1H_\mu B^2$ to $B^1B^2B^3$	170
B^2—B^3—B^4	98	$B^1B^3B^4$ to $B^1B^2B^3$	118

The structural relationship between the *closo-*, *nido-* and *arachno*-species containing seven skeletal bonding electron pairs is shown in Fig. 13.

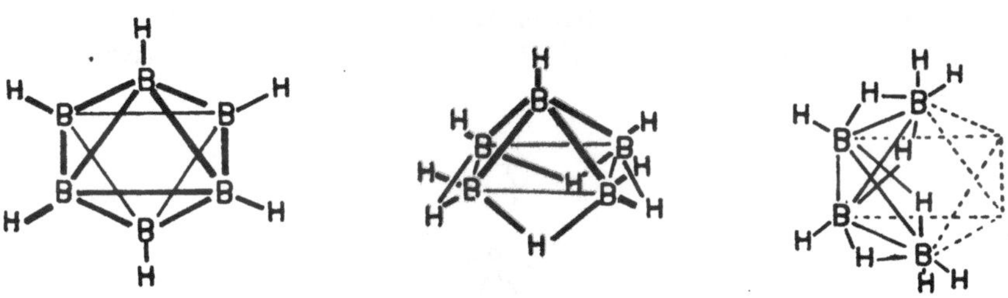

Fig. 13. Structural Relationship Between $[B_6H_6]^{2-}$, B_5H_9 and B_4H_{10}

A *hypho*-borane adduct, $B_3H_53P(CH_3)_3$ has been prepared recently which belongs to the N = 7 class [111)]. Boron-11 NMR data support the structure shown in Fig. 14.

4.3.3 The N = 8 Class; $[B_7H_7]^{2-}$, B_6H_{10}, B_5H_{11} and $B_4H_8L_2$

The species $[B_7H_7]^{2-}$ is well characterized and although an X-ray diffraction study has not been made, the D_5h pentagonal bipyramidal structure, shown in Fig. 15, is suggested by ^{11}B NMR data" [112)]. This structure is also suggested by the fact that the isoelectronic *closo*-carborane $C_2B_5H_7$, has been shown to have the pentagonal bipyramid structure [113)].

nido-Hexaborane(10) has the expected pentagonal-pyramidal structure obtained by the removal of an apical vertex from $[B_7H_7]^{2-}$. The structure is shown in Fig. 16.

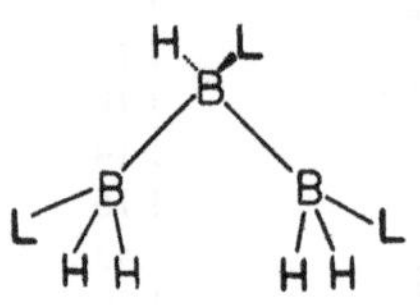

Fig. 14. The Proposed Structure of *hypho*-$B_3H_5 3P(CH_3)_3$

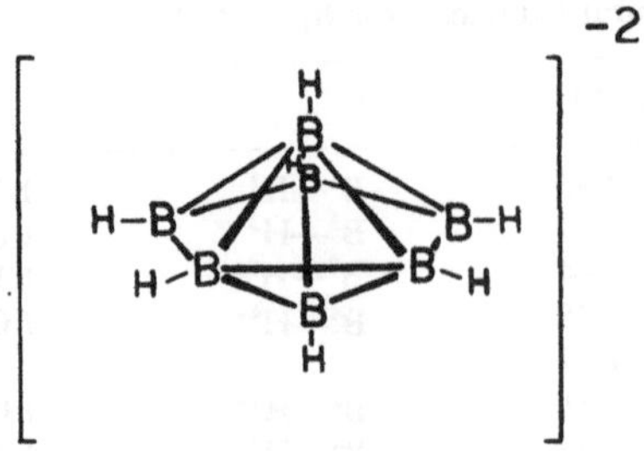

Fig. 15. Proposed Structure of the $[B_7H_7]^{2-}$ Ion

The molecule shows C_s symmetry with an unusually short non-bridged B—B bon in the basal pentagon; this latter distance, B^4—B^5, is 1.63 Å. The other interestin feature is that the hydrogen bridges are asymmetric, the average distances ar B^5—H^9 and B^2—H^{10} = 1.17 Å and B^6—H^{10} and B^6—H^9 = 1.30 Å. The B—H terminal distances are quite normal, the basal bridged B—B distances are ≈1.77 Å and the B_{apex}—B_{base} distances are ≈1.76 Å[114].

The structure of B_5H_{11} is the classical *arachno*-structure obtained by the removal of a basal boron atom from B_6H_{10}. The molecular structure is that of a shallow square pyramid with an open side. The structure, shown in Fig. 17, which was determined by low temperature X-ray diffraction, is that of an asymmetric molecule with C_1 symmetry[115]. The terminal hydrogen H^2 is actually bridging to one of the basals borons, B^2 or B^5. The distances, B^2—H^2 and B^5—H^2 are clearly asymmetric, being 1.55 Å and 1.83 Å, respectively.

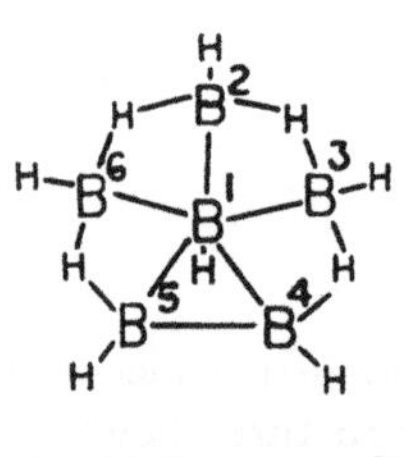

Fig. 16. Structure of B_6H_{10}

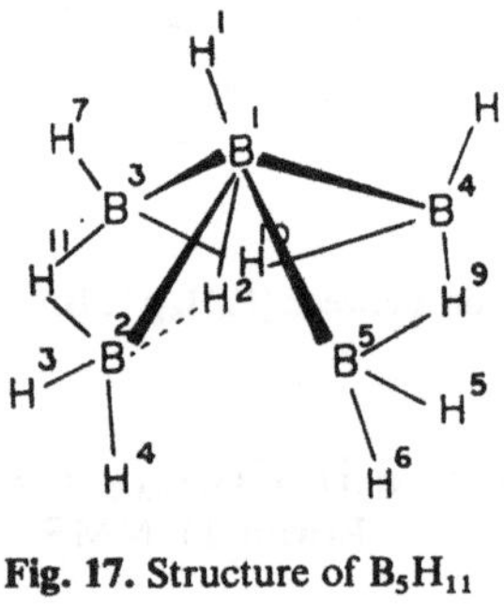

Fig. 17. Structure of B_5H_{11}

Table 5. Selected Bond Distances in B_5H_{11}

Atoms	Distance in Å	Atoms	Distance in Å
B^1—B^2	1.87	B^5—H^2	1.83
B^1—B^3	1.72	B^2—H^{11}	1.34
B^2—B^3	1.80	B^3—H^{11}	1.19
B^3—B^4	1.79	B^3—H^{10}	1.25
B^2—H^2	1.55	B^4—H^{10}	1.28

Some bond distances are given in Table 5. The B—$H_{terminal}$ distances not listed are normal. The structure, in solution, is assumed to be stereochemically non-rigid; a fluxional process occurs with the unique bridging hydrogen moving between B^2 and B^5. This assumption is based on the ^{11}B NMR spectrum which indicates Cs symmetry[116].

Base adduct analogues B_4H_82L of the *hypho*-borane $[B_4H_{10}]^{2-}$ have been known for some time[76] although only recently have NMR data been obtained which give an indication of structure. Kodama[76d] and Colquhoun[76e] have independently suggested structures and these are indicated in Fig. 18. The second structure in 18 (a) is essentially the same as that in (b). The cyclic structure (a) has been suggested for the unknown B_4H_{12} [117].

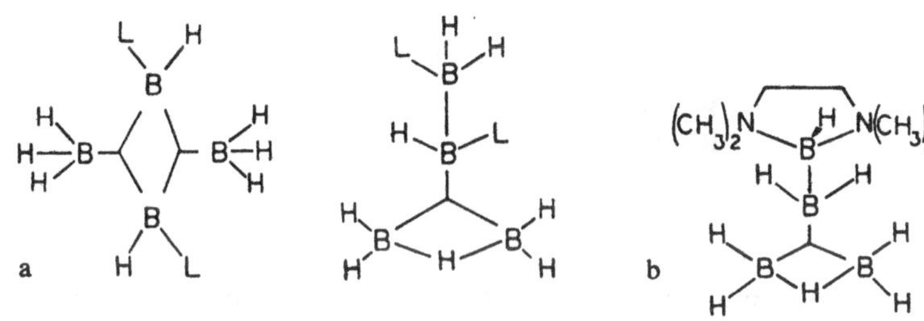

Fig. 18a. Suggested Structures for $B_4H_82P(CH_3)_3$; **b** Suggested Structure of $B_4H_8[(CH_3)_2NCH_2]_2$

4.3.4 The N = 9 Class; $[B_8H_8]^{2-}$, $B_7H_{11}Fe(CO)_4$, B_6H_{12} and $B_5H_9L_2$

The solid state structure of [*closo*-$B_8H_8]^{2-}$ is very similar to that of an idealized triangulated dodecahedron and is shown in Fig. 19[118]. Bond distances not related by molecular symmetry are given in Table 6.

The species is fluxional in solution[119]. The energy separation between the three related structures, dodecahedron D_{2d}, bicapped trigonal prism C_{2v} and square anti-prism D_{4d} has been shown to be very small by ^{11}B NMR spectroscopy.

Neither the *nido*-B_7H_{11} species nor any of its isoelectronic analogues have been observed. Of interest, however, is the species $B_7H_{11}Fe(CO)_4$ [120] which can formally be regarded as an adduct between the acid $Fe(CO)_4$ and a basic B—B bond in B_7H_{11}. The structure of $B_7H_{11}Fe(CO)_4$ is predicted to be related to that of $[B_7H_{12}Fe(CO)_4]^-$ [121] and $[B_7H_{12}]^-$ [122] which are discussed later. The structure predicted from first principles for B_7H_{11}, is that obtained by removal atoms B^3 or B^4 from the $[B_8H_8]^{2-}$ structure. However, as Williams[123] has pointed out, this structure involves bridge-hydrogen congestion, thus a structure which is effectively μ-BH_2—B_6H_9 is the most likely and this is discussed in the subsequent discussion of $[B_7H_{12}]^-$.

The structure of *arachno*-B_6H_{12} is that of an isosahedral belt which may be obtained by the removal of a boron atom from either of the predicted B_7H_{11} structures, discussed above to produce the most "open" structure. Study of B_6H_{12} by X-ray diffraction has proved unsuccessful owing to glass formation, however, the structure shown in Fig. 20 is consistent with ^1H and ^{11}B NMR measurements[124].

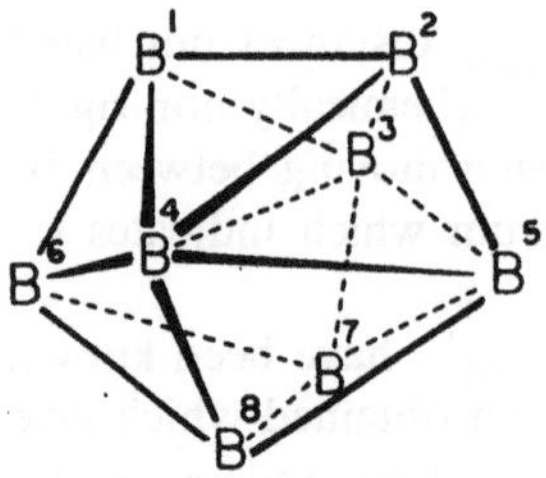

Fig. 19. Structure of the $[B_8H_8]^{2-}$ Ion (terminal H's omitted)

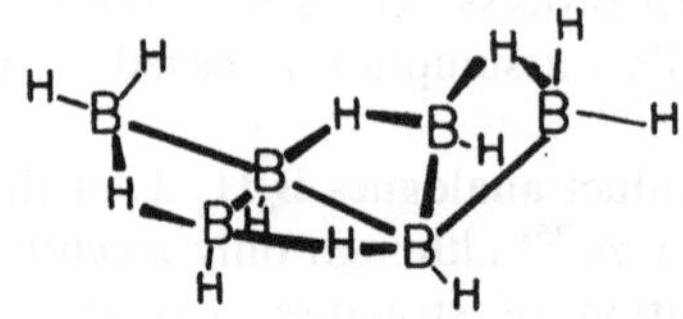

Fig. 20. Proposed Structure of the B_6H_{12}

Table 6. Bond distances in $[B_8H_8]^{2-}$

Atoms	Distance (Å)	Atoms	Distance (Å)
B^1—B^2	1.56	B^1—H^1	1.11
B^1—B^3	1.76	B^3—H^3	1.25
B^3—B^7	1.72		

The above structure is in contrast to that of the isoelectronic species $[B_6H_{11}]^-$ whose structure was predicted by Shore and coworkers from NMR data[125]. This structure is essentially that of B_5H_8 with a BH_3 inserted into the basal B—B bond, i.e., $[\mu\text{-}BH_3B_5H_8]^-$.

There are several examples of species isoelectronic with *hypho*-B_5H_{13}. Lipscomb[126], several years ago, predicted the valence structure of $[B_5H_{11}]^{2-}$ as that shown in Fig. 21.

The only binary *hypho*-borane thus far observed is the species $[B_5H_{12}]^-$ prepared by Shore and coworkers[122]. Although a crystal structure of the species has not been determined, NMR data suggest the structure shown in Fig. 22 which is similar to that proposed by Lipscomb.

Structural data are available for several *hypho*-pentaboranes which are $B_5H_9 2L$ adduct species[77]. Shore and coworkers have determined the crystal and molecular structure of $B_5H_9\,2\,P(CH_3)_2$ [77a]. The structure is a shallow pyramid (C_1 point symmetry) with basal borons distorted from a square pyramid. The two trimethyl-

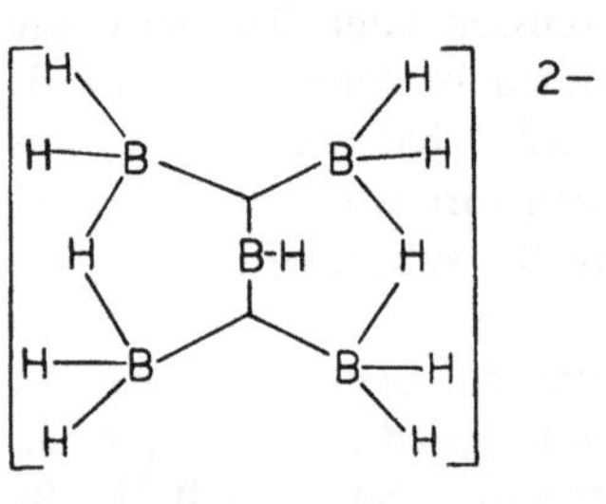

Fig. 21. Proposed Structure of the $[B_5H_{11}]^{2-}$ Ion

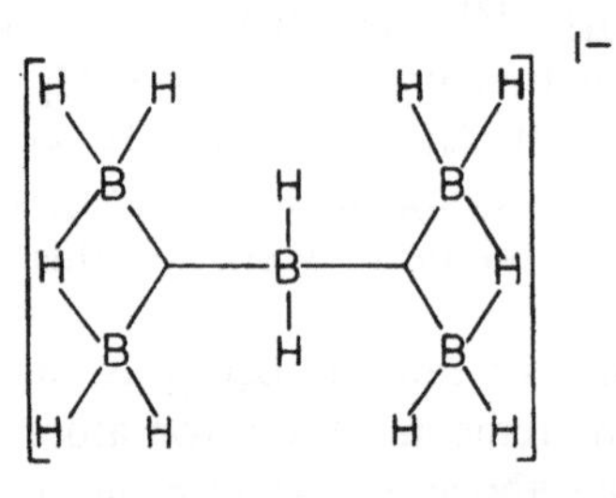

Fig. 22. Proposed Structure of the $[B_5H_{12}]^-$ Ion

phosphine molecules coordinate one to the apical boron and the other to a basal boron as shown in Fig. 23.

The structures of the bidentate ligand adducts $B_5H_9L_2$, where L_2 = $[(C_6H_5)_2P]_2CH_2$, $[(C_6H_5)_2PCH_2]_2$ or $[(CH_3)_2NCH_2]_2$, have been determined by Allcock and coworkers[77b]. The structures of the first two are analogous to that of B_5H_9 2 $P(CH_3)_3$ with the diphosphine ligand bridging between apical and basal borons whereas the structure of the 1,2-bis(dimethylamino)ethane adduct indicates coordination of the two nitrogen atoms to the same basal B atom as shown in Fig. 24.

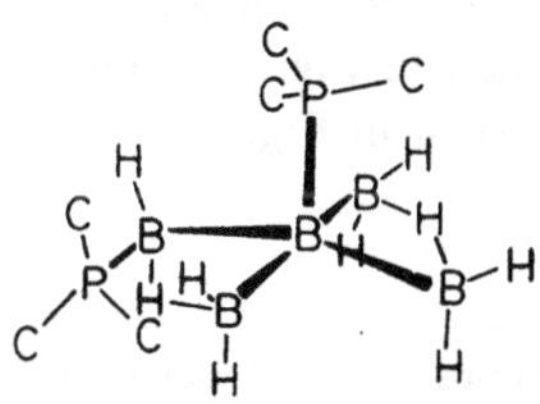

Fig. 23. Structure of $B_5H_92P(CH_3)_3$ (methyl H's omitted)

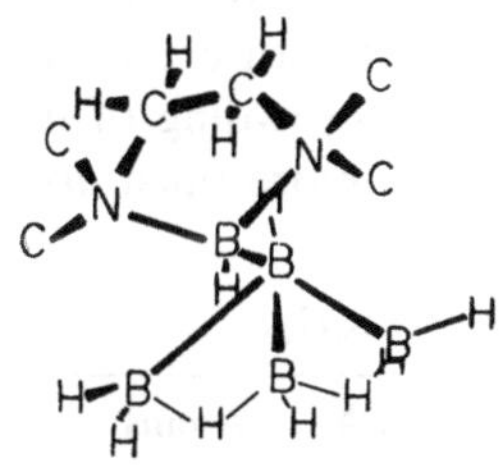

Fig. 24. Structure of $B_5H_9[(CH_3)_2NCH_2]_2$ (methyl H's omitted)

In both $B_5H_9[(C_6H_5)_2P]_2CH_2$ and $B_5H_9[(C_6H_5)_2PCH_2]_2$ the angles between opposite triangular B_3 faces are 130.4° and 126.5°, respectively, compared with 90 to 91° in B_5H_9. The structure ought to reflect that of the $[B_8H_8]^{2-}$ dodecahedron, i.e., the boron atoms should occupy all but three of the eight vertices in $[B_8H_8]^{2-}$. Allcock and coworkers suggest that these structures appear to be derived from a hexagonal bipyramid. In $B_5H_9[(C_6H_5)_2P]_2CH_2$ the dimensions of the hexagonal bipyramid are center to apex height 0.64 Å and edge lengths of 1.81 Å (equatorial) and 1.77 Å (apex to base) and in $B_5H_9[(C_6H_5)_2PCH_2]_2$ the corresponding lengths are 0.645, 1.82 and 1.78 Å. The non-bonded distances of adjacent borons in B_5H_9 2 $P(CH_3)_3$ are comparable to the corrresponding distances in B_5H_{11}.

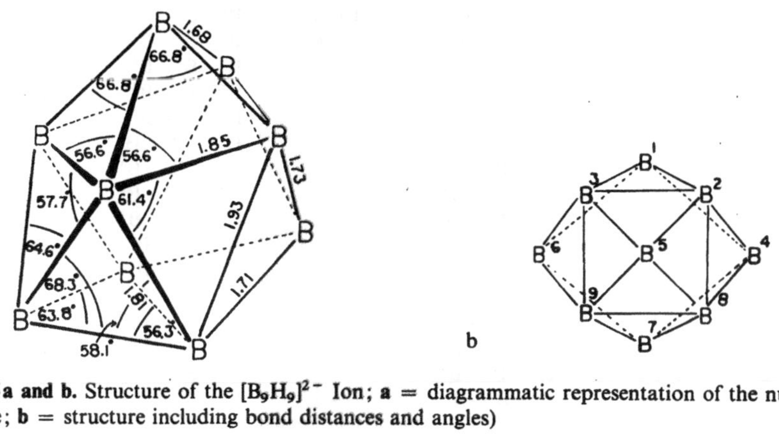

a b

Fig. 25a and b. Structure of the $[B_9H_9]^{2-}$ Ion; a = diagrammatic representation of the numbering scheme; b = structure including bond distances and angles)

4.3.5 The N = 10 Class; $[B_9H_9]^{2-}$, B_8H_{12}, $[B_7H_{12}]^-$ and $B_6H_{10}2L$

The structure of the $[B_9H_9]^{2-}$ ion is that of a tricapped trigonal prism which has C_{2v} symmetry exactly although it may be regarded to possess idealized D_{3h} symmetry[127]; Fig. 25 also shows the bond distances and angles.

The predicted structures for B_8H_{12}, a *nido*-borane in terms of electron counting, are obtained by the removal of a 5- or 6-coordinate vertex from $[B_9H_9]^{2-}$, i.e., (a) or (b) in Fig. 26; however, the observed structure of B_8H_{12} in the solid state is the *arachno*-structure shown in Fig. 27[128]. Some bond distances and angles are given in Tables 7 and 8.

It is noteworthy that the predicted *nido*-structures for B_8H_{12} are non-icosahedral whereas the observed structure may be considered to be derived from an icosahedron. Apparently the crowding of bridge hydrogens, as seen in Fig. 26a and b, mitigate in favor of the more open *arachno*-like structure shown in Fig. 27.

Table 7. Bond Distances in B_8H_{12}

Atoms	Distance in Å	Atoms	Distance in Å	Atoms	Distance in Å
B^1—H^1	1.18	B^3—H^{12}	1.35	B^5—H^{10}	1.28
B^1—B^3	1.71	B^3—B^8	1.68	B^5—H^9	1.51
B^1—B^7	1.79	B^3—B^4	1.83	B^5—B^6	1.72
B^1—B^2	1.84	B^3—H^9	2.01	B^8—H^{12}	1.32
B^2—H^2	1.13	B^4—H^4	1.12		
B^2—B^5	1.72	B^4—H^9	1.27		
B^2—B^4	1.81	B^4—B^5	1.80		
B^3—H^3	1.11	B^5—H^5	1.11		

Table 8. Selected Bond Angles in B_8H_{12}

Angle	Degrees	Angle	Degrees	Angle	Degrees
H^1—B^1—B^3	123.8	H^2—B^2—B^4	123.3	H^9—B^4—B^5	55.4
H^1—B^1—B^7	120.6	H^2—B^2—B^1	118.2	H^9—B^4—B^2	104.4
H^1—B^1—B^2	113.5	B^3—B^2—B^6	59.9	H^9—B^4—B^3	78.5
B^3—B^1—B^8	58.9	B^5—B^2—B^4	61.6	B^1—B^4—B^5	109.0
B^3—B^1—B^7	108.6	B^5—B^2—B^7	107.4	B^1—B^4—B^2	61.4
B^3—B^1—B^4	68.9	B^5—B^2—B^1	110.6	B^1—B^4—B^3	56.2
B^3—B^1—B^2	113.6	B^4—B^2—B^7	113.2	B^5—B^4—B^2	56.9
B^7—B^1—B^4	104.8	B^4—B^2—B^1	58.9	B^5—B^4—B^3	119
B^7—B^1—B^2	60	B^8—B^3—B^1	60.6	B^2—B^4—B^3	109.2
H^2—B^2—B^5	122.3	B^8—B^3—B^4	108.4	H^{10}—B^5—H^9	87.1
H^3—B^3—H^{12}	112	B^8—B^3—H^9	107.3	H^{10}—B^5—B^6	46.9
H^3—B^3—B^8	134.3	B^1—B^3—B^4	60.8	H^{10}—B^5—B^2	101.5
H^3—B^3—B^1	138.7	H^4—B^4—H^9	109.2	H^{10}—B^5—B^4	115.8
H^3—B^3—B^4	116.7	H^4—B^4—B^1	123.8	H^9—B^5—B^6	113.3
H^3—B^4—H^9	112.5	H^4—B^4—B^5	114	H^9—B^5—B^2	99.1
H^{12}—B^3—B^8	50.3	H^4—B^4—B^2	126	H^9—B^5—B^4	44.2
H^{12}—B^3—B^1	104.2	H^4—B^4—B^3	118	H^5—B^5—H^{10}	113
H^{12}—B^3—B^4	114.9	H^9—B^4—B^1	117	H^5—B^5—H^9	108.5
H^5—B^5—B^6	130.6	H^5—B^5—B^4	110	B^6—B^5—B^4	108.4
H^5—B^5—B^2	136	B^6—B^5—B^2	60.2	B^2—B^5—B^4	62.8

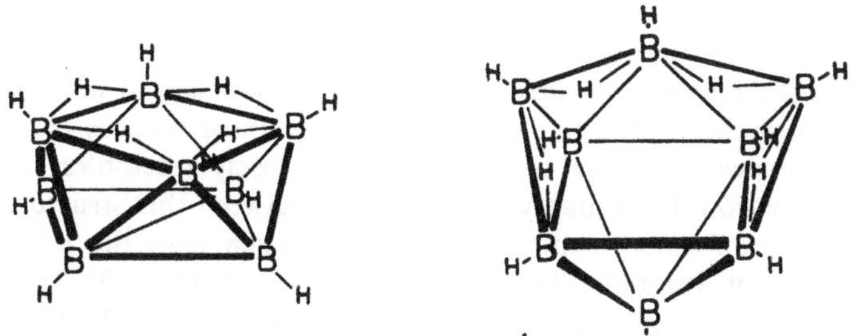

a b

Fig. 26a and b. Possible *nido*-Structures for B_8H_{12}

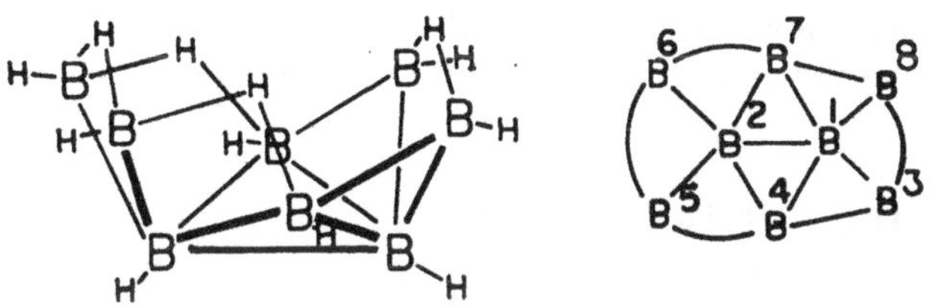

Fig. 27. Molecular Geometry and Numbering Scheme for B_8H_{12} (terminal hydrogens are numbered the same as their parent borons and the four bridges are numbered 9 through 12 starting with the 4,5-position and counting clockwise around the structure)

The parent *arachno*-B_7H_{13} species has never been observed although its conjugate base, $[B_7H_{12}]^-$, has been prepared by the addition of BH_3 to $[B_6H_9]^-$ by Shore and coworkers[129,122]. The proposed structure of the species is consistent with one obtained by the removal of atoms B^1 and B^2 from $[B_9H_9]^{2-}$ in Fig. 25 or by the insertion of a BH_3 group into a B—B bond of the $[B_6H_9]^-$ ion as shown in Fig. 28. The structure in Fig. 28 is consistent with boron-11 and proton NMR data[122] although the solid state structure is not yet available. The structure proposed is similar to that of $[(CO)_4FeB_7H_{12}]^-$ whose structure is known and is shown in Fig. 29[121].

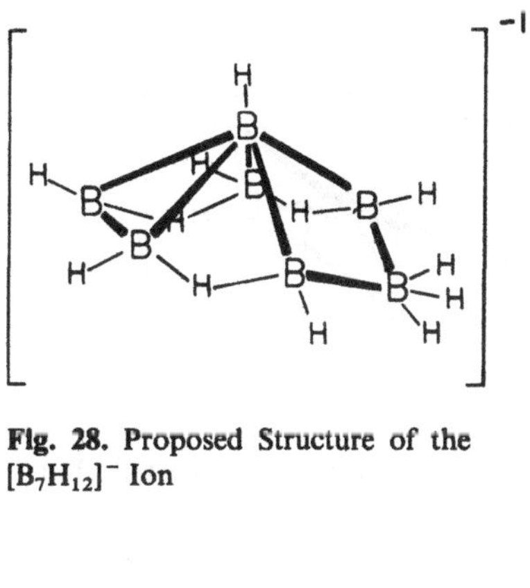

Fig. 28. Proposed Structure of the $[B_7H_{12}]^-$ Ion

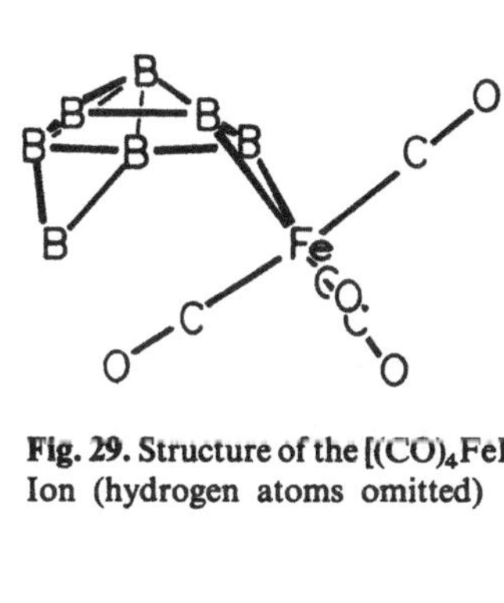

Fig. 29. Structure of the $[(CO)_4FeB_7H_{12}]^-$ Ion (hydrogen atoms omitted)

The species $B_6H_{10}[P(CH_3)_3]_2$, prepared by Shore and coworkers, has been characterized structurally by X-ray crystallography[78]. The structure, which is shown in Fig. 7, establishes the species as a derivative of the *hypho*-class of boranes and is consistent with a structure obtained by the removal of three vertices from the $[B_9H_9]^{2-}$ ion. The framework of the six boron atoms closely resembles the equatorial belt of an icosahedron. Presumably the difference between this structure and that of B_6H_{12} is that the former is more flattened and open than the latter whose detailed structure is not available. The bond distances in B_6H_{10} 2 $P(CH_3)_2$ are consistent with those in other boranes. The range of B—B distances is 1.745 to 1.841 Å and within the triangular faces, B—B—B angles vary from 57.4 to 62.8° with an average value of 61.6°. The B—H bonds vary between 0.99 and 1.19 Å, one hydrogen bridge is symmetric with B—H_μ distances of 1.23 Å while the other is asymmetric with B—H_μ distances of 1.05 and 1.29 Å. Dihedral angles between the triangular faces are not provided, however, the angles B^1—B^2—B^3 and B^6—B^5—B^4 which are 122.8° and 123.8°, respectively, give an indication of the flattened nature of this *hypho*-structure.

4.3.6 The N = 11 Class; $[B_{10}H_{10}]^{2-}$, $[B_9H_{12}]^-$ and B_8H_{14}

The structure of [*closo*-$B_{10}H_{10}]^{2-}$, which is shown in Fig. 30, is that of a bicapped Archimedian antiprism[30]. The boron framework may be inscribed in a prolate spheroid with semi-axes a = 1.88 and b = c = 1.43 Å [130]. The bond distances and angles are quite normal for this type of structure.

The species *nido*-B_9H_{13} is unknown and Williams[70] has pointed out this is not surprising when one studies the predicted structure. Removal of a 6-coordinate equatorial boron atom from $[B_{10}H_{10}]^{2-}$ generates a *nido*-structure with a puckered five-membered face in which the placement of four bridge hydrogens is difficult. The removal of a proton from this structure relieves this bridge crowding yielding an acceptable structure. Todd and coworkers[131] reported the ^{11}B NMR spectrum of $[B_9H_{12}]^-$ which is consistent with the predicted *nido*-structure for $[B_9H_{12}]^-$ which is shown in Fig. 31. The carborane isoelectronic with B_9H_{13}, i.e., $C_2B_7H_{11}$, does not have the problem of hydrogen bridge crowding and thus an X-ray diffraction

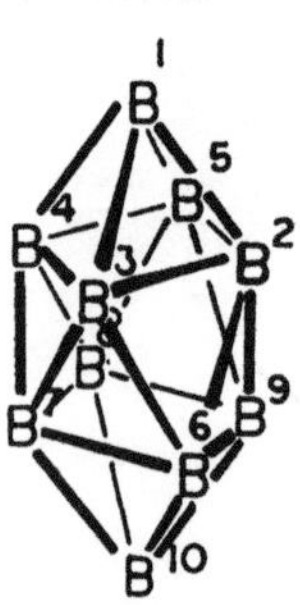

Fig. 30. Boron Atom Positions and Numbering Scheme for the $[B_{10}H_{10}]^{2-}$ Ion (hydrogens omitted)

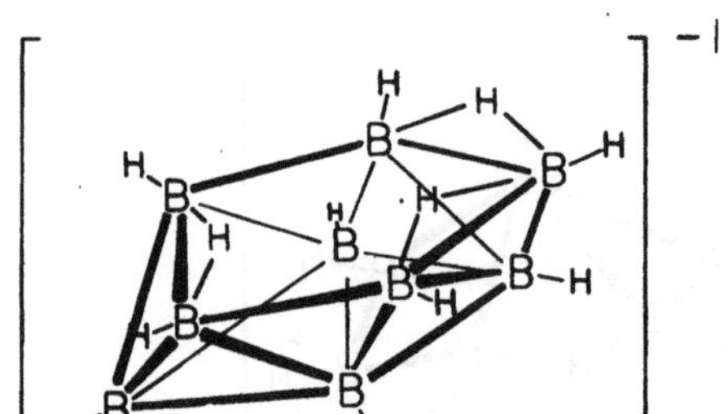

Fig. 31. Proposed Structure for the $[B_9H_{12}]^-$ Ion

study of $C_2B_7H_9(CH_3)_2$ indicates the expected bicapped Archimedian antiprism structure with the 6-position removed[132]. The structure shows very little deviation from the *closo*-structure, i.e., the remaining vertex positions occupy essentially the same positions as they would in $C_2B_7H_9$.

The unsubstituted *nido*-carborane $C_2B_7H_{11}$ exhibits an NMR spectrum indicating a structure similar to that of the dimethyl species, in terms of vertex positions, but the hydrogen positions are different[133]. Apparently there is a BH_2 group and only one bridging hydrogen.

The *arachno*-species B_8H_{14} is known although there is little structural information available. The only structural data available are those from NMR spectra[134] which suggest that species has the expected *arachno*-structure shown in Fig. 32.

Neither the *hypho*-B_7H_{15} species nor its isoelectronic analogues are known.

4.3.7 The N = 12 Class; $[B_{11}H_{11}]^{2-}$, $B_{10}H_{14}$, *n*-B_9H_{15}, *i*-B_9H_{15} and B_8H_{16}

The structure of [*closo*-$B_{11}H_{11}]^{2-}$ has not been determined however its structure is assumed to be the same as that of the isoelectronic species $B_9C_2H_{11}$ whose structure has been shown to be an octadecahedron with C_{2v} symmetry[136] as shown in Fig. 33. Bond distances are quite normal for carboranes. NMR spectra of solutions of $[B_{11}H_{11}]^{2-}$ indicate that the species is fluxional[137]; it is suggested that the species undergoes a low energy internal framework rearrangement involving a C_{5v} intermediate structure.

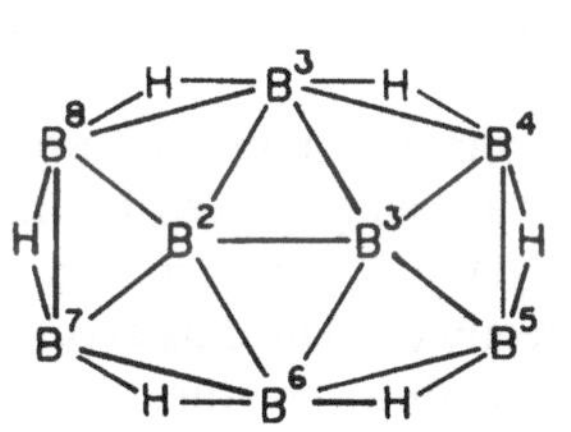

Fig. 32. Proposed Structure of B_8H_{14} (terminal hydrogens omitted)

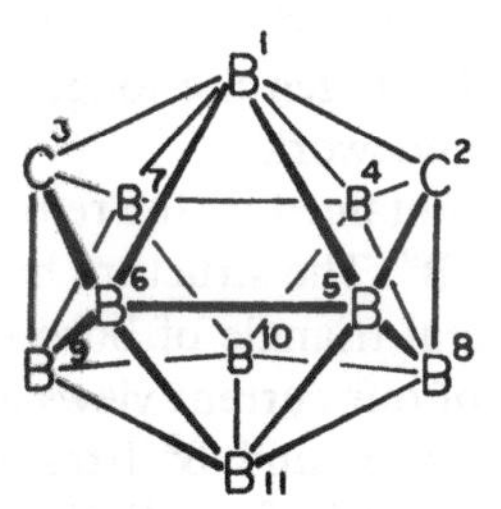

Fig. 33. Atom Positions in $B_9C_2H_9(CH_3)_2$ (vertex positions only are shown)

The structure of $B_{10}H_{14}$ is that predicted for the *nido*-species derived by removing the unique 7-coordinate vertex from $[B_{11}H_{11}]^{2-}$. The structure[138,139], shown in Fig. 34, has the four bridge hydrogen atoms symmetrically disposed on the open surface. Boron-boron and boron-hydrogen distances are normal, the hydrogen bridges are unsymmetrical, with shorter distances towards the center of the molecule.

arachno-B_9H_{15} exists as two isomers. Removal of a highest coordination vertex from the open-face of $B_{10}H_{14}$, i.e., B^5, or removal of B^1 and B^4 from $[B_{11}H_{11}]^{2-}$ yields *n*-B_9H_{15} which is shown in Fig. 35 [140].

The structure may also be considered to approximate a fragment of an icosahedron, derived by the removal of three connected boron atoms not forming an equilateral triangle, although this description is confusing in the context of this discussion.

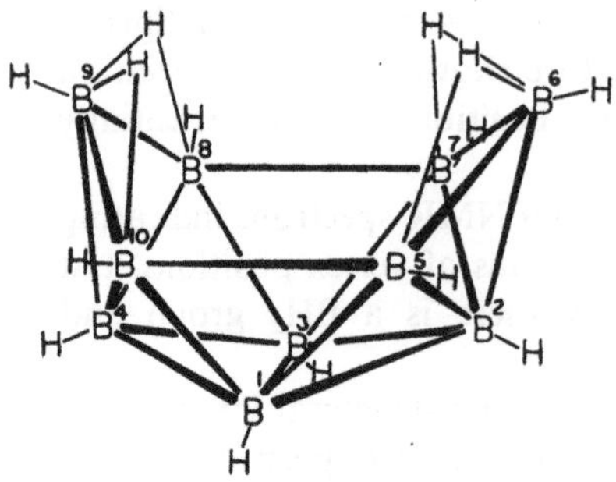

Fig. 34. Structure of $B_{10}H_{14}$

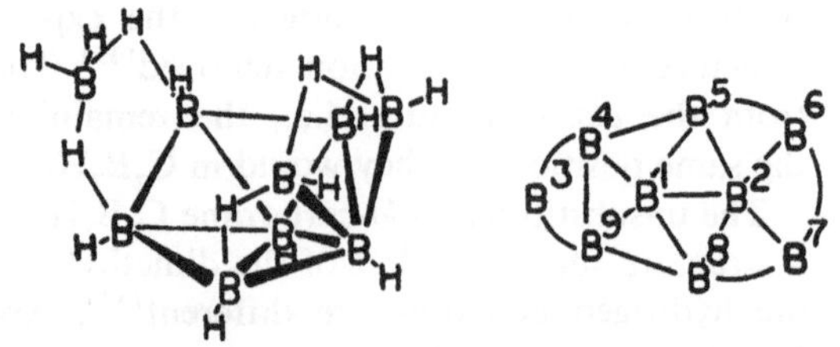

Fig. 35. Molecular Structure and Numbering Scheme for n-B_9H_{15} (terminal hydrogens bonded to B^1 through B^9 are similarly numbered; H^{10} is bonded to B^3 and the bridge hydrogens are numbered H^{11} to H^{15} starting at position B^3—B^4 and moving clockwise around the structure)

Again bond distances and angles are quite normal for boranes. The disposition of the BH_2 group relative to the remainder of the structure which resembles B_8H_{12} may be seen by comparing the bond angles B^5—B^4—B^3 (115.3°) and B^4—B^5—B^6 (122.1°)[141)]. The resemblence of the structure without the BH_2 group, to B_8H_{12} has chemical significance since n-B_9H_{12} is known to lose BH_3 to form B_8H_{12} [23)]. The bridge hydrogens between B^5 and B^6 and between B^7 and B^8 are almost symmetrically placed between the respective boron atoms whereas the bridge hydrogens between the B_8 unit and the BH_2 group are unsymmetrical. The respective distances are B^4—H^{11} = 1.17 Å and B^3—H^{11} = 1.39 Å, i.e., the bonds between the BH_2 group and the bridge hydrogen are longer than the bond between the same bridge hydrogen and the B_8 framework.

i-Nonaborane(15) is obtained by the removal of a low coordination number vertex from $B_{10}H_{14}$ i.e. B^6 or B^9. The structure has been described as being derived by the removal of an equilateral triangle of boron atoms from $[B_{12}H_{12}]^{2-}$ although this is inappropriate in view of our current views and is more correctly described as arising from the removal atoms B^1 and B^2 from $[B_{11}H_{11}]^{2-}$. Although a structure determination has not been made for i-B_9H_{15}, NMR data[142)] are consistent with the structure described above and shown in Fig. 36. This structure is very reasonable since i-B_9H_{15} is prepared by protonation of $[B_9H_{14}]^-$, and since the positions of the boron atoms in $[B_9H_{14}]^-$ are the same as those proposed for i-B_9H_{15}. The structure of $[B_9H_{14}]^-$ has been described by Greenwood and coworkers[81)].

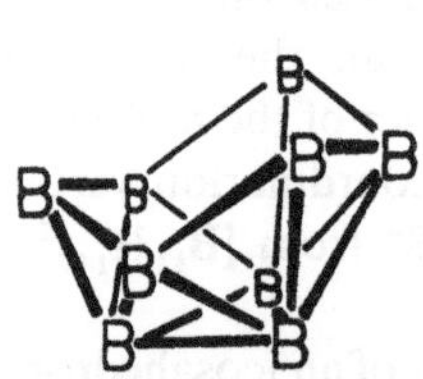

Fig. 36. Proposed Boron Positions in i-B_9H_{15}

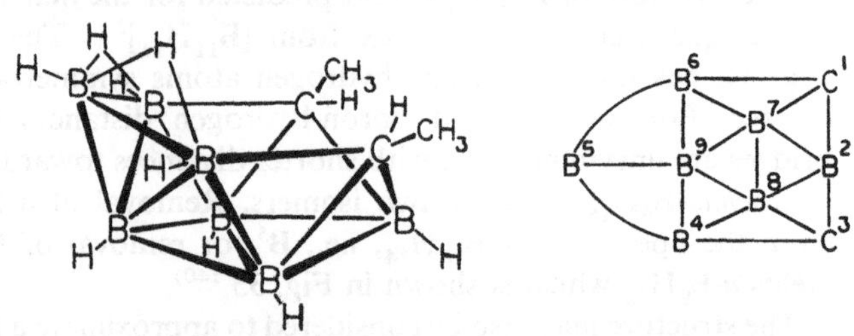

Fig. 37 a. Molecular Structure of 1,3-$(CH_3)_2C_2B_7H_{11}$; b Numbering Scheme

The structure of the carborane 1,3-$(CH_3)_2C_2B_7H_{11}$ has been determined[43] and is the same as that proposed for *i*-B_9H_{15}. The structure, shown in Fig. 37, again is derived from the formal removal of vertices numbered 1 and 2, in the $[B_{11}H_{11}]^{2-}$ *closo*-polyhedron.

A species with the formula B_8H_{16} has been prepared but structural information is unavailable[143].

4.3.8 The N = 13 Class; $[B_{12}H_{12}]^{2-}$, $[B_{11}H_{13}]^{2-}$ and $[B_{10}H_{14}]^{2-}$

The structure of [*closo*-$B_{12}H_{12}]^{2-}$ is that of a slightly distorted icosahedron as shown in Fig. 38. Several structural determinations using X-ray diffraction were performed between 1960 when Wunderlich and Lipscomb[31] first determined the structure and 1976[144]; however, the best data available are from a very recent structural determination by Shoham, Schomburg and Lipscomb[145]. This structure, which was determined in an X-ray diffraction study of $[B_{12}H_{12}]^{2-}$ $[N(C_2H_5)_3H^+]_2$, reveals a very small range of B—B distances with a mean value of 1.781 ± 0.002 Å. This is probably due to removal of distortion in the previously determined structures of alkali or alkaline earth metal salts by distribution of the positive charge over several atoms in the $[N(C_2H_5)_3H]^+$ cation.

Of the various possible *nido*-boranes containing eleven boron atoms, i.e., $B_{11}H_{15}$ and its derivatives, only $[B_{11}H_{14}]^-$ and $[B_{11}H_{13}]^{2-}$ have been positively identified. The species $B_{11}H_{15}$(1,4-dioxane)$_2$ has been prepared[146]; however, the correct electron count including two electrons per dioxane molecule places it in the *hypho*-class, isoelectronic with $B_{11}H_{19}$. Structural data are not available for $[B_{11}H_{14}]^-$; however, Lipscomb[74] suggested a structure based on replacement of a BH group in $[B_{12}H_{12}]^{2-}$ by a H_3^+ triangle perpendicular to the molecular axis. The structure of the isoelectronic $[B_{11}H_{13}]^{2-}$ has been determined and it is the expected *nido*-structure with the open face containing non-adjacent bridge hydrogens as shown in Fig. 39[147]. Bond distances are quite normal. The open face is not quite a regular pentagon as expected owing to the presence of the bridging hydrogens. Thus, bond distances are B^8—B^9 = 1.80 Å, B^7—B^8 = 1.89 Å and B^7—B^{11} = 1.82 Å and bond angles are B^8—B^9—B^{10} = 108° B^7—B^8—B^9 = 109° and B^8—B^7—B^{11} = 107°.

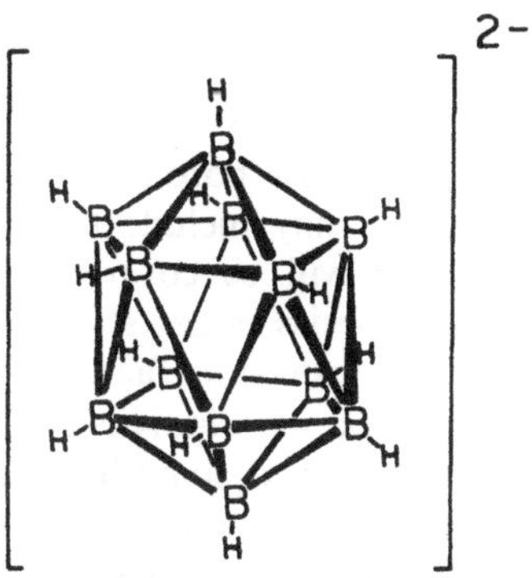

Fig. 38. Structure of the $[B_{12}H_{12}]^{2-}$ Ion

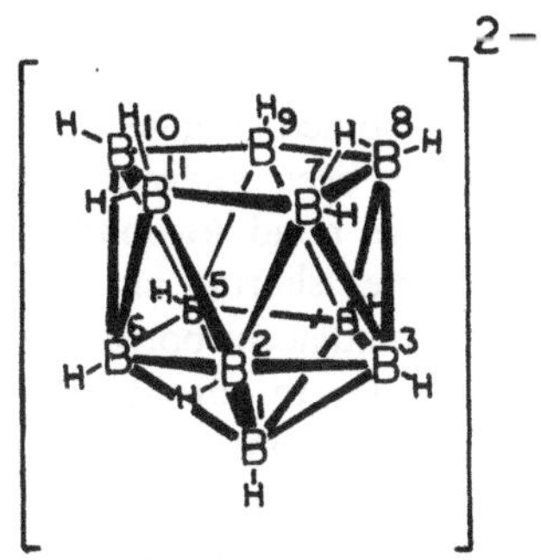

Fig. 39. The Structure of $[B_{11}H_{13}]^{2-}$

arachno-$B_{10}H_{16}$ is unknown; the well characterized $B_{10}H_{16}$ is the σ-bonded *conjuncto*-borane involving B_5H_8 units bonded as the 1,1-, 1,2- and 2,2-isomers[148]. The isoelectronic anions $[B_{10}H_{15}]^-$ [149] and $[B_{10}H_{14}]^{2-}$ are also known and since the latter is well characterized, it will be described. The structure of [*arachno*-$B_{10}H_{14}]^{2-}$ has been determined by X-ray diffraction[151] to be very similar to that of $B_{10}H_{14}$ as shown in Fig. 40.

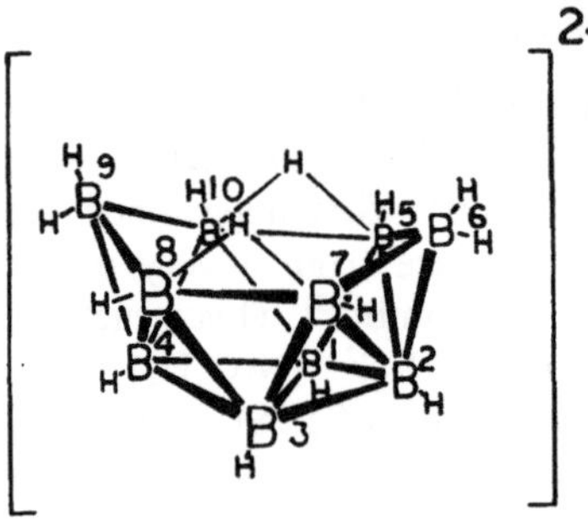

Fig. 40. The Structure of $[B_{10}H_{14}]^{2-}$

The structure indicates two hydrogen bridges disposed symmetrically between B^7 and B^8 and between B^5 and B^{10} and also endo-hydrogens on B^9 and B^6. It is not significantly different from $B_{10}H_{14}$. In view of the usual properties of related *nido*- and *arachno*-boranes this is surprising. Presumably the presence of the two bridge hydrogens and the endo-hydrogens serves to maintain the tight cluster almost as effectively as the four bridge hydrogens do in $B_{10}H_{14}$. The $[B_{10}H_{14}]^{2-}$ cluster is somewhat more open than $B_{10}H_{14}$ and this is indicated by the following comparison of the two structures. The bonds B^{10}—B^5, B^6—B^7 and B^2—B^6 tend to be shorter in $B_{10}H_{14}$ than in $B_{10}H_{14}^{2-}$, the distances being (in Å) 1.97, 1.88; 1.76, 187 and 1.72, 1.76, respectively. The angles at the 6.9 positions, i.e., B^5—B^6—B^7 are 105.3 and 103° respectively in $B_{10}H_{14}$ and $[B_{10}H_{14}]^{2-}$, again consistent with a slightly more open cluster for the latter:

4.4 Boranes Containing More than Twelve Boron Atoms

Almost all the known borane structures may be described in terms of fragments of polyhedra containing 5 to 12 vertices. This section deals with species which may be considered as joined fragments of boranes containing 12 or less boron atoms. Such species have been referred to as *conjuncto*-boranes and may include those joined by a single bond (σ-bonded *conjuncto*-boranes which will be covered in Section 4.5), species sharing an apex (*commo*-boranes) and species either sharing and edge (i.e., two adjacent boron atoms), a triangular face or a four-atom face.

It is not necessary to consider the icosahedron as the largest *closo*-polyhedron from which other boranes are derived. Lipscomb[55] has described hypothetical *closo*-boranes $[B_nH_n]^{2-}$ where n = 13 to 24. Furthermore, several examples of metallo-boranes and -carboranes are known. Callahan and Hawthorne[152] have described several *closo*-metallocarboranes containing 13 and 14 vertices and Maxwell and

Grimes[153] have prepared 11–13 atom *nido*-cages by the insertion of transition metal atom moieties into carboranes. Recently, 13 and 15-vertex *nido*-carboranes have been prepared by Hosmane and Grimes[154] by the oxidative fusion of carboranes and borane anions.

The material is arranged by increasing number of boron atoms in the species.

4.4.1 Tridecaborane(19), $B_{13}H_{19}$

The structure of $B_{13}H_{19}^{\bullet}$ [155] is that of a *conjuncto*-borane consisting of a hexaborane and a nonaborane fragment sharing two boron positions. The structure is shown in Fig. 41 a with the numbering scheme in the topological representation in Fig. 41 b. The structure involves the atoms B^3 and B^9 being shared between 6 boron and 9-boron units. The 6-boron unit, B^3—B^{13}—B^{12}—B^{11}—B^9—B^{10} with four bridging hydrogens is structurally identical to B_6H_{10} with the B—B bond of the latter forming part of a 9-boron species, B^3—B^4—B^5—B^6—B^7—B^8—B^9—B^1—B^2. The arrangement of the latter is similar to *n*-B_9H_5 except for the arrangement around the BH_2 group in *n*-B_9H_{15}. (An alternative view of the molecule, suggested by Lipscomb and associates[156], considers it to be a *conjuncto*-borane involving 8-boron and 7-boron units sharing a pair of boron atoms.) B^9, which is seven-coordinate, does not have a terminal hydrogen atom as is the case with other seven-coordinate borons in *conjuncto*-boranes.

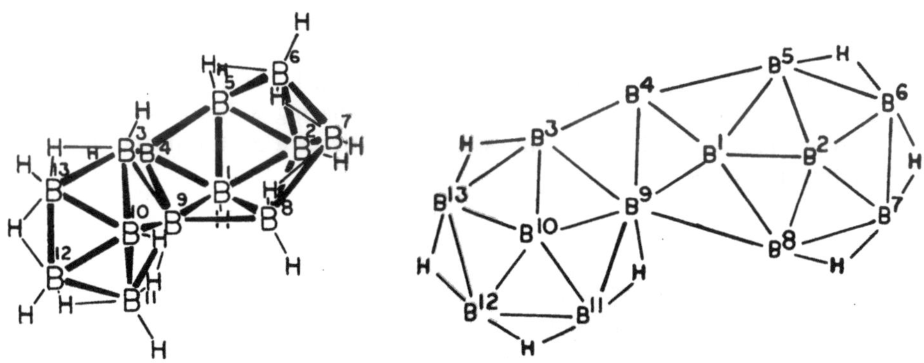

Fig. 41 a. The Structure of $B_{13}H_{19}$; **b** Topological Representation of $B_{13}H_{19}$

4.4.2 Tetradecaborane(18), $B_{14}H_{18}$

The structure of $B_{14}H_{18}$ has not been determined; however, it has been deduced from its NMR spectrum and from a consideration of its mode of formation from $B_{16}H_{20}$ and from analogy with $B_{18}H_{22}$ [157]. The proposed structure is that of a *conjuncto*-borane with $B_{10}H_{14}$ and B_6H_{10} cages sharing a pair of boron atoms (B^5 and B^6 in $B_{10}H_{14}$, B^2 and B^3 in B_6H_{10}). The structure is shown in Fig. 42. Note that the shared atoms do not have terminal H's.

4.4.3 Tetradecaborane(20), $B_{14}H_{20}$

The structure of $B_{14}H_{20}$, which is shown in Fig. 43, is that of two B_8H_{12} fragments fused at the B^3—B^8 position[158]. The bond distances and angles are normal and this is predicted in calculations reported by Lipscomb[156], which suggest that the fusion should not distort the bonding of the B_8 fragments. The two fragments have their open faces *cis* to each other giving the molecule virtual C_{2v} symmetry. The four boron atoms in the center portion of the molecule are almost planar.

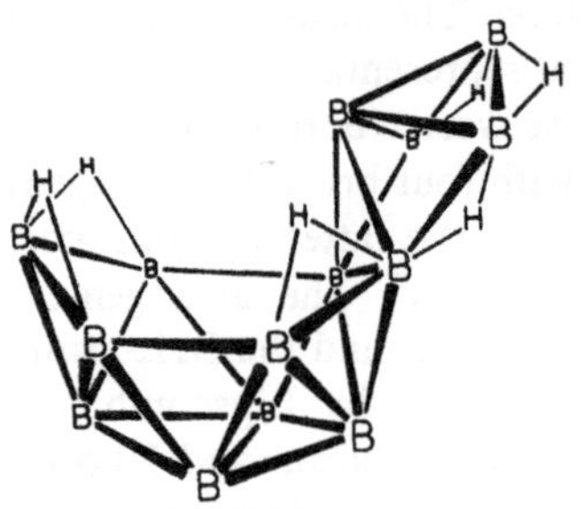

Fig. 42. Proposed Structure of $B_{14}H_{18}$

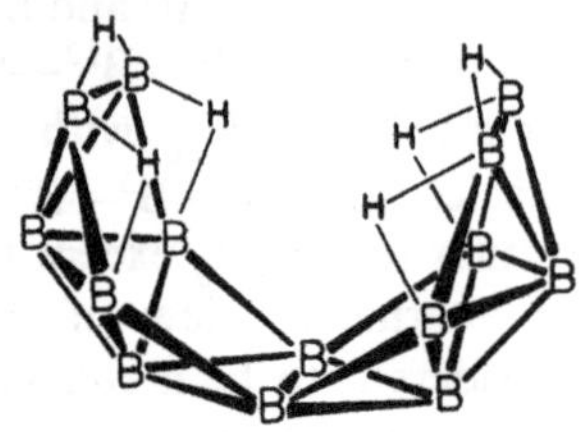

Fig. 43. The Structure of $B_{14}H_{20}$ (terminal hydrogens are not shown)

4.4.4 Pentadecaborane(23), $B_{15}H_{23}$

The structure of $B_{15}H_{23}$, as deduced from its boron-11 NMR spectrum[159], is that of a *commo*-borane involving a B_9H_{13} cage sharing a boron atom with a B_7H_{11} group. The shared boron atom essentially replaces a proton in the $[B_6H_{11}]^+$ ion as shown in Fig. 44.

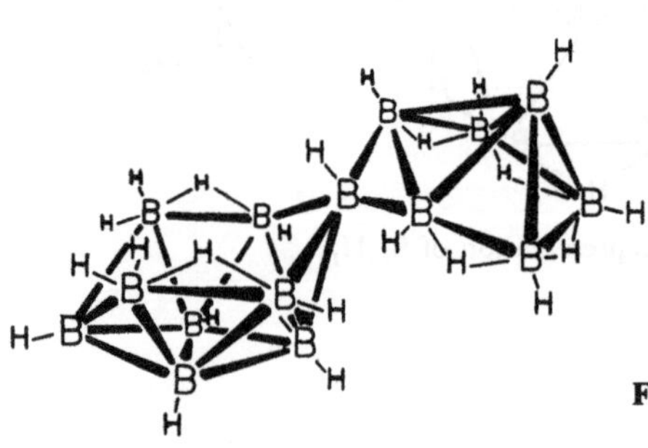

Fig. 44. Proposed Structure of $B_{15}H_{23}$

4.4.5 Hexadecaborane(20), $B_{16}H_{20}$

The structure of the $B_{16}H_{20}$ molecule has been determined by single crystal X-ray diffraction to be that of a *conjuncto*-borane involving the fusion of a $B_{10}H_{14}$ cage with a B_8H_{12} cage[160]. The formal fusion is along the B^5—B^6 edge of $B_{10}H_{14}$ and the B^3—B^8 edge of B_8H_{12} with the two fragments opening in opposite directions. The structure is shown in Fig. 45a and a topological representation in Fig. 45b. The molecule is not significantly distorted from the original structures $B_{10}H_{14}$ and B_8H_{12}.

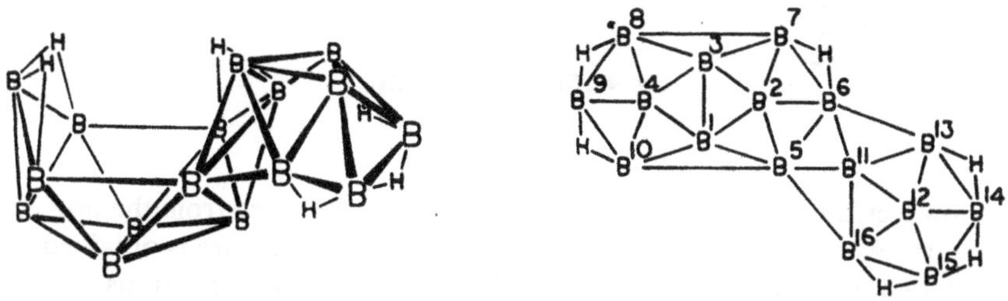

Fig. 45a. Structure of $B_{16}H_{20}$; **b** Topological Representation of $B_{16}H_{20}$ (terminal H's omitted)

4.4.6 *n*-Octadecaborane(22), *n*-$B_{18}H_{22}$

The *n*-$B_{18}H_{22}$ molecule, as determined by single crystal X-ray diffraction[161)], is centrosymmetric, possessing six bridging hydrogens, and has the geometry of two $B_{10}H_{14}$ cages sharing a common B^5—B^6; B^6—B^7 edge as indicated in Fig. 46.

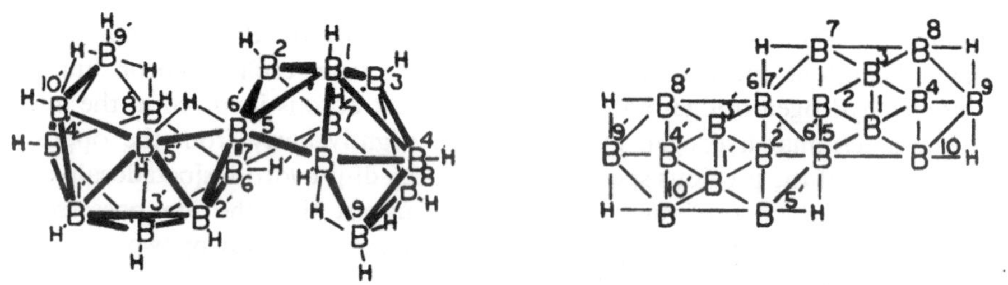

Fig. 46a and b. Molecular Structure; **a** and Topological Representation; **b** of *n*-$B_{18}H_{22}$

4.4.7 *i*-Octadecaborane(22), *i*-$B_{18}H_{22}$

The structure of the species *i*-$B_{18}H_{22}$ [163)] is very similar to that of *n*-$B_{18}H_{22}$ except the molecule possesses a two-fold axis and not a center of symmetry. The molecule has two 10-boron units from $B_{10}H_{14}$ sharing a common B^6—B^7 edge such that the two halves of the molecule open in opposite directions giving it a two-fold axis.

4.4.8 Icosaborane(16), $B_{20}H_{16}$

The structure of $B_{20}H_{16}$, as determined by single crystal X-ray diffraction and shown in Fig. 47, represents that of the the only well characterized derivative of a *closo*-borane with more than 12 vertices. The structure may be considered as two $B_{10}H_{14}$ cages fused at the open faces, with the loss of all bridge H atoms and the loss of the terminal H atoms on the four B atoms which join the cages. This is not so

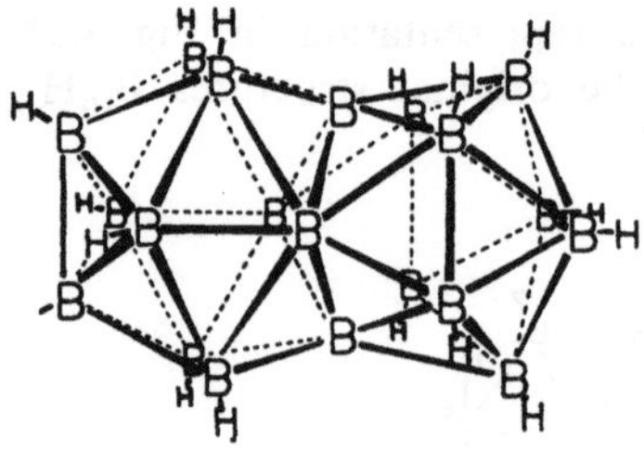

Fig. 47. Structure of $B_{20}H_{16}$

satifactory, however, since the 10-boron units are partially collapsed from the original $B_{10}H_{14}$ geometry. A more satisfactory description is that the structure is obtained by the fusion of two *nido*-twelve-boron cages sharing four common atoms or the fusion of two *closo*-thirteen-boron cages each missing a vertex atom, the four coordinate atom, and joined at that face. Lipscomb[55)] has described such *closo*-species for which, as mentioned above, there is much experimental precedent in metalloborane chemistry.

4.4.9 Oligomeric *closo*-Borane Anions

There are several species which are essentially formed from the coupling of $[B_{10}H_{10}]^{2-}$ cages sometimes followed by subsequent reduction of the coupled product. The octadecahydroicosaborate(2—) ion, $[B_{20}H_{18}]^{2-}$, consists of two 10-boron units, very similar to $[B_{10}H_{10}]^{2-}$ species, linked by B—B interaction among one pair of adjacent apical and equatorial B atoms from each 10-boron unit[163)]. There are no hydrogen bridges and each of the apical borons involved in the interaction has a terminal hydrogen atom displaced slightly away from the opposite cage. An isomer of the previous species exists, 2,2:6,6-di-μ-hydro-bis[octadecahydrodecaborate], for which preliminary X-ray data suggest that the ion possesses a plane of symmetry[164)]. Two $[B_{10}H_9]^-$ units are linked symmetrically by two hydrogen bridges between pairs of equatorial borons.

The species octadecahydroicosaborate(4—), $[B_{20}H_{18}]^{4-}$, yields NMR spectra which suggest that it exists as two $[B_{10}H_9]^{2-}$ units sigma bonded with boron-boron bonds yielding the 1,1-, 1,2- and 2,2-isomers [165)]. The ion nonadecahydroicosaborate(3—), $[B_{20}H_{19}]^{3-}$, gives infrared spectra and NMR data consistent with a structure involving two $[B_{10}H_9]^{2-}$ moieties bridged by a proton[165,166)].

Analogous species derived from the $[B_{12}H_{12}]^{2-}$ ions exist. $[B_{24}H_{23}]^{3-}$ is analogous to $[B_{20}H_{19}]^{3-}$, above[167)], and $[B_{48}H_{45}]^{2-}$ presumably exists as four B_{12} units linked by three hydrogen bridges[167c)].

4.5 Sigma-Bonded conjuncto-Boranes

4.5.1 Octaborane(18), B_8H_{18}

Low temperature X-ray data and NMR spectra for octaborane(18) suggest that the species exists as 2,2'-$(B_4H_9)_2$ [168,169)].

4.5.2 Pentaborane(9)yl-pentaborane(9), $(B_5H_8)_2$

The species $B_{10}H_{16}$ has been shown to be two B_5H_8 moieties bonded by a single boron-boron bond. All three possible isomers, 1,1′-$(B_5H_8)_2$, 1,2′-$(B_5H_8)_2$ [169,171] and 2,2′-$(B_5H_8)_2$ [169,172] have been identified by NMR spectrometry although detailed structural data are not available.

4.5.3 Bidecaboran(14)yl Species, $(B_{10}H_{13})_2$

The icosaboranes(26), $B_{20}H_{26}$, which have been known for several years, have been shown to be the bidecaborane(14)yl species, i.e., two $B_{10}H_{13}$ cages joined by a single boron-boron bond. There are eleven distinct geometric isomers of this species of which four should exist as enantiomeric pairs[173]. Using the numbering scheme for $B_{10}H_{14}$, the possible isomers include 1,1′, 1,2′, 1,6′, 2,2′, 2,6′, and 6,6′. The isomers involving monosubstitution at the 5, 7, 8 or 10 position on $B_{10}H_{14}$ represent enantiomeric pairs since these positions are chiral. This leads to geometrically distinct isomers 5,1′, 5,2′, and 5,6′ which are paired enantiomerically with the isomers 7,1′, 7,2′, and 7,6′, respectively. The 5,5′- and 7,7′-isomers constitute a geometrically distinct enantiomeric pair and the 5,7′-isomer is also geometrically distinct, being the unique mesodiastereoisomer of the 5,5′ and 7,7′ enantiomeric pair[174].

Four species have been identified. The 6,6′-species has been identified by detailed boron-11 and proton NMR spectral measurements[173]. The 2,2′, 2,6′ and 1,5′ isomers have been studied by single crystal X-ray diffraction and detailed structural data are available. The 1,5′-species is unsymmetrical[175]. The conformations of the 2,2′- and 2,6′-isomers may be defined in terms an eclipsed conformation of C_{2v} symmetry for the former and of Cs symmetry for the latter[176]. Thus, the actual conformations are defined by the torsion angle 118° for B^6—B^2:$B^{2'}$—$B^{6'}$ in the 2,2′-isomer and 92° for B^6—B^2:$B^{6'}$—$B^{2'}$ in the 2,6′-isomer. These torsion angles represent rotations from the eclipsed conformation about the boron-boron bond between the two $B_{10}H_{13}$ moieties. In all three molecules whose structure is known in detail, all boron-boron bonds involving one of the cage-linked atoms are slightly longer than the other boron-boron distances and are also slightly longer than the corresponding distances in $B_{10}H_{14}$, otherwise the bond lengths and angles are very similar to those in $B_{10}H_{14}$.

5 References

1. Lipscomb, W. N.: Science, *196*, 1047 (1977)
2. Brown, H. C.: Science, *210*, 485 (1980)
3. Stock, A.: Hydrides of Boron and Silicon, Cornell University Press, Ithaca, N.Y. 1933
4. Bartow, V.: Advan. Chem. Ser., *32*, 5 (1961)
5. Wiberg, E.: Pure Applied. Chem., *49*, 691 (1977)
6. Sidgewick, N. V.: The Chemical Elements and Their Compounds, Claredon, London, 1950, p. 338
7. Shriver, D. F.: The Manipulation of Air-sensitive Compounds, McGraw Hill, New York, 1969

8. Bauer, S. H.: J. Amer. Chem. Soc., *59*, 1096 (1937)
9. a) Stock, A., Massenez, C.: Ber., *45*, 3529 (1912). b) Core, A. F.: Chem. Ind., (London) *46*, 642 (1927). c) Pauling, L.: The Nature of the Chemical Bond, Ithaca, New York (1939). d) Bauer, S. H.: Chem. Rev., *31*, 43 (1942)
10. Dilthey, W.: Z. Angew. Chem., *34*, 596 (1921)
11. Bell, R. P., Longuet-Higgins, H. C.: Proc. Roy. Soc., (London) A. *183*, 357 (1945)
12. a) Price, W. C.: J. Chem. Phys., *15*, 614 (1947). b) Price, W. C.: J. Chem. Phys., *16*, 894 (1948)
13. Hedberg, K., Schomaker, V.: J. Am. Chem. Soc., *73*, 1482 (1951)
14. Lipscomb, W. N.: Boron Hydrides, W. A. Benjamin. New York 1963
15. Pauling, L.: J. Inorg. Nucl. Chem., *32*, 3745 (1970)
16. Reference 3, p. 17
17. Parry, R. W., Edwards, L. J.: J. Am. Chem. Soc., *81*, 3554 (1959)
18. Hawthorne, M. F., Pitochelli, A. R.: J. Am. Chem. Soc., *81*, 5519 (1959)
19. Burg, A. B.: J. Am. Chem. Soc., *79*, 2129 (1957)
20. Kodama, G.: J. Am. Chem. Soc., *92*, 3482 (1970)
21. a) Guter, G. A., Schaeffer, G. W.: J. Am. Chem. Soc., *78*, 3546 (1958). b) Hawthorne, M. F., Miller, J. J.: J. Am. Chem. Soc., *80*, 754 (1958). c) Siegel, B., Mack, J. L., Lowe, J. O., Gallaghan, J.: J. Am. Chem. Soc., *80*, 4523 (1958). d) Hough, W. V., Edwards, L. J.: Abstracts 133rd Nat. Meet. Amer. Chem. Soc., San Francisco, California, (1958) 28L. e) Miller, J. J., Hawthorne, M. F.: J. Am. Chem. Soc., *81*, 4501 (1959). f) Schaeffer, G. W., Burnes, J. J., Klingen, T. J., Martincheck, L. A., Rozett, R. W.: Abstr., 13th Nat. Meet. Amer. Chem. Soc., Boston, Mass, (1959) 44M. g) Hawthorne, M. F., Pitochelli, A. R., Strahm, R. D., Miller, J. J.: J. Am. Chem. Soc., *82*, 1825 (1960)
22. Johnson, H. D., Shore, S. G.: J. Am. Chem. Soc., *92*, 7586 (1970)
23. Ditter, J. F., Spielman, J. R.: Williams, R. E.: Inorg. Chem., *5*, 118 (1966)
24. Longuet-Higgins, H. C., Roberts, M. de V.: Proc. Roy. Soc., (London) *224A*, 336 (1954)
25. Longuet-Higgins, H. C., Roberts, M. de V.: Proc. Roy. Soc., (London) *230A*, 110 (1955)
26. Allard, G.: Bull, Soc. Chim., (France), *51*, 1213 (1932)
27. Pauling, L., Weinbanm, S.: Z. Kristallog., *87*, 181 (1934)
28. Hoard, J. L., Geller, S., Hughes, R. E.: J. Am. Chem. Soc., *73*, 1892 (1951)
29. Hawthorne, M. F., Pitochelli, A. R.: J. Am. Chem. Soc., *82*, 3328 (1960)
30. a) Kaczmarczyk, A., Dobrott, R. D., Lipscomb, W. N.: Proc. Nat. Acad. Sci., U.S., *48*, 729 (1962). b) Dobrott, R. D., Lipscomb, W. N.: J. Chem. Phys., *37*, 1779 (1962)
31. Wunderlich, J., Lipscomb, W. N.: J. Am. Chem. Soc., *82*, 4427 (1960)
32. Klanberg, F., Muetterties, E. L.: Inorg. Chem., *5*, 1955 (1966)
33. Klanberg, F., Eaton, D. R., Guggenberger, L. J., Muetterties, E. L.: Inorg. Chem., *6*, 1271 (1967)
34. Boone, J. L.: J. Am. Chem. Soc., *86*, 5036 (1964)
35. a) Williams, R. E., Good, C. D., Shapiro, I.: Abstr., 140th Nat. Meet., Amer. Chem. Soc., Chicago (1961) 14N. b) Shapiro, I., Good, C. D., Williams, R. E.: J. Am. Chem. Soc., *84*, 3837 (1962). c) Shapiro, I., Keilin, B., Williams, R. E., Good, C. D.: J. Am. Chem. Soc., *85*, 3167 (1963)
36. a) Heying, T. L., Ager, J. W., Clark, S. L., Mangold, D. J., Goldstein, H. L., Hillman, M., Polak, R. J., Symanski, J. W.: Inorg. Chem., *2*, 1089 (1963). b) Fein, M. M., Grafstein, D., Paustain, J. E., Bobinski, J., Lichstein, B. M., Mayes, N., Schwartz, N. N., Cohen, M. S.: Inorg. Chem., *2*, 1115 (1963)
37. a) Potenza, J. A., Lipscomb, W. N.: Inorg. Chem., *3* 1673 (1964). b) Voet, D., Lipscomb, W. N.: Inorg. Chem., *3*, 1679 (1964). c) Potenza, J. A., Lipscomb, W. N.: J. Am. Chem. Soc., *86*, 1874 (1964). d) Stanko, V. I., Struchkov, Yu. T.: Zh. Obshch. Khim., *35*, 930 (1965). e) Potenza, J. A., Lipscomb, W. N.: Inorg. Chem., *5*, 1471 (1966). f) Potenza, J. A., Lipscomb, W. N.: *5*, Inorg. Chem. 1478 (1966). g) Potenza, J. A., Lipscomb, W. N.: Inorg. Chem., *5*, 1483 (1966). h) Andrianov, V. G., Stanko, V. I., Struchkov, Yu. T., Klimova, A. L.: Zh. Strukt. Khim., *8*, 707 (1967)
38. a) Onak, T., Drake, R., Dunks, G.: Inorg. Chem., *3*, 1686 (1964). b) Weiss, H. G., Shapiro, I.: U.S. Patent, 3,086,996 (1963). c) Shapiro, I., footnote in Onak, T., Gehard, F. J., Williams, R. E.: J. Am. Chem. Soc., *85*, 3378 (1963). d) Onak, T., Williams, R. E., Weiss, H. G.: J. Am. Chem. Soc., *84*, 2830 (1962)

39. a) Streib, W. D., Boer, F. P., Lipscomb, W. N.: J. Am. Chem. Soc., *85*, 2331 (1963). b) Boer, F. P., Streib, W. E., Lipscomb, W. N.: Inorg. Chem., *3*, 1666 (1964)
40. a) The terms *closo-* and *nido-* were adopted for carboranes by the Boron Nomenclature Committee of the Amer. Chem. Soc. in Sept. 1966. b) Williams, R. E.: in Progress in Boron Chemistry, Vol. 2, (Steinberg, H., McCloskey, A. L. ed.), Macmillan, New York, 1970, CL2
41. a) Williams, R.: Prog. Boron. Chem., *2*, 47 (1970) and references therein; b) Jutsi, P., Seufert, A., Buchner, W.: Chem. Ber., *112*, 2488 (1979)
42. a) Tebbe, F. N., Garrett, P. M., Hawthorne, M. F.: J. Am. Chem. Soc., *88*, 667 (1966). b) Tebbe, F. N., Garrett, P. M., Hawthorne, M. F.: J. Am. Chem. Soc., *90*, 869 (1968). c) Garrett, P. M., George, T. A., Hawthorne, M. F.: Inorg. Chem., *8*, 2008 (1969)
43. Voet, D., Lipscomb, W. N.: Inorg. Chem., *6*, 113 (1967)
44. Longuet-Higgins, H. C.: J. Chim. Phys., *46*, 268 (1949)
45. Longuet-Higgins, H. C.: Quart, Rev., *11*, 121 (1957)
46. Eberhardt, W. H., Crawford, B., Lipscomb, W. N.: J. Chem. Phys., *22*, 989 (1954)
47. Lipscomb, W. N.: J. Phys. Chem., *62*, 381 (L958)
48. Lipscomb, W. N.: Adv. Inorg. Chem. Radiochem., *1*, 118 (1959)
49. Dickerson, R. E., Lipscomb, W. N.: J. Chem. Phys., *27*, 212 (1957)
50. Lipscomb, W. N.: Inorg. Chem., *3*, 1683 (1964)
51. Epstein, I. R., Lipscomb, W. N.: Inorg. Chem., *10*, 1921 (1971)
52. Epstein, I. R.: Inorg. Chem., *12*, 709 (1973)
53. Rudolph, R. W., Thompson, D. W.: Inorg. Chem., *13*, 2780 (1974)
54. Switkes, E., Lipscomb, W. N., Newton, M. D.: J. Am. Chem. Soc., *92*, 3847 (1970)
55. Brown, L. D., Lipscomb, W. N.: Inorg. Chem., *16*, 2989 (1979)
56. Lipscomb, W. N.: Pure, Appl. Chem., *29*, 493 (1972)
57. Lipscomb, W. N.: in "Boron Hydride Chemistry", Muetterties, E. L.: ed., Academic Press, New York (1975), ch. 2
58. Lipscomb, W. N.: Acc. Chem. Res., *6*, 257 (1973)
59. Lipscomb, W. N.: Pure Appl. Chem., *49*, 701 (1977)
60. Lipscomb, W. N.: in "Boron Chemistry 4" Parry, R. W., Kodama, G.: editors, Pergamon Press, London (1980)
61. Wade, K.: J. Chem. Soc., Chem. Commun, 792 (1971)
62. Williams, R. E.: Inorg. Chem., *10*, 210 (1970)
63. Rudolph, R. W., Pretzer, W. R.: Inorg. Chem., *11*, 1974 (1972)
64. a) Wade, K.: Nature (London), Phys. Sci., *71*, 240 (1972). b) Wade, K.: Inorg. Nucl. Chem. Lett., *8*, 823 (1972). c) Wade, K.: Inorg. Nucl. Chem. Lett., *8*, 559 (1972). d) Wade, K.: Chem. Br., 177 (1974). e) Housecraft, C. E., Wade, K.: Tetrahedron Lett., 3175 (1979)
65. Mingos, D. M. P.: Nature, (London) Phys. Sci., 236, *99* (1972)
66. Bannister, A. J.: Nature (London) Phys. Sci., 239, *69* (1972)
67. Jones, C. J., Evans, W. J., Hawthorne, M. F.: J. Chem. Soc., Chem. Commun., 543 (1973)
68. Wade, K.: Adv. Inorg. Chem. Radiochem., *18*, 1 (1976)
69. Rudolph, R. W.: Acc. Chem. Res., *9*, 446 (1976)
70. Williams, R. E.: Adv. Inorg. Chem. Radiochem., *18*, 67 (1976)
71. a) Grimes, R. N.: Ann. N.Y. Acad. Sci., 239, 180 (1974). b) Eady, C. R., Johnson, B. F. G., Lewis, J.: J. Chem. Soc., Dalton, 2606 (1975)
72. a) Muetterties, E. L., Knoth, W. H.: Chem. Eng. News., May 9 (1966), p. 88. b) Muetterties, E. L., Knoth, W. H.: Polyhedral Boranes, Marcel Dekker, New York (1968). c) Todd, L. J.: in Progress in Boron Chemistry, Vol. 2 (Steinberg, H., McCloskey, A. L. ed.), Macmillan, New York 1970, Ch. 1
73. Lipscomb, W. N., Pitochelli, A. R., Hawthorne, M. F.: J. Am. Chem. Soc., *81*, 5833 (1959)
74. Moore, E. B., Lohr, L. L., Lipscomb, W. N.: J. Chem. Phys., *35*, 1329 (1961).
75. Williams, R. E.: "1970 McGraw-Hill Year book of Science and Technology" McGraw-Hill, New York 1970, p. 33
76. a) Miller, N. E., Miller, H. C., Muetterties, E. L.: Inorg. Chem., *3*, 860 (1964). b) Kodama, G., Kameda, M.: Inorg. Chem. *18*, 3302 (1979). c) Kodama, G., Dodds, A. R.: IMEBORON III, Ettal, Germany, July (1976). d) Colquhoun, H. M.: J. Chem. Research (5), 451 (1978). e) Dodds, A. R., Kodama, G.: Inorg. Chem., *18*, 1465 (1979)

77. a) Hough, W. V.: U.S. 3,167,559; Chem. Abs., *62*, 9009g (1965). b) Zhigach, A. F., Kazakova, E. B., Antonov, I. S.: Zh. Obs. Khim., *27*, 1655 (1967). c) Kodama, G., Engelhardt, V., Lafrenz, C., Parry, R. W.: J. Am. Chem. Soc., *94*, 407 (1972). d) Savory, G. C., Wallbridge, M. G. H.: J. Chem. Soc., Dalton 179 (1973). e) Fratini, A. V., Sullivan, G. W., Denniston, M. L., Hertz, R. K., Shore, S. G.: J. Am. Chem. Soc., *96*, 3013 (1974). f) Alcock, N. W., Colquhonn, H. M., Haran, G., Sawyer, T. F., Wallbridge, M. G. H.: J. Chem. Soc., Chem. Commun. 368 (1977)
78. a) Brubaker, G. L., Denniston, M. L., Shore, S. G., Carter, J. C., Swicker, F.: J. Am. Chem. Soc., *92*, 7216 (1976). b) Mangion, M., Hertz, R. K.: Denniston, M. L., Long, J. R., Clayton, W. R., Shore, S. G.: J. Am. Chem. Soc., *98*, 449 (1976). c) Duben, J., Hermanek, S., Stibr, B.: J. Chem. Soc. Chem. Commun., 287 (1978)
79. Guggenberger, L. J.: Inorg. Chem., *7*, 2260 (1968)
80. Reference 14, p. 53
81. Greenwood, N. N., Gysling, H. J., McGinnety, J. A., Owen, J. D.: J. Chem. Soc., Chem. Commun., 505 (1970)
82. Mingos, D. M. P.: J. Chem. Soc., Dalton, 133 (1974)
83. Mingos, D. M. P.: Pure Appl. Chem., *52*, 705, (1980)
84. Corbett, J. D.: Inorg. Chem., *7*, 198 (1968)
85. Corbett, J. D.: Prog. Inorg. Chem., *21*, 129 (1975)
86. Cragg, R. H., Smith, J. D., Toogood, G. E.: Ann. Rep., Prog. Chem., The Chemical Society, *74A*, 144 (1977) and references therein
87. Rudolph, R. W.: Inorg. Chem., *17*, 1097 (1978)
88. Lipscomb, W. N.: Inorg. Chem., *18*, 2328 (1979)
89. a) Adams, R. M.: Boron, Metallo-boron Compounds and Boranes. John Wiley, New York, 1964 ch. 7. b) Campbell, G. W.: in Progress in Boron Chemistry Vol. I, (Steinberg, H., McCloskey, A. L., ed.) Macmillan, New York, 1964. ch. 3. c) Holtzman, R. T.: Production of Boranes and Related Research, Academic Press, New York, 1967. d) Hawthorne, M. F.: in The Chemistry of Boron and Its Compounds (Muetterties, E. L., ed.) John Wiley, New York, 1967 ch. 5. e) Wade, K.: Electron-deficient Compounds, Nelson, London (1971). f) Shore, S. G.: in Boron Hydride Chemistry, (Muetterties, E. L., ed.) Academic Press, New York, 1975 ch. 3.
90. Fehlner, T. P.: in Boron Hydride Chemistry (Muetterties, E. L., ed.) Academic Press, New York, 1975, ch. 4. and references therein
91. Schwartz, M. E., Allen, L. C.: J. Am. Chem. Soc., *92*, 1466 (1970)
92. Kaldor, A., Porter, R. F.: J. Am. Chem. Soc., *93*, 2140 (1971)
93. Venkatachar, A. C., Taylor, R. C., Kuczkowski, R. L.: J. Mol. Struct., *38*, 17 (1977)
94. Rathke, J., Schaeffer, R.: Inorg. Chem., *13*, 760 (1974)
95. Bartell, L. S., Carroll, B. L.: J. Chem. Phys. *42*, 1135 (1965)
96. Kuchitsu, K.: J. Chem. Phys., *49*, 4456 (1968)
97. Smith, H. W., Lipscomb, W. N.: J. Chem. Phys., *43*, 1060 (1965)
98. Jones, D. S., Lipscomb, W. N.: J. Chem. Phys., *51*, 3133 (1969)
99. McNeill, E. A., Gallaher, K. L., Scholer, F. R., Bauer, S. H.: Inorg. Chem., *12*, 2108 (1973)
100. Grimes, R. N., Bramlett, C. L., Vance, R. L.: Inorg. Chem., *8*, 55 (1969)
101. Graham, G. D., Marynick, D. S., Lipscomb, W. N.: J. Am. Chem. Soc., *102*, 2939 (1980)
102. Hollins, R. E., Stafford, F. E.: Inorg. Chem., *9*, 877 (1970)
103. Kodama, G., Engelhardt, V., Lafreze, C., Parry, R. W.: J. Am. Chem. Soc., *94*, 407 (1972)
104. Peters, C. R., Nordman, C. E.: J. Am. Chem. Soc., *82*, 5758 (1960)
105. Muetterties, E. L., Phillips, W. D.: Advan. Inorg. Chem. Radiochem., *4*, 265 (1962)
106. Schaeffer, R., Johnson, Q., Smith, G. S.: Inorg. Chem., *4*, 917 (1965)
107. Wirth, H. E., Slick, P. I.: J. Phys. Chem., *65*, 1447 (1969)
108. Hedberg, K., Jones, M. E., Schomaker, V.: Proc. Nat. Acad. Sci. U.S., *38*, 679 (1952)
109. a) Dulmage, W. J., Lipscomb, W. N.: Acta, Cryst., *5*, 260 (1952). b) Huffman, J. C., Thesis, Indiana Univ., Michigan (1974), p. 194
110. Nordman, C. E., Lipscomb, W. N.: J. Chem. Phys., *21*, 1856 (1953)
111. Kameda, M., Kodama, G.: Inorg. Chem., *19*, 2288 (1980)
112. Klanberg, F., Eaton, D. R., Guggenberger, L. J., Muetterties, E. L.: Inorg. Chem., *6*, 1271 (1967)
113. Beaudet, R. A., Poynter, R. L.: J. Am. Chem. Soc., *86*, 1258 (1964)

114. Huffman, J. C.: Ph. D. Thesis, Indiana University (1974), reported by Barton, L., Onak, T., Shore, S. G.: Gmelin, Erg-werk Vol. *54*, 65 (1979)
115. Onak, T. P., Shore, S. G., Yamouchi, M.: Gmelin, Handbuch der Anorganischen Chemie, Erg-werk, Vol. *54*, 1–9 (1979). Calculated from positional parameters in: Huffman, J. C.: Ph. D. Thesis, Indiana University (1974)
116. Rietz, R. R., Schaeffer, R., Sneddon, L. G.: J. Am. Chem. Soc., *92*, 3514 (1970)
117. Pepperberg, I. M., Halgren, T. A., Lipscomb, W. N.: Inorg. Chem., *16*, 363 (1977)
118. Guggenberger, L. J.: Inorg. Chem., *8*, 2771 (1969)
119. a) Muetterties, E. L.: Tetrahedron, *30*, 1595 (1974). b) Muetterties, E. L., Hoel, E. L., Salentine, C. G., Hawthorne, M. F.: Inorg. Chem., *14*, 950 (1975). c) Muetterties, E. L., Beier, B. F.: Bull. Soc. Chim. Belges, *84*, 397 (1975)
120. Hollander, O., Clayton, W. R., Shore, S. G.: J. Chem. Soc., Chem. Commun., 604 (1974)
121. Mangion, M., Clayton, W. R., Hollander, O., Shore, S. G.: Inorg. Chem., *16*, 2110 (1977)
122. Remmel, R. J.: Johnson, H. D., Jaworisky, I. S., Shore, S. G.: J. Am. Chem. Soc., *97*, 5395 (1975)
123. Reference 70, p. 109
124. Leach, J. B., Onak, T., Spielman, J., Rietz, R. R., Schaeffer, R., Sneddon, L. G.: Inorg. Chem., *9*, 2170 (1970)
125. Johnson, H. D., Shore, S. G.: J. Am. Chem. Soc., *93*, 3798 (1971)
126. Lipscomb, W. N.: Boron Hydrides, W. A., Benjamin, New York, 1963, p. 221
127. Guggenberger, L. J.: Inorg. Chem., *7*, 2260 (1968)
128. a) Enrione, R. E., Boer, F. P., Lipscomb, W. N.: J. Am. Chem. Soc., *86*, 1451 (1964). b) Enrione, R. E., Boer, F. P., Lipscomb, W. N.: Inorg. Chem., *3*, 1659 (1964)
129. Johnson, H. D., Shore, S. G.: J. Am. Chem. Soc., *93*, 3798 (1971)
130. Kaczmarczyk, A., Kolski, G. B.: Inorg. Chem., *4*, 665 (1965)
131. Siedle, A. R., Bodner, G. M., Garber, A. R., Todd, L. J.: Inorg. Chem., *13*, 1750 (1974)
132. Huffman, J. C., Streib, W. E.: J. Chem. Soc., 665 (1972)
133. Rietz, R. R., Schaeffer, R.: J. Am. Chem. Soc., *95*, 6254 (1973)
134. Moody, D. C., Schaeffer, R.: Inorg. Chem., *15*, 233 (1976)
135. Berry, T. E., Tebbe, F. N., Hawthorne, M. F.: Tetrahedron Lett., 715 (1963)
136. Tsai, C., Streib, W. E.: J. Am. Chem. Soc., *88*, 4513 (1966)
137. Tolpin, E. I., Lipscomb, W. N.: J. Am. Chem. Soc., *95*, 2384 (1973)
138. Moore, E. B., Dickerson, R. E., Lipscomb, W. N.: J. Chem. Phys., *27*, 209 (1957)
139. Tippe, A., Hamilton, W.: Inorg. Chem., *8*, 464 (1969)
140. a) Simpson, P. G., Lipscomb, W. N.: J. Chem. Phys. *35*, 1340 (1961). b) Dickerson, R. E., Wheatley, P. J., Howell, P. A., Lipscomb, W. N., Schaeffer, R.: J. Chem. Phys., *25*, 606 (1956). c) Dickerson, R. E., Wheatley, P. J., Howell, P. A., Lipscomb, W. N.: J. Chem. Phys., *27*, 200 (1957)
141. Huffman, J. C.: Ph. D. Thesis, Indiana University, 1974, p. 771
142. a) Dobson, J., Keller, P. C., Schaeffer, R.: J. Am. Chem. Soc., *87*, 3522 (1965). b) Keller, P. C.: Inorg. Chem., *9*, 75 (1970). c) Moody, D. C., Schaeffer, R.: Inorg. Chem., *15*, 233 (1976)
143. Dobson, J., Maruca, R., Schaeffer, R.: Inorg. Chem., *9*, 2161 (1970)
144. a) Uspenskaya, S. I., Solntsev, K. A., Kuznetsov, N. T.: J. Struct. Chem., USSR, 149 (1973). b) Uspenskaya, S. I., Solntsev, K. A., Kuznetsov, N. T.: J. Struct. Chem., 450 (1973). c) Uspenskaya, S. I., Sontsev, K. A., Kuznetsov, N. T.: Kristallografiya, 1072 (1976)
145. Shoham, G., Schomburg, D., Lipscomb, W. N.: Cryst. Struct. Comm., *9*, 429 (1980)
146. Edwards, L. J., Makhouf, J. M.: J. Am. Chem. Soc., *88*, 4728 (1966)
147. Frischie, C. J.: Inorg. Chem., *6*, 1199 (1967)
148. See Gmelin, Erg-werk, Vol. *54*, 123 (1979)
149. Reddy, J., Lipscomb, W. N.: J. Chem. Phys., *31*, 610 (1959)
150. Muetterties, E. L.: Inorg. Chem., *2*, 647 (1963)
151. Kendall, D. S., Lipscomb, W. N.: Inorg. Chem., *12*, 546 (1973)
152. Callahan, K. P., Hawthorne, M. F.: Pure Appl. Chem., *39*, 475 (1974) and references therein
153. Maxwell, W. M., Grimes, R. N.: Inorg. Chem., *18*, 2174 (1979)
154. Hosmane, N. S., Grimes, R. N.: Inorg. Chem., *19*, 3482 (1980)
155. a) Huffman, J. C., Moody, D. C., Rathke, J. W., Schaeffer, R.: J. Chem. Soc., Chem. Commun., 308 (1973). b) Huffman, J. C., Moody, D. C., Schaeffer, R.: Inorg. Chem., *15*, 227 (1976)

156. Dixon, D. A., Kleier, D. A., Halgren, T. A., Lipscomb, W. N.: J. Am. Chem. Soc., *98*, 2086 (1976)
157. Hermanek, S., Fetter, K., Plesek, J., Todd, L. J., Garber, A. R.: Inorg. Chem. *14*, 2250 (1975)
158. Huffman, J. C., Moody, D. C., Schaeffer, R.: J. Am. Chem. Soc., *97*, 1621 (1970)
159. Rathke, J., Schaeffer, R.: Inorg. Chem., *13*, 3008 (1974)
160. a) Friedman, L. B., Coop, R. E., Glick, M. D.: J. Am. Chem. Soc., *90*, 6802 (1968). b) Friedman, L. B., Cook, R. E., Glick, M. D.: Inorg. Chem., *9*, 1452 (1970)
161. a) Simpson, P. G., Lipscomb, W. N.: Proc. Natl. Acad. Sci., U.S. *48*, 1490 (1962). b) Simpson, P. G., Lipscomb, W. N.: J. Chem. Phys., *39*, 26 (1963)
162. a) Simpson, P. G., Folting, K., Lipscomb, W. N.: J. Am. Chem. Soc., *85*, 1879 (1963). b) Simpson, P. G., Folting, K., Dobrott, R. D., Lipscomb, W. N.: J. Chem. Phys., *30*, 2339 (1963)
163. Schwalbe, C. H., Lipscomb, W. N.: Inorg. Chem., *10*, 151 (1971)
164. Hawthorne, M. F., Pilling, R. L.: J. Am. Chem. Soc., *88*, 3873 (1966)
165. Hawthorne, M. F., Pilling, R. L., Stokely, P. F.: J. Am. Chem. Soc., *87*, 1893 (1965)
166. a) Chamberland, B. L., Muetterties, E. L.: Inorg. Chem., *3*, 1450 (1964). b) Middaugh, R. L., Farha, F.: J. Am. Chem. Soc., *88*, 4147 (1966)
167. a) Wiersema, R. J., Middaugh, R. L.: J. Am. Chem. Soc., *89*, 5078 (1967). b) Wiersema, R. J., Middaugh, R. L.: Inorg. Chem., *8*, 2074 (1969). c) Bechtold, R., Kaczmarczyk, A.: J. Am. Chem. Soc., *96*, 5953 (1974)
168. Beall, H. A.: Diss. Harvard Univ., cited in: Rietz, R. R., Schaeffer, R., Sneddon, L. G.: Inorg. Chem., *11*, 1242 (1972)
169. Rietz, R. R., Schaeffer, R., Sneddon, L. G.: Inorg. Chem. *11*, 1242 (1972)
170. Grimes, R. N., Diss. Univ. of Minnesota, Minneapolis (1962); cited by Shore, S. G., in "Boron Hydride Chemistry", Muetterties, E. L.: editor, ch. 3. Academic Press, New York (1975), p. 103
171. Gaines, D. F., Iorns, T. V., Clevenger, E. N.: Inorg. Chem., *10*, 1096 (1971)
172. Rietz, R. R., Schaeffer, R.: J. Am. Chem. Soc., *95*, 4580 (1973)
173. Boockock, S. K., Greenwood, N. N., Kennedy, J. D., Taylorson, D.: J. Chem. Soc., Chem. Commun., *6* (1979)
174. Greenwood, N. N., Kennedy, J. D., Spalding, T. R., Taylorson, D.: J. Chem. Soc., Dalton Trans., 840 (1979)
175. Brown, G. M., Pinson, J. W., Ingram, L. L.: Inorg. Chem., *18*, 1951 (1979)
176. Boocock, S. K., Greenwood, N. N., Kennedy, J. D., McDonald, W. S., Staves, J.: J. Chem. Soc., Dalton Trans., 790 (1980)

Bei Fragen zur Produktsicherheit wenden Sie sich bitte an:
If you have any questions regarding product safety,
please contact:

Walter de Gruyter GmbH
Genthiner Straße 13
10785 Berlin
productsafety@degruyterbrill.com